T0252460

HIPPOCRATES
IX

LCL 509

HIPPOCRATES

VOLUME IX

EDITED AND TRANSLATED BY
PAUL POTTER

HARVARD UNIVERSITY PRESS
CAMBRIDGE, MASSACHUSETTS
LONDON, ENGLAND
2010

Library of Congress Control Number 2009935559
CIP data available from the Library of Congress

ISBN 978-0-674-99640-3

*Composed in ZephGreek and ZephText by
Technologies 'N Typography, Merrimac, Massachusetts.
Printed on acid-free paper and bound by
The Maple-Vail Book Manufacturing Group*

CONTENTS

INTRODUCTION

The eleven Hippocratic works in this volume include a range of medical genres: monographs on the form and function of various parts of the human body; collections of prognostic observations; practical clinical manuals.[1]

While *Anatomy* presents a brief rehearsal of the main thoracic and abdominal viscera, *Nature of Bones* is devoted mainly and *Heart* completely to expounding the cardiovascular system, the former describing the distribution of vessels through the body and explaining their roles in health and disease, the latter examining the parts of the heart in terms of structure and purpose. The level of anatomical knowledge attained in *Nature of Bones* and *Heart* suggests the employment of more active and experimental methods of investigation (e.g. human dissection), not evidenced in other works of the Collection, which limit themselves on the whole to opinions derived from clinical observation and philosophical speculation. The treatise *Eight Months' Child* seeks, by combining various *a priori* numerological theories with many accurate clinical observations, to establish that children born during a forty-day

[1] The individual works are analysed in more detail in their particular introductions.

period centred on the eighth month of pregnancy cannot survive.

With its 640 chapters, *Coan Prenotions* represents the largest single collection of prognostic aphorisms in the Hippocratic Collection. These statements, many excerpted from other Hippocratic works, are organized by subject, and describe a wide range of medical signs and conditions. Two other semeiotic works, *Crises* and *Critical Days*, are limited in subject matter to the phenomenon of crisis in diseases; the former consists of aphorisms similar to those that make up *Coan Prenotions*, *Prorrhetic I*, and *Aphorisms*, while the latter contains eleven longer passages on the topic taken mainly from other extant Hippocratic works.

The remaining four treatises in the volume are devoted to specialty practice. *Superfetation* is a manual of obstetrical knowledge, loosely organized by theme and focused on prognosis and therapy. *Girls* is a fragment describing the untoward mental effects that can result in girls at puberty from increased blood production, and *Excision of the Fetus* is a short collection of miscellaneous notes on embryotomy and other obstetrical subjects. *Sight* is the fragmentary remains of a handbook of ophthalmology arranged by specific disorders of the eyes: surgical treatments predominate.

Manuscript Tradition

M = Marcianus Venetus Graecus 269	X/XI c.
A = Parisinus Graecus 2253	XI c.
V = Vaticanus Graecus 276	XII c.
I = Parisinus Graecus 2140	XIII c.

H = Parisinus Graecus 2142
     Ha (older part)[2]   XII/XIII c.
     Hb (newer part)   XIV c.
R = Vaticanus Graecus 277   XIV c.
Recentiores = approximately 20 manuscripts XV/XVI c.

The stemma codicum appearing as Fig. 1 provides an overview of the interdependencies among the manuscripts containing the eleven treatises in this volume. The particular treatises are transmitted in the following independent witnesses:

Anatomy	V
Nature of Bones	M
Heart	V
Eight Months' Child	M V
Coan Prenotions	A I
Crises	V
Critical Days	M
Superfetation	M Va Vb
Girls	M V
Excision of the Fetus	M^{I} V I^{II}
Sight	M

Superfetation appears in two different versions in V (Va and Vb). *Excision of the Fetus* was also once contained twice in M (M^{I} and M^{II}); however, the second text (M^{II}) was lost in the fourteenth century, after this version had been copied into the manuscripts H and I (H^{II} and I^{II}).

In both cases where the M text is lost and must be reconstructed from its copies—*Coan Prenotions* and the

---

2 Folios 46, 49, 55–78, and 80–308.

Fig. 1: *Stemma Codicum*

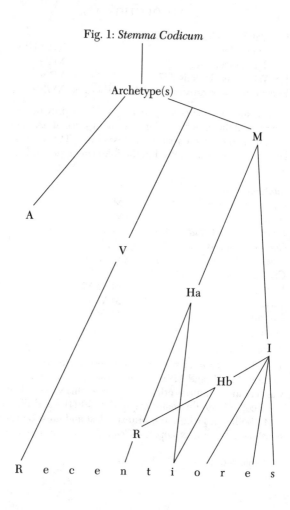

second version of *Excision of the Fetus*—the text in Parisinius Graecus 2142 is contained in the newer part of that manuscript, Hb, (fol. 459v.-466r. and 441r.-v.), and thus of no independent textual authority,[3] leaving I as M's sole representative.

BIBLIOGRAPHY

Editions, Translations, and Commentaries

Hippocratic Collection

Hippocratis Coi . . . octoginta volumina . . . per M. Fabium Calvum, Rhavennatem . . . latinitate donata . . . , Rome, 1525. (=Calvus)

Omnia opera Hippocratis . . . in aedibus Aldi & Andreae Asulani soceri, Venice, 1526. (=Aldina); marginal notes by Ianus Cornarius in a copy of this edition presently kept in the Göttingen University Library. (=Cornarius in marg.)

Hippocratis Coi . . . libri omnes . . . [per Ianum Cornarium], Basel, 1538. (=Froben)

Hippocratis Coi . . . opera . . . per Ianum Cornarium . . . Latina lingua conscripta, Venice, 1546. (=Cornarius)

[3] See J. Jouanna, *Hippocrate. La nature de l'homme*, Corpus Medicorum Graecorum I 1, 3, Berlin, 1975, pp. 85f., 131; H. Grensemann, *Hippokratische Gynäkologie*, Wiesbaden, 1982, pp. 69–76.

BIBLIOGRAPHY

Hippocratis Coi . . . viginti duo commentarii . . . Theod. Zvingeri studio & conatu, Basel, 1579. (=Zwinger)

Hippocratis Coi Iusjurandum, Aphorismorum Sectiones VIII, Prognostica, Prorrheticorum libri II, Coaca praesagia. Graecus et Latinus contextus accurate renovatus . . . studio Joannis Opsopoei, Frankfurt, 1587. (=Opsopoeus)

Magni Hippocratis . . . opera omnia . . . latina interpretatione & annotationibus illustrata Anutio Foesio . . ., Geneva, 1657–62. (=Foes)

Magni Hippocratis Coi opera omnia graece & latine edita . . . industria & diligentia Joan. A. Vander Linden, Leiden, 1665. (=Linden)

J. F. K. Grimm, *Hippokrates Werke aus dem griechischen übersetzt . . .*, Altenburg, 1781–92. (=Grimm)

E. Littré, *Oeuvres complètes d' Hippocrate*, Paris, 1839–61. (=Littré)

F. Z. Ermerins, *Hippocratis . . . reliquiae*, Utrecht, 1859–64. (=Ermerins)

R. Fuchs, *Hippokrates, sämmtliche Werke. Ins Deutsche übersetzt . . .*, Munich, 1895–1900. (=Fuchs)

R. Kapferer and G. Sticker, *Die Werke des Hippokrates . . . in neuer deutscher Übersetzung*, Stuttgart, 1933–40. (=Kapferer / Sticker)

H. Grensemann, *Hippokrates Über Achtmonatskinder*, Corpus Medicorum Graecorum I 2, 1, Berlin, 1968. (=Grensemann)

R. Joly, *Hippocrate, . . . Du Foetus de huit mois*, Budé XI, Paris, 1970. (=Joly)

C. Lienau, *Hippokrates über Nachempfängnis, Geburtshilfe und Schwangerschaftsleiden*, Corpus Medicorum Graecorum I 2, 2, Berlin, 1973. (=Lienau)

R. Joly, *Hippocrate, . . . De la vision . . .* , Budé XIII, Paris, 1978. (=Joly)

C. García Gual et al., *Tratados hipocráticos . . . , introducciones, traducciones y notas*, Madrid, 1983–2003.

J. Jouanna, *Hippocrate, Airs, Eaux, Lieux*, Budé II . . . , Paris, 1996.

A. Roselli, *Ippocrate, La malattia sacra*, Venice, 1996.

E. M. Craik, *Hippocrates, Places in Man*, Oxford, 1998.

M.-P. Duminil, *Hippocrate, Plaies, Nature des os, Coeur, Anatomie*, Budé VIII, Paris, 1998. (=Duminil)

J. Jouanna and M. D. Grmek, *Hippocrate, Epidémies V et VII*, Budé IV (3), Paris, 2000.

J. Jouanna, *Hippocrate, La maladie sacré*, Budé II (3), Paris, 2003.

E. M. Craik, *Two Hippocratic Treatises*, On Sight *and* On Anatomy, Leiden, 2006. (=Craik)

F. Giorgianni, *Hippokrates, Über die Natur des Kindes* (De genitura *und* De natura pueri), Wiesbaden, 2006.

F. Bourbon, *Hippocrate, Nature de la femme*, Budé XII (1), Paris, 2008.

Other Authors

F. Marx, A. *Cornelii Celsi quae supersunt*, Corpus Medicorum Latinorum I, Leipzig and Berlin, 1915. (= Celsus)

O. Stählin, *Clemens Alexandrinus . . .* , Leipzig, 1905–36. (=Clement of Alexandria)

E. Nachmanson, *Erotiani Vocum hippocraticarum collectio*, Gothenburg, 1918. (=Erotian)

E. Nachmanson, *Erotianstudien*, Uppsala, 1917. (=Nachmanson)

BIBLIOGRAPHY

C. G. Kühn, *Claudii Galeni Opera omnia* . . . , Leipzig, 1825–33. (=Galen)

K. Latte, *Hesychii Alexandrini Lexicon*, Copenhagen, 1953–66. (=Hesychius)

Ch. Daremberg and Ch. E. Ruelle, *Oeuvres de Rufus d'Ephèse*, Paris, 1879. (=Rufus)

M. Wellmann, *Die Fragmente der sikelischen Ärzte* . . . , Berlin, 1901, «Der Tractat des Vindicianus», pp. 208–34. (=Vindicianus)

General Works

A. Anastassiou and D. Irmer, *Testimonien zum Corpus Hippocraticum*, Göttingen, 1997–2006. (=Anastassiou / Irmer)

L. A. Dean-Jones, *Women's Bodies in Classical Greek Science*, Oxford, 1994.

N. Demand, *Birth, Death and Motherhood in Classical Greece*, Baltimore, 1994.

P. Diepgen, *Die Frauenheilkunde der alten Welt*, Munich, 1937.

Ph. J. van der Eijk, H. F. J. Horstmanshoff, and P. H. Schrijvers (edd.), *Ancient Medicine in its Socio-cultural Context*, Amsterdam, 1995.

Ph. J. van der Eijk (ed.), *Hippocrates in Context. Papers read at the XIth International Hippocrates Colloquium*, Leiden, 2005. (=Eijk)

Ph. J. van der Eijk, *Medicine and Philosophy in Classical Antiquity*, Cambridge, 2005.

H. Fasbender, *Entwickelungslehre, Geburtshülfe und Gynäkologie in den hippokratischen Schriften*, Stuttgart, 1897.

BIBLIOGRAPHY

K.-D. Fischer, D. Nickel, and P. Potter (edd.), *Text and Tradition, Studies in Ancient Medicine and its Transmission*, Leiden, 1998.

H. Flashar and J. Jouanna (edd.), *Médecine et morale dans l'antiquité*, Geneva, 1997.

S. Föllinger, *Differenz und Gleichheit. Das Geschlechterverhältnis in der Sicht griechischer Philosophen des 4. bis 1. Jahrhunderts v. Chr.*, Stuttgart, 1996.

I. Garofalo, A. Lami, D. Manetti, and A. Roselli (edd.), *Aspetti della terapia nel Corpus Hippocraticum. Atti del IXe Colloque International Hippocratique*, Florence, 1999.

G. Harig, *Aufsätze zur Medizin- und Wissenschaftsgeschichte*, Marburg, 2007.

J. Jouanna, *Hippocrates*, trans. M. B. DeBevoise, Baltimore, 1999.

H. King, *Hippocrates' Woman*, London, 1998.

K.-H. Leven, *Antike Medizin. Ein Lexikon*, Munich, 2005.

C. W. Müller, C. Brockmann, and C. W. Brunschön (edd.), *Ärtze und ihre Interpreten. Medizinische Fachtexte der Antike als Forschungsgegenstand der Klassischen Philologie*, Munich, 2005.

V. Nutton, *Ancient Medicine*, London, 2004.

C. M. Oser-Grote, *Aristoteles und das Corpus Hippocraticum. Die Anatomie und Physiologie des Menschen*, Stuttgart, 2004.

A. Thivel and A. Zucker (edd.), *Le normal et le pathologique dans la Collection hippocratique. Actes du Xème colloque international hippocratique*, Nice, 2002.

R. Wittern and P. Pellegrin (edd.), *Hippokratische Medizin und antike Philosophie. Verhandlungen des VIII Internationalen Hippokrates-Kolloquiums*, Hildesheim, 1996.

ANATOMY

INTRODUCTION

The Greek text of this short account of the internal parts is transmitted only in the manuscript V and its descendents. The work is mentioned in no ancient text, but verbal echos may be present in Celsus' *De medicina* (e.g. 4, 1, 3 *constat ex circulis quibusdam* ~ κρίκοις ξυγκειμένη ὁμορρύσμοις; 4, 1, 8 *in sinus vehementer inplicitum* ~ ἑλικηδὸν ἐν κόλποις ἐνειλούμενον) and pseudo-Rufus' *Anatomy of the Parts of Man* (e.g. Daremberg-Ruelle p. 175, 4 τὴν χροιὰν τεφρὸς καὶ ὑπόλευκος ~ τεφρίνης χροιῆς; p. 176, 9f. σπλὴν ... ἀνθρωπίνῳ ἴχνει ~ σπλὴν ... ὁμοιόρρυσμος ἴχνει ποδός). Many expressions in the text are rare or unique in Greek usage, some with parallels in the fragments of the pre-Socratic philosophers (e.g. the Democritean ῥυσμός ~ ὁμ(οι)όρρυσμος).

Anatomy is included in all the collected editions and translations of Hippocrates, and recently two scholars have independently published new editions of the treatise accompanied by valuable analyses of the text and its place in the history of anatomy:

> E. M. Craik, "The Hippocratic Treatise *On Anatomy*," *Classical Quarterly* 48 (1998), 135–67; reprinted with revisions and additions in Craik, pp. 115–70.
>
> M.-P. Duminil, *Hippocrate ... Anatomie*, Budé VIII,

Paris, 1998, 197–209, 258–9, and 289–91. (= Duminil)

The present edition is based on a collation of the manuscript V from microfilm.

ΠΕΡΙ ΑΝΑΤΟΜΗΣ

VIII 538
Littré

1. Ἀρτηρίη ἐξ ἑκατέρου φαρυγγέθρου τὴν ἔκφυσιν ποιευμένη ἐς ἄκρον πνεύμονος τελευτᾷ, κρίκοις ξυγκειμένη ὁμορρύσμοις, τῶν περιηγέων ἁπτομένων[1] κατ' ἐπίπεδον ἀλλήλων. αὐτὸς δὲ ὁ πνεύμων συνεξαναπληροῖ τὴν χέλυν, τετραμμένος ἐς τὰ ἀριστερά, πέντε ὑπερκορυφώσιας ἔχων, ἃς δὴ καλέουσι λοβούς, τεφρίνης χροιῆς τυχὼν στίγμασιν [σφραγῖσι][2] κεκεντημένος, φύσει ἐὼν τενθρηνιώδης.[3] μέσῳ δ' αὐτῷ ἡ καρδίη ἐγκαθίδρυται, στρογγυλωτέρη καθεστεῶσα πάντων ζῴων. ἀπὸ δὲ καρδίης ἐς ἧπαρ βρογχίη πολλὴ καθήκει, καὶ μετὰ βρογχίης φλὲψ μεγάλη καλευμένη, δι' ἧς οὖλον τὸ σκῆνος τρέφεται. τὸ δὲ ἧπαρ ὁμορρυσμίην μὲν ἔχει τοῖς ἄλλοις ἅπασιν, αἱμορρωδέστερον δέ ἐστι τῶν ἄλλων, ὑπερκορυφώσιας ἔχον δύο, ἃς καλέουσι πύλας, ἐν δεξιοῖς τόποις κειμένας· ἀπὸ δὲ τούτου, σκαληνὴ φλὲψ ἐπὶ τὰ κάτω νεφρὸν ἀποτείνουσα. νεφροὶ δὲ ὁμοιόρρυσμοι, τὴν χροιὴν δὲ ἐναλίγκιοι μήλοισιν· ἀπὸ δὲ τούτων, ὀχετοὶ σκαληνοειδέες ἄκρην κορυφὴν κύστιος κεῖνται. κύστις δὲ νευρώδης οὔλη καὶ μεγάλη· ἐκ δὲ τῆς κύστιος μετοχέ-

ANATOMY

1. An air pipe (*arteria*) growing out of the throat on each side ends at the apex of the lung; it is composed of symmetrical rings, which in their circular course meet one another in a plane. The lung itself occupies the chest, facing towards the left, and possesses five prominences called lobes; it is of ashen colour, marked with spots, and in structure like honeycomb. In the centre of the lung the heart is set, being more spherical than in all other animals. From the heart a large tube (*bronchia*) descends to the liver; running with this tube is what is called the great vessel, through which the whole frame is nourished. The liver has a symmetry with all others, but more blood-flow than the others; it possesses two prominences called "gates" which lie on the right side. From the liver a vessel slanting downwards reaches the kidney. The kidneys are symmetrical, and in colour are like apples.[1] From the kidneys two oblique ducts reach the topmost apex of the bladder. The bladder is large and entirely sinewy, and out of it grows a

---

1 Or "like sheep" (Calvus: *ovillis similes*).

---

1 Ermerins: ἁπτομένη V.
2 Del. Potter as gloss on στίγμασιν: ὀφροναγεσι V.
3 Foes in note 6: τὲ θρηνιώδης V.

τευσις ἔξω[4] πέφυκε. καὶ τὰ μὲν ἐξ ἀνὰ μέσον ἐντὸς
φύσις ἐκόσμει.

Οἰσοφάγος δὲ ἀπὸ γλώσσης τὴν ἀρχὴν ποιεύμενος
540 ἐς κοιλίην τελευτᾷ, ὃν δὴ καὶ ἐπὶ σηπτικῆς κοιλίης
στόμαχον καλέουσι. πρὸς δὲ ἀκάνθης ὄπισθεν ἥπατος
φρένες πεφύκασι. ἐκ δὲ πλευρῆς νόθης, λέγω δὲ ἀριστερῆς, σπλὴν ἀρξάμενος ἐκτέταται ὁμοιόρρυσμος
ἴχνει ποδός. κοιλίη δὲ ἥπατι παρακειμένη κατ᾽ εὐώ-
νυμον μέρος, οὐλομελής ἐστι νευρώδης. ἀπὸ δὲ κοιλίης πέφυκεν ἔντερον, ὁμοιόρρυσμον μικρόν, πηχέων
οὐκ ἔλασσον δώδεκα, ἑλικηδὸν ἐν κόλποις ἐνειλούμενον, ὃ καλέουσιν ἔνιοι κόλον, δι᾽ οὗ ἡ παραφορὰ τῆς
τροφῆς γίνεται. ἀπὸ δὲ κόλου πέφυκεν ἀρχὸς λοίσθι-
ος, σάρκα περιπληθέα ἔχων, ἐς ἄκρον δακτυλίου τελευτῶν.

Τὰ δὲ ἄλλα ἡ φύσις διετάξατο.

---

[4] Ermerins: ἔκαθε κύστιος, μεσοχὴ εἶσα V.

channel. These six parts nature has arranged in the interior around the mid-line.

The oesophagus takes its origin from the tongue and ends at the cavity,[2] which is also called the orifice (*stomachos*) next the digestive cavity. Against the back-bone behind the liver the diaphragm is attached. Out of the side by the false ribs—I mean on the left—the spleen has its origin; it spreads out symmetrical with a foot print. The cavity lying beside the liver on the left side is all sinewy. The intestine grows out of the cavity, small and symmetrical, not less than twelve cubits wound in a spiral in the lap; this some people call the colon, and through it the transport of nutrients occurs. Out of the colon grows the anus, the final part; it has very full flesh, and ends at the outer margin of the (sc. muscular) ring.

The rest nature has arranged.

[2] I.e. the abdominal part of the gastro-intestinal tract.

NATURE OF BONES

INTRODUCTION

Although the title *Nature of Bones*[1] is never mentioned
in ancient literature, the text of this work apparently be-
longed to the Hippocratic Collection at the time of Bacchi-
us of Tanagra in the third century B.C., who according to
Erotian[2] glossed the term ἐνεφλεβοτόμησε (ch.18) in his
third book. Erotian himself, who makes no reference to
the treatise in his introductory census of Hippocratic writings, includes sixteen terms from it in his Hippocratic glossary.[3] A century later Galen discusses fourteen terms from
Nature of Bones among the Hippocratic words he ex-
plains,[4] referring to the source once as "the texts appended
to *Instruments of Reduction*"[5] and once as "*On Vessels*
(περὶ φλεβῶν), which is appended to *Instruments of Reduction*."[6] Finally, the fifth century A.D. lexicographer
Hesychius of Alexandria takes over five terms directly or

---

[1] The transmitted title is derived from the first paragraph of
the text, and applies but poorly to the work as a whole, which is
centred on angiology rather than osteology.

[2] Erotian E39, p. 38.

[3] See Nachmanson, pp. 346–7 and 354–8.

[4] See Duminil, pp. 131–4.

[5] S.v. κοτυληδόνα, Galen vol. 19, 114.

[6] S.v. παραστάτας, Galen vol. 19, 128.

11

indirectly from *Nature of Bones* for comment in his *Lexicon*.[7]

Considerable portions of the text of *Nature of Bones* are also transmitted in other extant Greek writings:

(a) A sentence near the end of chapter 1 comparing the size of the colon in humans and dogs appears in very similar form as *Epidemics VI* 4,6 (Loeb *Hippocrates* vol. 7, 248–9).

(b) The first six lines of chapter 8 are quoted almost verbatim by Aristotle in *History of Animals* 511b23–30, where he attributes them to an otherwise unknown Syennesis of Cyprus.

(c) The whole of chapter 9, which is also quoted by Aristotle (*History of Animals* 512b11–513a7) in a somewhat shortened and reworked form and attributed to Polybus, originated as *Nature of Man* 11 (Loeb *Hippocrates* vol. 4, 30–3).

(d) The text of chapter 10 is present in virtually identical wording as *Epidemics II* 4,1 (Loeb *Hippocrates* vol. 7, 66–71).

These recurrences as well as the treatise's lack of coherence have led to much scholarly discussion about its composition and authorship.[8] Generally it seems clear that what unifies the work is not a single origin, but the focus of its contents—although not exclusively—on human angiology. The resulting account, however, is neither well inte-

---

[7] See Duminil, p. 134.

[8] For a detailed account of these discussions cf. Duminil, pp. 75–115.

grated nor doctrinally consistent: the same questions are sometimes dealt with more than once,[9] no overarching architecture shapes the work, contradictions of terminology and fact jar the reader.

Still, *Nature of Bones* represents the most comprehensive and accurate Hippocratic account of the human vascular system we have, in many instances revealing a knowledge of the structure of vessels, and their paths through the body which is unattested in the rest of the Collection. Among the signs of this hightened anatomical awareness are:

(i) An incipient separation of "artery" (ἀρτηρίη) from its erstwhile synonym "bronchus" (βρόγχος) to signify a kind of vessel (φλέψ) (ch. 7 and 10).

(ii) A differentiation of the category "vessel" (φλέψ) into "blood-vessel" (αἱμόρρους or ἔναιμος φλέψ) (ch. 7, 12, 16 and 17) and "artery" (ἀρτηρίη) (ch. 7 and 10).

(iii) A widening of the range of meaning of "band" (τόνος) beyond its former synonym "cord" (νεῦρον) to include in addition the category of "vessel" (φλέψ) (ch. 7 and 10).

(iv) An interest in the positional relationships of the large abdominal vessels (ch. 7 and 10).

(v) A description and naming of hitherto unrecorded parts of the male genital apparatus such as the "seminal vesicle" (σπέρμα) (ch. 1), the "epididymus"

[9] E.g. the four pairs of large vessels described as descending from the head in chapter 9 are echoed by several vessels mentioned in subsequent chapters of the treatise. Furthermore, the hepatic vessels depicted in ch. 10 parallel to a degree the vessels described in ch. 4–7. Cf. Harris, pp. 72f.

($\pi\alpha\rho\alpha\sigma\tau\acute{\alpha}\tau\eta\varsigma$) (ch. 14), and the vessels of the penis "which are curved and run close together" (ch. 15), a possible reference to the *corpus spongiosum* and the *corpora cavernosa*.

(vi) Reference for the first time to the intercostal vessels (ch. 5 and 10).

(vii) Some appreciation of the complexity of the vascular connections between the lungs and the heart (ch. 19).

The chapters of *Nature of Bones* are organized as follows:

    1:   Osteology and miscellaneous other anatomy.
    2:   The paths of two vessels from the heart.
    3:   The paths of cords (*neura*) through the body.
    4:   Structure and function of the kidneys.
  5–6:   The paths of the thoracic vessels.
    7:   The paths of the veins and arteries of the trunk.
  8–9:   Two accounts of the "wide vessels" (*venae cavae*).
   10:   An account of the vessels and cords of the trunk.
11–19:   A detailed human angiology.

Generally, *Nature of Bones* attracted little attention from scholars until about the middle of the twentieth century, when it became the centre of an extended controversy concerning Hippocratic knowledge of the vessels and their functions. In his 1938 German translation of the treatise, Richard Kapferer attempted—partly by rearranging and, on occasion, even rewriting sections of the work—to extract from the text a consistent, anatomically correct account of the human circulatory system. In response to the outspoken criticism this questioning of William Harvey's priority as the discoverer of the circulation of the blood evoked, Kapferer published, in collaboration with A. Fingerle and F. Lommer, *Die anatomischen Schriften die Anatomie, das Herz, die Adern in der hippokratischen Sammlung* (Stuttgart, 1951), in which he attempted to buttress his case with additional arguments, explanations, and illustrations. Kapferer's position has found little support among subsequent historians, such as C. R. S. Harris[10] and M.-P. Duminil,[11] who have tended, rightly in my opinion, to see more confusion and less hidden meaning in the text than Kapferer did.

Professor Duminil's recent Budé edition, *Hippocrate, . . . Nature des os*, Paris, 1998 (=Duminil), is an important contribution to the study of *Nature of Bones*, for which all subsequent workers owe a sincere debt of gratitude.

My text is based on a collation of the sole independent Greek witness to the text, M, from microfilm.

---

[10] *The Heart and the Vascular System in Ancient Greek Medicine*, Oxford, 1973, pp. 50–73. (= Harris)

[11] *Le Sang, les vaisseaux, le coeur dans la Collection hippocratique*, Paris, 1983, pp. 281–7.

ΠΕΡΙ ΟΣΤΕΩΝ ΦΥΣΙΟΣ

1. Ὀστέα χειρὸς εἰκοσιεπτὰ καὶ ποδὸς εἰκοσιτέσσαρα· τραχήλου ἐς τὸν μέγαν ἑπτά, ὀσφύος πέντε, ῥάχιος εἴκοσι· κεφαλῆς ξὺν ὀπωπίοις ὀκτώ· ξύμπαντα ὀγδοήκοντα ὀκτώ, σὺν ὄνυξιν ἑκατὸν ὀκτώ. ἃ δ᾽ ἡμεῖς αὐτοὶ ἐξ ἀνθρώπου ὀστέων κατεμάθομεν, σφόνδυλοι οἱ ἄνω τῆς κληῖδος ξὺν τῷ μεγάλῳ ἑπτά· οἱ δὲ κατὰ τὰς πλευρὰς ὅσαιπερ αἱ πλευραὶ δώδεκα· οἱ δὲ κατὰ κενεῶνας ἐκτός ⟨εʹ⟩·[1] ἐν ᾧ τὰ ἰσχία, ἐν τῇ ὀσφύϊ πέντε.

Τὸ δὲ σπέρμα οἷον κηρίον ἑκατέρωθεν τῆς κύστιος· ἐκ δ᾽ αὐτῶν φλέβες ἑκατέρωθεν τοῦ οὐρητῆρος ἐς τὸ αἰδοῖον τείνουσι. ποτὸν διὰ φάρυγγος καὶ διὰ στομάχου· λάρυγξ ἐς πλεύμονα καὶ ἀρτηρίην· ἀπὸ δὲ τούτων ἐς ἄκρην κύστιν. ἥπατος πέντε λοβοί· ἐπὶ τοῦ τετάρτου λοβοῦ ἡ χολή, ἡ στόμα ἐπὶ φρένας, κραδίην, πλεύμονα φέρει. κραδίην ὑμὴν περίεστιν. τὰ κόλα ἔχει κυνὸς[2] μείζω· ἤρτηται δ᾽ ἐκ τῶν μεσοκόλων·

---

[1] Potter.
[2] Foes in note 6, after Cornarius' *quam canis*: κοινὸς M.

---

[1] ὀσφῦς here means "sacrum," as earlier in the chapter.

NATURE OF BONES

1. The bones of the arm are twenty-seven in number, of the leg, twenty-four; of the neck up to the great vertebra, seven; of the sacrum, five; of the spine, twenty; of the head including those of the eyes, eight: altogether eighty-eight, and with the nails (sc. of the fingers and toes) one hundred and eight. As for what we ourselves have observed of human bones: the vertebrae from the collar-bone up, including the great vertebra, seven; those in the region of the ribs, twelve, the same as the ribs themselves; those in the region where the flanks are on the outside, ⟨five;⟩ those where the hip bones lie next the sacrum, five.[1]

A honeycombed seminal vesicle[2] is situated on each side of the bladder: from these arise vessels which pass on each side of the urethra to the penis. Drink moves through the throat and oesophagus, the larynx leads to the lung and the artery, and from these the course is to the apex of the bladder. Five lobes of the liver: on the fourth lobe is the gall bladder, which turns its orifice to the diaphragm, the heart, and the lung. A membrane encloses the heart. He (sc. a person) has intestines larger than a dog; they are suspended from mesocolons, which are attached to cords

[2] Cf. Theophrastus *History of Plants* 6, 4, 3 for σπερματικόν as the seed vessel in plants.

ταῦτα δ' ἐκ νεύρων ἀπὸ τῆς ῥάχιος ὑπὸ τὴν γαστέρα. νεφροὶ ἐκ νεύρων ἀπὸ ῥάχιος καὶ ἀρτηρίης.

2. Καρδίης πηγὴ ξυγγενής· φλὲψ τείνει διὰ φρενῶν, ἤπατος, σπληνός, νεφρῶν ἐς ἰσχίον· περὶ γαστροκνημίην ἐπὶ τὸν ταρσόν. ἑτέρη δ' ἐκ καρδίης ὑπὸ 170 μασχάλας, κληῖδας, σφαγάς, κεφαλήν, | ῥῖνα, μέτωπον, παρὰ τὰ ὦτα, ὤμους, μετάφρενον, στήθεα, γαστέρα, διὰ πήχεως· ἡ δὲ διὰ μασχαλέων ἐπὶ πῆχυν, ταρσόν.

3. Νεύρων ἔκφυσις ἀπὸ τοῦ ἰνίου ἄχρι παρὰ ῥάχιν, παρὰ ἰσχίον, ἐς αἰδοῖα, ἐς μηρούς, πόδας, κνήμας, ἐς χεῖρας. ἄλλ' ἐς βραχίονας, τὰ μὲν ἐς σάρκας, τὰ δὲ παρὰ τὴν περόνην ἐς τὸν μέγαν δάκτυλον, τὰ δ' ἐκ τῶν σαρκῶν ἐπὶ τοὺς ἄλλους δακτύλους. ἀλλ' ⟨ἐς⟩[3] ὠμοπλάτην, στῆθος, γαστέρα † ὀστέοισι συνδέσμοις † ἀπὸ δὲ αἰδοίου παρ' ἀρχόν, κοτυληδόνα· τὸ μὲν ἄνωθεν μηροῦ, τὸ δὲ κάτωθεν ἐπὶ τὰ γούνατα, ἐντεῦθεν γούνατι ξυντάθεν, ἐπὶ τένοντα, πτέρναν, πόδας· τὸ δ' ἐς περόνην, ἐς τὸν ἀστράγαλον.[4]

4. . . . . [ἢ][5] ἐς τοὺς νεφρούς· αὗται δ' αἱ φλέβες ἐφ' ἑκάτερα διχῇ τὰ μέγιστα σχίζονται, τὰ μὲν ἔνθεν τοῦ νεφροῦ ἑκατέρου, τὰ δ' ἔνθεν, καὶ διατέτρηνται ἐς τοὺς νεφρούς. καὶ εἶδος καρδίης οἱ νεφροὶ ἔχουσι· καὶ οὗτοι κοιλιώδεες· ὁ δὲ νεφρὸς τὰ κοῖλα ἑωυτοῦ πρὸς τὰς

---

3 Foes.   4 τὸν ἀστράγαλον Potter: τὴν οὐκ εἰς **
ἄλλαι M in an erasure.   5 The text is unclear; I suggest assuming a lacuna and deleting M's ἤ.

from the spine beneath the stomach.[3] The kidneys are attached to cords from the spine and from the artery.

2. The congenital fountain-head of the heart: one vessel passes through the diaphragm, the liver, the spleen, and the kidneys to the pelvis; and around the calf to the flat of the foot. A second vessel out of the heart passes to the axillae, the collar-bones, the jugulars, the head, the nose, the forehead, past the ears, the shoulders, the back, the chest, the stomach, and through the forearms; this vessel also passes through the axillae to the forearm and the palm of the hand.

3. Cords grow out from the occiput along the spine and the hip-bone to the genital parts, the thighs, the legs and the calves, as well as to the arms. Other cords grow to the upper arms, some into the muscles, others along the process (radius) into the thumb, and others out of the muscles into the other fingers. Other cords grow to the shoulder blade, the chest and the belly † by means of bones and ligaments †, and from the genital part past the anus and the acetabulum; an upper cord occupies the thigh, a lower one extends to the knees, and from there connects the knee to the Achilles tendon, the heel and the foot; a third one passes to the process (fibula) and the ankle.

4. . . . to the kidneys: these vessels each divide to send off two large branches in opposite directions which pass in a pair on each side to one of the kidneys or the other, and pierce through into the kidneys. The kidneys are heart-shaped and each has a hollow place. A kidney lies with this

[3] Cf. *Epidemics VI* 4,6 where this sentence is also present.

19

φλέβας ἔχων κεῖται τὰς μεγάλας· ὅθεν ἐκπεφύκασιν
ἐξ αὐτοῦ αἱ φλέβες αἱ ἐς κύστιν, ᾗ εἵλκετο τὸ ποτὸν
διὰ τῶν φλεβῶν ἐς τοὺς νεφρούς· ἔπειθ' ὥσπερ καὶ διὰ
τῶν νεφρῶν διηθεῖται τὸ ὕδωρ καὶ δι' αὐτῶν τούτων
τῶν ἐντέρων ὧν ξυνεπακολουθεῖ. σπογγοειδὲς γάρ
ἐστι τὸ ἀπ' αὐτῶν ἐς τὴν κύστιν, καὶ ἐνταῦθα διηθού-
μενον καὶ ἀποκρινόμενον ἀπὸ τοῦ αἵματος τὸ οὖρον,
διὸ δὴ ἴσως ἐρυθρόν ἐστι. οὐδὲ γὰρ ἐς τοὺς νεφροὺς
οὐκ ἦσαν ἄλλαι φλέβες ἢ αἱ εἴρηνται, οὐδ' ὅποι ἂν τὸ
ποτὸν ξυντήκοιτο, ὅσον ἐγὼ οἶδα.

5. Αἱ παρὰ τὰς πλευρὰς κατατείνουσαι κάτωθέν
εἰσιν ἑκάστης τῶν πλευρέων, οὐ πρὸς κεφαλῆς, κατω-
τέρω δὲ καὶ ἀπὸ ἀρτηρίης· ἀρτηρίη μὲν οὖν εἶθ'
ὑπορρεύσασα διαδιδοῖ τῇσι πλευρῇσι. ἀπὸ δὲ τῆς
παχείης ἀπὸ καρδίης παλινδρομέει μία ἐς τὰ ἀριστερὰ ἐγκεκλεισμένη. ἔπειτα ἡ μὲν διὰ μέσων σφονδύλων μέχρι ἄκρων πλευρέων πορεύεται, πλευρῇσιν οὐκ
172   ἐξ ἴσου διαδιδοῦσα τοῖσι δεξιοῖσι καὶ | ἀριστεροῖσι
διασχίδας· ἀλλ' ἴσας μέν, ἀνωτέρωθεν δὲ τοῖσι δεξιοῖς ἀποσχίζεται.

6. Παρὰ δὲ κληῖδας ἑκατέρης τῶν φλεβῶν δύο μὲν
ἄνω, δύο δὲ ὑπὸ στῆθος, αἱ μὲν ἐς δεξιά, αἱ δ' ἐς
ἀριστερὰ ἀπεσχίσθησαν ἀποσχίδες· ἐὰν πρὸς αὐχένος μὲν μᾶλλον αὗται, δύο δὲ πρὸς καρδίην μᾶλλον,
αἱ μὲν ἐπὶ δεξιά, αἱ δ' ἐπ' ἀριστερά· ἀφ' ἑκατέρης
παρὰ τὰς πλευράς· καὶ ἀπ' αὐτῶν ὥσπερ αἱ κάτω
ἐσχίζοντο, μέχρι ὅτου ξυνέμιξαν τῇ κάτω παλινδρο-
μησάσῃ ἀπὸ καρδίης.

20

hollow part facing the large vessels, and from this part also grow out the vessels (ureters) that pass to the bladder. It is here that drink is drawn through the vessels into the kidneys, and the fluid is then, as it were, filtered through the kidneys and also through the internal parts themselves (ureters) into which it follows; for the vessel from the kidney to the bladder is spongy, and urine is filtered out there and separated from the blood, for which reason, I imagine, it is reddish. Indeed, the only vessels entering the kidney are the ones I have described, nor is there any other place where drink might liquify, at least as far as I am aware.

5. The vessels extending down along the ribs lie beneath each of the ribs, not on the side towards the head, but lower down and away from the artery. Now the artery, when it has descended, distributes to the ribs. From the wide vessel out of the heart, a vessel runs back hemmed in on the left, and then proceeds by the middle of the vertebrae as far as the last ribs, distributing branches unequally to the ribs on the right and the left; but it then gives off equal branches higher up on the right side.

6. Two branches from each of the vessels split off upwards along the collar-bones, and two beneath the sternum, on both the right and the left; if these are more on the side of the neck, two others are more against the heart, on both the right and the left. Vessels arising from each of these extend along the ribs, and from those in turn other vessels branch off just like the ones below, and continue until they join the vessel coming back below from the heart.[4]

---

[4] As described in chapter 5 above.

7. Ἡ δ' αἱμόρρους ἀπὸ τῆς ἀρτηρίης ταύτης διὰ
τοῦτο ἐσχίσθη ὅτι μετέωρος ἐνταῦθά ἐστι διὰ καρδίης
πορευομένη. τὰ δὲ κάτω πλευρέων, ἡ αἱμόρρους ἡ
παχείη καλεομένη φλὲψ τοῖσι σφονδύλοισι αὐτὴ[6] ἐφ'
ἑωυτῆς διαδιδοῖ, καὶ ἐνταῦθα προσέχεται, καὶ οὐκέτι
κρέμαται ὥσπερ ἄνω δι' ἥπατος ἰοῦσα. ἔστι δὲ κατὰ
μὲν ὀσφὺν ἄνω ἡ ἀρτηρίη, ὑποκάτω δὲ ἡ αἱμόρρους ἡ
ἀπὸ τοῦ ἥπατος διὰ φρενῶν ἐλθοῦσα μετέωρος, παρὰ
τὰ ἐπὶ δεξιὰ τῆς καρδίης φέρει ἄχρι κληΐδων, ἁπλῆ
πλὴν ὅσον αὐτῇ τῇ καρδίῃ κοινωνέει. τὰ μὲν κατ'
αὐτὴν σχιζόμενα ἐπιπολαιότερα, τὰ δὲ τὴν κοιλίην
τῆς καρδίης διέχοντα. ἔπειτα ἀπὸ τῆς καρδίης τὸ ἐπ'
ἀριστερὰ κάθηται ἁπλῆ πρὸς ῥάχιν· παλινδρομέει ἐς
μὲν τὸ ἄνω μέρος τοῦ σώματος ἄχρι τῶν ἀνωτάτω
πλευρέων· καὶ ἀποσχίδας ἀφ' ἑαυτῆς ἔχει παρ'
ἑκάστην πλευρὴν παρατεταμένας κατὰ φύσιν ἄχρι
στήθεος συνοκωχῆς καὶ ἐπ' ἀριστερὰ καὶ ἐπὶ δεξιά·
καὶ τὸ ἰθὺ αὐτῆς πρὸς σφονδύλων μᾶλλόν ἐστιν ἢ ὁ
τῆς ἀρτηρίης τόνος καὶ ὁ τῆς ἀπὸ τοῦ ἥπατος φλεβός.
πρὸς δὲ τὸ κάτω μέρος τῆς καρδίης ὁ μὲν ἰθὺς τόνος
ἀπ' αὐτῆς πρὸς σφονδύλων μᾶλλόν ἐστιν ἢ ὁ τῆς
ἀρτηρίης, ὁ δ' ἕτερος ὁ παρὰ καρδίην καὶ ἐς τὰ κάτω
174 μέρη | φρενῶν ἐτράπετο, τὰ πρὸς ῥάχιος ἠρτημένα·

___

[6] Potter: αὖθις M.

7. The blood-vessel is separated from this artery because here it is suspended[5] as it passes through the heart.
In the region below the ribs, the blood-vessel called the
wide vessel distributes by itself to the vertebrae, being attached there and no longer suspended as it is above where
it goes through the liver. At the sacrum the artery is above
(anterior) and the blood-vessel below (posterior); the latter comes out of the liver, is suspended through the diaphragm, and passes on the right side of the heart to the collar-bones, being alone except where it communicates with
the heart. (Some branches split off more superficially at
the level of the heart, while others go through the heart's
cavity.) Then, leaving the heart on the left side, the blood-
vessel[6] resides alone next the spine, and runs back to the
upper part of the body up as far as the highest ribs; here it
sends off branches from itself which extend along each rib
in a systematic way, right through to their junction at the
sternum on the left and right. And the straight part of it is
closer to the vertebrae than are the band[7] of the artery and
the band of the vessel from the liver. Towards the lower
part of the heart, the straight band growing out of it is
closer to the vertebrae than is the band of the artery; the
second band is the one next the heart, and it turns to the
lower parts of the diaphragm which are attached to the

5 For this meaning of μετέωρος based on Galen see Duminil,
p. 223, n. 22.

6 Although grammatically the feminine relative pronoun (ἥ)
could refer equally well to any one of the three possible feminine
antecedents "artery," "vessel" or "blood-vessel," the sense of the
passage seems to militate for blood-vessel, i.e the hollow vessel.

7 See n. 12 below.

ἐντεῦθεν δὲ ἀποσχίδες ἐς ἰθὺ ἕκαστον ἐπιφέρονται, δι᾿ ὀστέων καὶ σαρκῶν περαιωθεῖσαι ἀλλήλαις.

8. Αἱ φλέβες δὲ αἱ παχεῖαι ὧδε πεφύκασιν· ἐκ τοῦ ὀφθαλμοῦ παρὰ τὴν ὀφρύν· διὰ τοῦ νώτου παρὰ τὸν πλεύμονα ὑπὸ τοῦ στήθους, ἡ μὲν ἐκ τοῦ δεξιοῦ ἐς τὸ ἀριστερόν, ἡ δὲ ἐκ τοῦ ἀριστεροῦ ἐς τὸ δεξιόν. ἡ μὲν οὖν ἐκ τοῦ ἀριστεροῦ διὰ τοῦ ἥπατος ἐς τὸν νεφρὸν καὶ τὸν ὄρχιν, ἡ δὲ ἐκ τοῦ δεξιοῦ ἐς τὸν σπλῆνα καὶ νεφρὸν καὶ ὄρχιν· ταύτῃσι δὲ στόμα αἰδοῖον. ἀπὸ δὲ τοῦ δεξιοῦ τιτθοῦ ἐς τὸ ἀριστερὸν ἰσχίον καὶ ἐς τὸ σκέλος· καὶ ἀπὸ τοῦ ἀριστεροῦ ἐς τὰ δεξιά. ὁ δὲ ὀφθαλμὸς ὁ δεξιὸς ἐκ τοῦ ἀριστεροῦ καὶ ὁ ὄρχις, κατὰ ταὐτὰ ἐκ τοῦ δεξιοῦ ὁ ἀριστερός.

9. Αἱ παχύταται τῶν φλεβῶν ὧδε πεφύκασιν· τέσσαρα ζεύγεά ἐστιν ἐν τῷ σώματι. καὶ αἱ μὲν αὐτῶν ἀπὸ τῆς κεφαλῆς ὄπισθεν διὰ τοῦ αὐχένος, ἔξωθεν παρὰ τὴν ῥάχιν ἔνθεν καὶ ἔνθεν ἐς τὰ ἰσχία ἀφικνέεται καὶ ἐς τὰ σκέλεα· ἔπειτα διὰ τῶν κνημέων ἐπὶ τῶν σφυρῶν τὰ ἔξω καὶ ἐς τοὺς πόδας ἀφήκε. δεῖ οὖν τὰς φλεβοτομίας ἐπὶ τῶν ἀλγημάτων τῶν ἐν τῷ νώτῳ καὶ ἐν τοῖς ἰσχίοισιν ἀπὸ τῶν ἰγνύων ποιέεσθαι καὶ ἀπὸ τῶν σφυρῶν ἔξωθεν. αἱ δ᾿ ἕτεραι φλέβες ἐκ τῆς κεφαλῆς παρὰ τὰ ὦτα διὰ τοῦ αὐχένος, σφαγίτιδες καλεόμεναι, ἔσωθεν παρὰ τὴν ῥάχιν ἑκατέρωθεν φέρουσι παρὰ τὰς ψόας ἐς τοὺς ὄρχιας καὶ ἐς τοὺς

---

8 See Aristotle, *History of Animals* 511b24–30 for a quotation of this same passage.

spine; from there branches extend straight in each direction, passing through the bones and the muscles to meet one another.

8.[8] The wide vessels are disposed as follows: out of the eye, along the eyebrow,[9] along the spine, past the lung, under the breasts: the one, from the right to the left; the other, from the left to the right. The one from the left, through the liver to the kidney and the testicle; the one from the right to the spleen, the kidney and the testicle: for these the genital part is the outlet. From the right breast to the left hip and leg, and from the left breast to the right parts. The right eye from the left eye; and the testicle in the same way: the left one from the right one.

9.[10] The widest of the vessels are disposed as follows: there are four pairs in the body. The first of these go posteriorly from the head through the neck, and, passing on the outside along the spine on the left and the right, arrive at the hips and legs; then each passes through the calves to the ankle on the outside, and arrives at the foot. Thus for pains in the back and hips one must practice phlebotomies from the hams and from the outer part of the ankles. The second pair of vessels from the head, past the ears and through the neck—called the jugulars—pass inside along the spine on each side past the psoas muscles to the testi-

---

[9] Some manuscripts of the Aristotelean tradition read in place of "out of the eye, along the eyebrow" the alternative text "from the navel, across the loins "(ἐκ τοῦ ὀμφαλοῦ παρὰ τὴν ὀσφύν); cf. J. A. Smith and W. D. Ross (edd.), *The Works of Aristotle translated into English* vol. 4, *Historia Animalium*, by D. W. Thompson, Oxford, 1910, ad loc., n. 2.

[10] This chapter is taken from *Nature of Man* 11.

μηρούς, καὶ ⟨διὰ⟩[7] ἰγνύων ἐκ τοῦ ἔσωθεν μέρεος·
ἔπειτα διὰ τῶν κνημέων ἐπὶ τὰ σφυρὰ τὰ ἔσωθεν καὶ
τοὺς πόδας. δεῖ οὖν τὰς φλεβοτομίας ποιέεσθαι πρὸς
τὰς ὀδύνας τὰς ἀπὸ τῶν ψοῶν καὶ τῶν ὀρχίων, ἀπὸ
τῶν ἰγνύων καὶ ἀπὸ τῶν σφυρῶν ἔσωθεν. αἱ δὲ τρίται
φλέβες ἐκ τῶν κροτάφων διὰ τοῦ αὐχένος ἐπὶ τὰς
176   ὠμο|πλάτας, ἔπειτα ξυμφέρονται ἐς τὸν πλεύμονα, καὶ
ἀφικνέονται ἡ μὲν ἀπὸ τῶν δεξιῶν ἐς τὰ ἀριστερὰ ὑπὸ
τὸν μαζὸν καὶ ἐς τὸν σπλῆνα καὶ ἐς τὸν νεφρόν, ἡ δ᾽
ἀπὸ τῶν ἀριστερῶν ἐς τὰ δεξιὰ ἐκ τοῦ πλεύμονος ὑπὸ
τὸν μαζὸν καὶ ἐς τὸ ἧπαρ καὶ ἐς τὸν νεφρόν· τελευτῶσι
δὲ ἐς τὸν ἀρχὸν αὗται ἀμφότεραι. αἱ δὲ τέταρται ἀπὸ
τοῦ ἔμπροσθεν τῆς κεφαλῆς καὶ τῶν ὀφθαλμῶν ὑπὸ
τὸν αὐχένα καὶ ὑπὸ τὰς κληῖδας· ἔπειτα ἀπὸ τῶν
βραχιόνων ἄνωθεν ὑπὸ τὰς συγκαμπάς· ἔπειτα διὰ
τῶν πήχεων ἐς τοὺς καρποὺς καὶ τοὺς δακτύλους·
ἔπειτα ἀπὸ τῶν δακτύλων πάλιν διὰ τῶν στηθέων καὶ
τῶν πήχεων τῶν χειρῶν ᾧ ἐς τὰς ξυγκαμπὰς καὶ διὰ
τῶν βραχιόνων τοῦ κάτωθεν μέρεος ἐς τὰς μασχάλας·
καὶ ἐκ τῶν πλευρῶν ἄνωθεν ἡ μὲν ἐς τὸν σπλῆνα
ἀφικνέεται, ἡ δὲ ἐς τὸ ἧπαρ· ἔπειτα ὑπὲρ τῆς γαστρὸς
ἐς τὸ αἰδοῖον τελευτῶσιν ἀμφότεραι. καὶ αἱ μὲν παχεῖαι τῶν φλεβῶν οὕτω πεφύκασιν.

Εἰσὶ δὲ καὶ ἀπὸ τῆς κοιλίης φλέβες ἀνὰ τὸ σῶμα
πολλαί τε καὶ παντοῖαι, δι᾽ ὧν ἡ τροφὴ τῷ σώματι
ἔρχεται. φέρουσι δὲ καὶ ἀπὸ τῶν παχειῶν φλεβῶν ἐς
τὴν κοιλίην καὶ ἐς τὸ ἄλλο σῶμα καὶ ἀπὸ τῶν ἐξωτάτω
καὶ ὑπὸ τῶν ἔσω, καὶ ἐς ἀλλήλας διαδιδόασι, αἵ τε

cles and the thighs, and through the hams from the interior part; and then through the calves to the ankle on the inside and the feet. Thus for pains arising from the psoas muscles and the testicles one must practice phlebotomies from the hams and the inside of the ankles. The third pair of vessels pass out of the temples through the neck to the shoulder-blades, and then meet at the lung; the one from the right side arrives on the left under the breast and comes to the spleen and the kidney, while the one from the left side arrives on the right side coming out of the lung under the breast to the liver and the kidney; both of these terminate at the anus. The fourth pair of vessels arising from the front of the head and the eyes pass beneath the neck and the collar-bones; then from the upper arms down under the elbow joints; and then through the forearms to the wrists and the fingers; then from the fingers back through the balls of the hand and the forearms to the elbow joints; and through the lower part of the upper arm into the axillae. Coming from the ribs above the one vessel arrives at the spleen, and the other at the liver; and then passing over the belly they both terminate in the genital part. This is how the wide vessels are disposed.

There are also vessels passing from the cavity through the body; these are numerous and sundry, and through them nutriment comes to the body. Vessels also pass from the wide vessels to the cavity and the rest of the body, and also from the most outwardly parts and from under the inwardly parts; these communicate with one another, the

---

[7] L. Servin in Foes' *Variae Lectiones*.

ἔσωθεν ἔξω καὶ αἱ ἔξωθεν ἔσω. τὰς οὖν φλεβοτομίας
ποιέεσθαι κατὰ τούτους τοὺς τρόπους· ἐπιτηδεύειν δὲ
χρὴ τὰς τομὰς ὡς προσωτάτω ταμεῖν ἀπὸ χωρίων,
ἔνθα ἂν αἱ ὀδύναι μεμαθήκωσι γίνεσθαι καὶ τὸ αἷμα
συλλέγεσθαι· οὕτω γὰρ ἂν ἥκιστα ἥ τε μεταβολὴ
178 γίνοιτο μεγάλη ἐξαπίνης, καὶ | τὸ ἔθος μεταστήσειας
ἄν, ὥστε μηκέτι ἐς τὸ αὐτὸ χωρίον ξυλλέγεσθαι.

10. Ἡ δ' ἡπατῖτις ἐν ὀσφύϊ μέχρι τοῦ μεγάλου
σφονδύλου κάτωθεν, καὶ σφονδύλοισι προσδιδοῖ, ἐντεῦθεν μετέωρος καὶ διὰ φρενῶν ἐς καρδίην. καὶ ἤει[8]
μὲν εὐθεῖα ἐς κλεῖδας· ἐντεῦθεν δὲ αἱ μὲν ἐς τράχηλον,
αἱ δ' ἐπ' ὠμοπλάτας, αἱ δὲ ἀποκαμφθεῖσαι κάτω, παρὰ
σφονδύλους καὶ πλευρὰς ἀποκλίνουσιν. ἐξ ἀριστερῶν
μὲν μία ἐγγὺς κληΐδων, ἐκ δεξιῶν δ' ἐπί τι αὐτῆς
χωρίον. ἄλλη δ' ἑκατέρωθεν ἀποκαμφθεῖσα, ἄλλη δὲ
σμικρὸν κατώτερον ἀποκαμφθεῖσα, ὅθεν μὲν ἐκείνη
ἀπέλιπε, προσέδωκε τῇσι πλευρῇσι, ἔστ' ἂν τῇ ἐπ'
αὐτῆς τῆς καρδίης προστύχῃ ἐπικαμπτομένη ἐς τὰ
ἀριστερά· ἀποκαμφθεῖσα δὲ κάτω ἐπὶ σφονδύλους
καταβαίνει, ἔστ' ἂν ἀφίκηται ὅθεν ἤρξατο μετεω-
ρίζεσθαι, ἀποδιδοῦσα τῇσι πλευρῇσι τῇσιν ἐπιλοί-
ποις ἁπάσαις, καὶ ἔνθεν καὶ ἔνθεν ἀποσχίδας παρ'
ἑκάστην διδοῦσα μία ἐοῦσα, ἀπὸ μὲν τῆς καρδίης ἐπί
τι χωρίον ἐν τοῖς ἀριστεροῖσι μᾶλλον ἐοῦσα, ἔπειτα
ὑποκάτω τῆς ἀρτηρίης, ἔστ' ἂν καταναλωθῇ ὅθεν ἡ
ἡπατῖτις ἐμετεωρίσθη. πρότερον δὲ πρὶν ἐνταῦθα ἐλ-

---

[8] Littré from *Epidemics* 2 4,1: ἡ M.

ones from inside going to the exterior, and the ones from outside inward. Thus phlebotomies must be practiced in the following way: you must manage the incisions so as to cut as far away as possible from the sites where the pains have been wont to arise and the blood to be collected. For in this way the shock caused will be the least great and sudden, and you will have the effect of altering the habitual state so that blood will no longer be collected in the same place.

10.[11] The hepatic vessel in the loin passes to the great vertebra (sacrum) from below and distributes to the vertebrae; running from there it is suspended and passes through the diaphragm into the heart, and its straight continuation proceeds to the collar-bones. From there some vessels extend to the neck, others to the shoulder blades, and others, turning off downward, incline along the vertebrae and the ribs, one from the left being close to the collar-bones, and one from the right sharing its space. Another vessel turns off on both sides, and yet another a little lower down; from the point where the former splits off, it distributes branches to the ribs until, turning to the left, it meets the one on the heart itself; then, turning away downward, it descends to the vertebrae, until it arrives where it began to be suspended, sending off branches to all the remaining ribs. It gives off branches here and there to each rib, being a single vessel as it first leaves the heart toward a certain place more on the left side, and then it passes behind the artery until it is consumed and arrives where the hepatic vessel was suspended. But first, before it arrives

[11] This chapter also appears as *Epidemics II* 4,1.

θεῖν, παρὰ τὰς ἐσχάτας δύο πλευρὰς ἐδικραιώθη·[9] καὶ
ἡ μὲν ἔνθα, ἡ δ' ἔνθα τῶν σφονδύλων ἐλθοῦσα κατ-
αναλώθη· ἡ δ' εὐθεῖα ἀπὸ καρδίης πρὸς κληῖδας
τείνουσα ἄνωθεν τῆς ἀρτηρίης ἐστί, [καὶ ἀπὸ ταύ-
της][10] ὥσπερ καὶ παρ' ὀσφὺν κάτωθεν τῆς ἀρτηρίης,
180 καὶ ἀπὸ ταύτης ἀΐσσει ἐς | τὸ ἧπαρ ἡ μὲν ἐπὶ πύλας
καὶ λοβόν, ἡ δ' ἐς τὸ ἄλλο ἑξῆς ἀφωρμήκει σμικρὸν
κάτωθεν φρενῶν. φρένες δὲ προσπεφύκασι τῷ ἥπατι,
ἃς οὐ ῥᾴδιον χωρίσαι. δισσαὶ δ' ἀπὸ κληῖδων, αἱ μὲν
ἔνθεν, αἱ δ' ἔνθεν ὑπὸ στῆθος ἐς ἦτρον· ὅπη δ' ἐντεῦθεν, οὔπω οἶδα. φρένες δὲ κατὰ τὸν σφόνδυλον τὸν
κάτω τῶν πλευρέων, ᾗ νεφρὸς ἐξ ἀρτηρίης, ταύτῃ
ἀμφιβεβηκυῖαι. ἀρτηρίαι δ' ἐκ τούτου ἐκπεφύκασιν
ἔνθεν καὶ ἔνθεν, ἀρτηρίης τρόπον ἔχουσαι. ταύτῃ πῃ
παλινδρομήσασα ἀπὸ καρδίης ἡ ἡπατῖτις ἔληγεν.
ἀπὸ δὲ τῆς ἡπατίτιδος διὰ τῶν νεφρῶν αἱ μέγισται
δύο, ἡ μὲν ἔνθεν, ἡ δ' ἔνθεν φέρονται μετέωροι,
πολυσχιδεῖς δὲ διὰ τῶν φρενῶν εἰσιν ἀμφὶ ταύταις,
καὶ πεφύκασιν ἄνωθεν δὲ φρενῶν, αὗται δὲ μᾶλλόν τι
ἐμφανέες.

Δύο δὲ παχεῖς τόνοι ἀπ' ἐγκεφάλου ὑπὸ τὸ ὀστέον
τοῦ μεγάλου σφονδύλου ἄνωθεν, καὶ πρὸς τοῦ στομάχου μᾶλλον ἑκατέρωθεν τῆς ἀρτηρίης παρελθὼν
ἑκάτερος εἰς ἑαυτὸν ἦλθεν ἴκελος ἑνί· ἔπειτα ᾗ σφόνδυλοι καὶ φρένες προσπεφύκασιν, ἐνταῦθα ἐτελεύτων·

9 Duminil after Erotian's gloss Δ37: ἐδιχώθη Μ.
10 Del. Littré as an intrusion from the following line.

there, it divides in two in the region of the last two ribs, one branch passing along one side of the vertebrae before being consumed, the other on the other side. The straight vessel that leads from the heart to the collar-bones is above (i.e. anterior to) the artery, just as in the loins it is beneath (i.e. posterior to) the artery; from this vessel one branch starts up towards the liver, to the *porta* and the lobe, and another starts off directly to the other part (sc. of the liver), a little beneath the diaphragm. The diaphragm is closely attached to the liver, and not easy to separate. Pairs (sc. of vessels) lead from the clavicles on both sides under the sternum into the abdomen. Where they go from there, I do not yet know. The diaphragm, located next the first vertebra below the ribs, curves around where the kidney comes off the artery; tubes (ureters) growing out of the kidneys on both sides have the structure of arteries. It is somewhere there that the hepatic vessel running back from the heart ends. From the hepatic vessel the two greatest branches are suspended through the diaphragm, one on each side; there are branches through the diaphragm around these which grow above the diaphragm, being somewhat more visible there.

Two wide bands[12] passing down from the brain beneath the bone of the great vertebrae and then closely along the oesophagus on each side of the artery, meet one another as if they were one, and then where the vertebrae and the diaphragm are attached, they end; other somewhat doubtful

---

[12] Cf. Erotian (T5) on this passage: "bands ($\tau \acute{o} \nu o \iota$): bodies stretching all through the tissues, such as vessels ($\phi \lambda \acute{\epsilon} \beta a \varsigma$), cords ($\nu \epsilon \hat{v} \rho a$), and the like."

καί τινες ἐνδοιαστοὶ πρὸς ἧπαρ καὶ σπλῆνα ἀπὸ
τούτου τοῦ κοινωνήματος ἐδόκεον τείνειν. ἄλλος τόνος
ἑκατέρωθεν ἐκ τῶν κατὰ κληΐδων σφονδύλων παρὰ
ῥάχιν παρέτεινεν, ἐκ πλαγίου σφονδύλου, καὶ τῆσι
πλευρῇσιν ἀπένεμεν ὥσπερ αἱ φλέβες· οὗτοι διὰ φρενῶν ἐς μεσεντέριόν μοι δοκέουσι τείνειν, ὅθεν δὲ αὗται
ἐξέλιπον. αὖτις ἔνθεν φρένες ἐξεπεφύκεσαν ἀπ᾽ ὅτου[11]
ξυνεχέες ἐόντες, κατὰ μέσον κάτωθεν ἀρτηρίης· τὸ
ἐπίλοιπον παρὰ σφονδύλους ἀπεδίδου, ὥσπερ αἱ φλέβες, μέχρι καταναλώθησαν πᾶν διελθόντες τὸ ἱερὸν
ὀστέον. |

182   11. Τὰ ὀστέα τῷ σώματι στάσιν καὶ ὀρθότητα καὶ
εἶδος παρέχονται· τὰ δὲ νεῦρα κάμψιν καὶ ξύντασιν
καὶ ἔκτασιν· αἱ δὲ σάρκες καὶ τὸ δέρμα πάντων
ξύνδεσιν καὶ ξύνταξιν. αἱ φλέβες διὰ τοῦ σώματος
κεχυμέναι πνεῦμα καὶ ῥεῦμα καὶ κίνησιν παρέχονται,
ἀπὸ μιῆς πολλαὶ διαβλαστέουσαι, καὶ αὕτη μὲν ἡ μία
ὅθεν ἦρκται[12] καὶ ᾗ τετελεύτηκεν οὐκ οἶδα· κύκλου γὰρ
γεγενημένου ἀρχὴ οὐχ εὑρέθη. τὰς δ᾽ ἀποφυάδας
αὐτῆς, ὅθεν ἤρτηνται καὶ ᾗ παύονται τοῦ σώματος,
καὶ ὡς ἡ μία ταύτῃσιν ὁμολογέει, καὶ ἐν ὁποίοις
τόποις τέτακται τοῦ σώματος, δηλώσω.

12. Περὶ μὲν γὰρ τὴν κεφαλὴν[13] κατὰ τὸ μέσον ἐκ
πλαγίου περίκειται ἡ φλέψ, αὐτὴ πλατεῖα καὶ λεπτή,
οὐ πολύαιμος· τῷ γὰρ ἐγκεφάλῳ κατὰ τὰς ἁρμονίας
ἐνερρίζωκε πολλὰ καὶ λεπτὰ φλέβια, καὶ περὶ τὴν
ὅλην κεφαλὴν τετάρσωται μέχρι τοῦ μετώπου καὶ τῶν
κροτάφων. αὐτὴ δὲ ἀπιθύνεται ἐς τοὔπισθεν τῆς κεφα-

ones seem to pass from this conjunction to the liver and the spleen. Another band emerging from the vertebrae at the level of the collar-bones runs along the spine on both sides, over the transverse processes of the vertebrae, and sends branches to the ribs like vessels; these seem to me to go through the diaphragm to the mesentery, where the vessels end. The diaphragm grows out again from where the bands, being continuous in the centre beneath the artery, continue on to disappear by giving off branches to the vertebrae, like vessels, until they are all consumed as they go through the sacrum.

11. The bones give the body rigidity, straightness and form; the cords allow it to bend, to contract and to extend; the muscles and the skin maintain everything's connection and order. Vessels coursing through the body provide breath, fluid and movement, many branching off from one: where this single vessel arises and where it ends, I do not know, for, just as in a circle, no beginning point may be found. About these branches of the single vessel—where they are attached, which part of the body they end in, how the single vessel relates to them, and in which parts of the body it is distributed—I will now explain.

12. Around the head in the middle a vessel runs at an angle; it is flat, narrow, and does not contain much blood; then in the brain next the sutures many narrow little vessels implant, and these form a network around the whole head as far as the forehead and the temples. The vessel itself goes straight to the back of the head on the outside

---

11 Duminil: ἀπὸ τοῦ M.

12 HIR: ἤρται M.

13 Ambrosianus Gr. C 85 sup. (XVI c.): τῆς -λῆς M.

λῆς ἐκτὸς παρὰ τῆς ἀκάνθης τὸ δέρμα· ἐντεῦθεν δὲ
καθεῖται παρὰ τὴν ἔξωθεν καὶ τὴν ἐντὸς φλέβα τῶν ἐν
τῆσι σφαγῆσι. πέρην δὲ τῆς ἀκοῆς ὑποσχισθεῖσα
ἀπὸ τῆς γένυος ἔξωθεν τείνει παχείη· ἀπὸ δὲ ταύτης ἐς
τὴν γλῶσσαν πολλαὶ καὶ λεπταί· πλὴν ἢ ὑπὸ τὴν
γλῶσσαν ἢ ὑπὸ τοὺς γομφίους. αὐτὴ δὲ παχείη διὰ
τῆς κληῖδος καθήκει ὑπὸ τὴν ὠμοπλάτην· καὶ ταύτῃ
ἀπ᾽ αὐτῆς βεβλάστηκε φλὲψ διὰ τοῦ νεύρου τοῦ ὑπὸ
τὴν ἐπωμίδα ἡ ἐπωμιδίη ὀνομαζομένη.[14] αὐτὴ δὲ αἱμόρρους καὶ αἱματώδης καὶ δυσίητος, ἢν ῥαγῇ ἢ
184 σπάσθῃ· τῇ μὲν γὰρ | αὐτὴν νεῦρον περιέχει πλατύ,
τῇ δὲ χόνδρος· τὸ δὲ μεταξὺ τῶν αὐτή τε ξυνέχει καὶ
ὑμὴν ἀφρώδης· ἀσάρκου οὖν ἐόντος τοῦ τόπου, ῥηϊδίως ῥήγνυται, οὐκ ἔχουσα περιφύεσθαι σάρκας· ἤν τε
ὑποδράμῃ τὸ αἷμα ἐς[15] τοῦτο τὸ μέρος, ἐπιτυχὸν
εὐρυχωρίης, οὐκ ἔχει ἀπαλλαγήν, ἀλλὰ σκληροῦται·
σκληρυνθὲν δὲ νοῦσον καὶ πόνον παρέχει· αὐτὴ μὲν
περαίνει ᾗ πρότερον εἶπον. ἡ δ᾽ ὑπὸ τὴν ὠμοπλάτην
ἀποβεβλάστηκεν ὑπὸ τοῖσι μαζοῖσι πυκνῇσι καὶ λεπτῇσι καὶ ἐπηλλαγμέναις φλεψί· καὶ διὰ τῆς ἐπωμίδος
παραλλάσσουσα τὸν χόνδρον, αὐτὴ νέρθεν ὑπονεμο-
μένη ἐς τὸν βραχίονα τείνει, τὸν μῦν ἐν ἀριστερᾷ
ἔχουσα. ἡ δὲ δεξιὴ σχίζεται αὐτὴ περὶ τὸν ὦμον καὶ
τοῦ ἀγκῶνος τὴν ἄνω μοῖραν· τὸ δ᾽ ἐντεῦθεν διαπέφυκε
τοῦ ἀγκῶνος ἑκατέρωθεν· ἔπειτα αὖθις παρὰ τὸν καρ-

[14] Littré after Cornarius' *vena humeralis appellata*: τῆς -ιδίης
-ομένης Μ.   [15] τ. α. ἐ. Linden: ἐς τὸ αἷμα Μ.

along the skin of the spine; from there it descends along the external and the internal vessel among the jugulars. Opposite the auditory meatus a wide vessel branching off from the direction of the jaw comes outward, and from it many narrow vessels lead to the tongue, as well as beneath the tongue and the molars. Another wide vessel descends along the collar-bone to beneath the shoulder blade; and at that point yet another vessel, called the acromial vessel, branches off from the one before and passes along the cord under the tip of the shoulder (acromion). This blood-vessel has a tendency to haemorrhage and to be difficult to treat if it ruptures or tears; for on the one side it is bordered by a flat cord, and on the other by a cartilage, the space between the two structures being occupied together by the vessel and a delicate membrane. Now since this region lacks any fleshy structure, it is easily torn, because it does not have muscles growing around it. And if blood runs down into this region, happening upon a wide open space, there is no opportunity for it to escape, but an induration arises, and as this hardening develops, it provokes disease and pain. This vessel continues where I have indicated above; the section going beneath the shoulder blade sends off numerous fine, intricate vessels beneath the breast; then running beyond the cartilage under the tip of the shoulder, it passes downward on the left into the upper arm together with its muscle. On the right the vessel divides in the area of the shoulder and the upper part of the elbow; from there it separates to pass on both sides of the elbow, and then continues on to the wrist of the hand; from

πὸν τῆς χειρός· ἐντεῦθεν δὲ ἤδη ἀπορρέουσα δι' ὅλου
ἀνὰ τὴν χεῖρα πολυπλανῶς ἐρρίζωται.

13. Ἡ δὲ παχείη[16] φλέψ, ἡ νεμομένη παρὰ τὴν
ἄκανθαν, διὰ δὲ τοῦ μεταφρένου, τῆς σφαγῆς καὶ τοῦ
βρόγχου, ἐμπέφυκεν ἐς τὴν καρδίην ἀφ' ἑωυτῆς φλέβα εὐμεγέθεα πολύστομον κατὰ τὴν καρδίην· ἐντεῦθεν
δὲ ἐς τὸ στόμα ἐσυρίγγωκεν, ἥπερ ἀρτηρίη διὰ τοῦ
πλεύμονος ὀνομάζεται, ὀλίγαιμός τε καὶ πνευματώδης. ἐν γὰρ εὐρυχωρίη καὶ ἀραιώσει σπλάγχνου πολ-
λαχῇ μὲν τοῦ πλεύμονος ὀχετεύεται, χονδρώδεις δὲ
τοὺς ἄλλους πεποίηται. διὸ καὶ ἤν τις ἐς ταύτας |
186 κατηνέχθη τὰς διόδους τοῦ πλεύμονος τῶν ἀηθῶν, ἢ
ἐν τῷ ποτῷ ἢ ἐν τῇ τοῦ πνεύματος καὶ αἵματος[17] διόδῳ,
ἅτε τῶν φλεβῶν τοιούτων ἐουσέων, καὶ τοῦ σπλάγχνου σπογγοειδέος πολύ τε ὑγρὸν δυναμένου δέξα-
σθαι ἄνω τε πεφυκότος· τῶν γὰρ ἐσιόντων ὑγρῶν
νόμος καθέστηκεν. ἔτι τε τὸ αἷμα διὰ τῶν φλεβῶν
τούτων οὐ πολὺ [καὶ][18] περισφίγγεται· καὶ οὐ ταχέως
χωρέον οὐκ ἐξάγει τὰ ἐμπίπτοντα· οὐχ ὑπεξαγομένων
δὲ αὐτῶν, ἀλλ' ἐμμενόντων, γίνεται πῶρος. οὗτος δὲ
ἀπελαύνεται[19] τὸ πλησιάζον τῆς τροφῆς, ταύτῃ[20] ἐούσης τῆς ἐσαγωγῆς τοῦ λάρυγγος καὶ πρὸς τὰ ἔξω.
ἐγκαταλαμβανομένων δὲ τῶν διόδων ὑπὸ τοῦ πώρου,
ταχύπνοιά τε καὶ δύσπνοια ἴσχει, τῶνδε μὴ δυνα-
μένων τὴν φύσιν ἐξεῖναι, τῇδε οὐκ εὐπόρως ἐχόντων

---

16 Harris: ἀρχαίη M.   17 καὶ αἵματος om. M in text,
add. in marg.   18 Om. HIR.   19 Potter: ἀπολύεται M.
20 HIR: ταύτης M.

there it breaks up entirely to send branches along the hand in all directions, which implant there.

13. The wide vessel[13] which is distributed beside the spine and along the back, the throat and the wind pipe sends off from itself into the heart a good sized vessel with many orifices at the heart. From there a pipe goes to the mouth which in its course through the lung is called "artery"; this contains little blood and much breath; indeed, in the spaciousness and porosity of the viscus it forms other cartilagenous conduits to many parts of the lung. Therefore, if something unusual is carried into these paths through the lung, either in the drink or in the passage of breath and blood, inasmuch as the vessels have the form they do and the viscus is very spongy and absorbtive, besides being situated in a high position (for the distribution of the liquids that enter takes place there; and furthermore the blood in these vessels is not very compressed and does not flow away quickly enough to lead off what arrives anew), these things are not disposed of but rather become impacted, so that a concretion is formed. This concretion excludes[14] the part of the nourishment that comes to it, since it is located at the entrance of the larynx and faces outward. As the passages are impeded by the concretion, rapid and difficult breathing comes on, and the passages are unable to expel the breath towards the outside, just as

[13] Harris's (p. 63) conjecture gives a better sense than M's "original vessel."

[14] Cf. Duminil, p. 233, n. 73: "Le sens de $\dot{\alpha}\pi o\lambda\acute{v}\epsilon\tau\alpha\iota$ n'est pas facile à déterminer."

κατασπᾶν. ἐκ δὴ τοιούτων αἱ τοιαῦται νοῦσοι γίνονται, οἷον ἄσθματα καὶ ξηραὶ φθινάδες. ἢν δὲ καὶ ἐν
αὐτοῖσι συνιστάμενον πλέον τὸ ὑγρὸν κρατήσῃ, ὥστε
μὴ δύνασθαι παχυνθὲν παγῆναι, καὶ σαπρὸν τὸν
πλεύμονα ποιέει καὶ τὰ πλησιάζοντα, καὶ γίνονται
ἔμπυοί τε καὶ φθινώδεες· γίνεται δὲ τὰ νοσήματα
ταῦτα καὶ δι' ἄλλας αἰτίας.

14. Ἐντεῦθεν δὲ ἡ φλὲψ αὕτη κατέχει τὸν πλεύμονα, καὶ διὰ τῶν λοβῶν τῶν δύο τῶν μεγάλων τῶν
ἔσω τετραμμένων ὑπὸ τὰς φρένας ἐπιτέταται τῇ ἀκάν-
θῃ λευκὴ καὶ νευρώδης, διαπέμπουσα φλέβια διὰ τοῦ
ἄλλου σώματος πυκνούμενα,[21] ἔντονα δέ, διά τε τῶν
σφονδύλων πυκνοῖσι φλεβίοισι ἐς τὸν νωτιαῖον μυε-
λὸν ἐγκισσεύεται. καὶ αἱ μὲν ἄλλαι φλέβες ἐν τῷ
σώματι τεταγμέναι, ἐκ πάντων τῶν μερῶν συντεί-
νουσαι ἐς τὴν ἄκανθαν, τὸ λεπτότατον καὶ εἰλικρι-
νέστατον ἑκάστη ξυνάγουσα, ἐνταῦθα ἐξερεύγεται.
αὕτη δὲ ἡ ἐπιτεταμένη διὰ | τῶν καθειμένων πλεκτα-
νέων ἐς ταὐτὸ ξυνάγει· ἐντεῦθεν δ' ἐς τοὺς νεφροὺς
ἀπερρίζωται παρὰ τὴν νόθην πλευρὴν λεπτῇσι καὶ
ἰνώδεσι φλεψί, καὶ τὸ ἐντεῦθεν συντείνουσα συμπε-
πύκνωται, ἔπειτα καὶ νενεύρωται πρὸς τὸν ἀρχόν,
πιέσασά τε τοὺς ξυναγωγέας ἐμπέφυκεν αὐτῷ· καὶ τήν
τε κύστιν καὶ τοὺς ὄρχιας καὶ τοὺς παραστάτας ἐρρί-
ζωκε πολυπλόκοισι λεπτῇσί τε καὶ στερεῇσι καὶ ἰνώδεσι φλεψί.

15. Ἐντεῦθεν αὐτῆς τὸ παχύτατον καὶ ἰθύτατον
ἀνάπαλιν τραπέν, προσκεκαύληκεν ὅπερ ἐστὶν αἰδοῖ-

they are not able easily to draw it down. Indeed, from such conditions diseases like asthmas and dry consumptions arise. And if yet more liquid collects in the passages and gains the upper hand, so that it cannot be thickened and congealed, it causes the lung and the neighbouring parts to putrefy, so that internal suppurations and consumptive states arise. These diseases can, however, also arise from other causes.

14. From there this same vessel enters the lung, and, passing through the two large lobes which face inward, reaches beneath the diaphragm to the spine, white and cord-like. It sends off small vessels in various directions pressed closely through the rest of the body, which are sinewy, and these pass along the vertebrae to form a complex of closely knit vessels by the spinal marrow. Other vessels which are spread out over the body run together from all its parts into the spine, where they discharge the finest and purest fluid which they have all collected. The vessel that passes through the complexes extending down the spine also comes together at the same place. From there it continues on to implant into the kidneys at the level of the false ribs in the form of fine, sinewy vessels, which, from that point onward join together and are compressed into a vessel; this vessel then becomes cord-like as it approaches the anus, where it presses on the sphincter muscles and implants. The bladder, testicles and *prostates* (i.e. epididymi) are all implanted with tangled, fine, solid, sinewy vessels.

15. From there the widest and straightest part of this vessel turns back and sends off a stalk in the form of the pe-

---

21 Potter: πυκναμένου M: πεπυκνωμένα Ermerins.

ον· ἐν δὲ τῇ ἀνακάμψει ἐνήρτηται[22] ἐς τὰ αὐτὰ ταῦτα,
καὶ διὰ τοῦ κτενὸς ἄνω ὑπὸ τὸ δέρμα τῆς γαστρὸς ἐκ[23]
τῆς φλεβὸς αὐτῆς ὡρμήκασι πρὸς τὰς κάτω φέρου-
σας, αἳ ἐς ἀλλήλας ἐσωχέτευνται· διαπεφύκασι δὲ καὶ
διὰ τοῦ αἰδοίου φλέβες παχεῖαι καὶ λεπταὶ καὶ πυκναὶ
καὶ καμπύλαι. τῇσι δὲ θηλείῃσιν αὐτὴ συντείνει ἔς τε
τὰς μήτρας καὶ ἐς τὴν κύστιν καὶ οὐρήθρην· ἐντεῦθεν
δὲ ἰθυπόρηκε, καὶ τῇσι γυναιξὶ μὲν περὶ τὰς μήτρας
ἤρτηται, τοῖσι δὲ ἄρσεσι περὶ τοὺς ὄρχιας ἐσπείρωται. διὰ ταύτην τὴν φύσιν αὐτὴ ἡ φλὲψ καὶ τὰ
γόνιμα πλεῖστα ξυλλαμβάνει· ἀπὸ γὰρ τῶν πλείστων
καὶ εἰλικρινεστάτων μερῶν τρεφομένη, ὀλίγαιμός τε
οὖσα καὶ κοίλη καὶ νευρόπαχυς καὶ πνευματώδης,
ἐντεινομένη τε ὑπὸ τοῦ αἰδοίου, τὰ καθειμένα ἐς τὴν
190   ἄκανθαν φλέβια βιάζεται, τὰ δὲ βιαζόμενα | ὥσπερ
σικύη ἐς ἑωυτὰ πάντα ἐκδιδοῖ ἐς τὴν ἄνω φλέβα·
συλλείβεται δὲ καὶ ἐκ τῶν ἄλλων μελῶν τοῦ σώματος
ἐς ταύτην· τὸ δὲ πλεῖστον, ὥσπερ εἴρηται, ἀπὸ τοῦ
μυελοῦ τοῦτο συναλίζεται. ἡ δὲ ἡδονὴ ἐν τῷ καιρῷ
τούτῳ παραγίνεται τῆς φλεβὸς ταύτης πληρευμένης
τῆς γονῆς· ἐωθυίης οὖν τὸν ἄλλον χρόνον ἐφαίμου τε
εἶναι καὶ πνευματώδεος, πληρευμένης τε καὶ θερμαι-
νομένης, καὶ ξυρρέοντος κάτω τοῦ σπέρματος,
περισφίγγει τὰ ἐν ἑωυτῇ. τὸ δὲ πνεῦμα τὸ ἐνεὸν καὶ ἡ
παροῦσα βίη καὶ ἡ θερμότης καὶ τῶν φλεβίων παντα-
χόθεν ἡ ξυντονίη γαργαλισμὸν ἐμποιέει.

16. Ἐκείνη δὲ ἡ ἀφ' ἑωυτῆς διέβλαστε, διά τε τοῦ

nis, also attaching, in the course of its curve, to the same parts. Out of this vessel other vessels branch off upwards through the pubic region beneath the skin of the abdomen to meet the vessels coming down, and these anastomose with one another. In the penis, too, there are vessels, both wide and narrow, that are curved and run close together. In the female, this (sc. main) vessel runs to the uterus, the bladder and the urethra. From that point it goes straight on, in women to be suspended around the uterus, in men to be coiled around the testicles. Because of this structural arrangement, it is this vessel that collects most of the seed: for being nourished by the most copious and purest components of the body, while it itself is bloodless, hollow, thick-corded and filled with breath, when it is pulled tight by the penis, the small vessels which branch off into the spine are compressed, and as they are all compressed in the manner of a cupping glass they secrete into the vessel lying above them; an influx into the vessel also occurs from various other parts of the body, the largest amount, as has been indicated, being collected from the marrow. The pleasure felt at this time arises from the vessel—used at other times to contain some blood and breath-like material —being filled with seed. When the vessel becomes full and warm, as the semen flows down and collects together in it compressing its contents, the breath in it, being subjected to the force present, the warmth, and the tension of the small vessels on all sides, produces a titillation.

16. The vessel which sends off branches from itself[15]

[15] See chapter 13 above.

---

[22] Duminil: ἐνῆρται M.   [23] ἐκ Littré: καὶ M.

μεταφρένου καὶ τῆς σφαγῆς παρὰ τὴν ἄκανθαν νεμομένη, πολλοῖσιν φλεβίοισι τὰς πλευρὰς διαπέπλοχε·
καὶ τοὺς σφονδύλους διὰ τῶν σαρκῶν ἐπηλλαγμένως
ξυμπεπύκνωκεν, ὥστε τρόφιμός τε καὶ ἔναιμος εἶναι.
αὐτὴ δὲ παρὰ τὸν γλουτὸν ἵεται, διὰ τοῦ μυός, ὑπὸ τῷ
μηρῷ ὑποβρυχίη· πρὸς δὲ τοῦ γλουτοῦ τῇ κοτυληδόνι
τοῦ μηροῦ παρὰ τὴν κεφαλὴν ἐστετρύπηκεν φλεβί,
ἥπερ ἀναπνοὴν τῷ μηρῷ παρέχει· ἐκπερᾷ τοῦ μηροῦ
παρὰ τὴν πρὸς τὸ γόνυ καμπήν· ἑτέρην δὲ παρὰ τὸν
βουβῶνα καθῆκεν πυκινόρριζον καὶ δυστράπελον. ἡ
δὲ διὰ τοῦ μυὸς τείνουσα περί τε τὸ γόνυ ἐσπείρωται,
καὶ διὰ τοῦ ὀστέου τοῦ κνημιαίου ἄκρου σεσυρίγγωκε
φλέβα, ἣ τρέφει τὸν μυελόν, καὶ ἐξωχέτευται διὰ τοῦ
νερτάτου τοῦ κνημιαίου, παρὰ τὴν ἔνδεσιν τοῦ ποδός.
αὐτὴ δὲ διὰ τῆς ἐπιγουνίδος ἐς τὸ ἐντὸς διὰ τῆς
κνήμης τοῦ μυὸς βρυχίη τέταται, καὶ ἐμπέπλεχε διὰ
192 τοῦ σφυροῦ ἐντὸς παχέη | καὶ ἔναιμος, καὶ ἐνταῦθα
περὶ τὸ σφυρὸν καὶ τὸν τένοντα δυσκρίτους φλέβας
μεμήρυκεν.

17. Αὐτὴ δὲ ὑποδεδράμηκε κάτωθεν τοῦ ποδὸς ὑπὸ
τὸν ταρσόν. καὶ ἐνταῦθα διαπλέξασα καὶ ἐς τὸν μέγαν
δάκτυλον ἐνερείσασα διπλὴν ἔναιμον φλέβα, ἄνωθεν
ὑπὸ τὸ δέρμα ἐκ τοῦ ταρσοῦ ἀνακέκαμπται, καὶ πέ-
φανται παχυνθεῖσα παρὰ τὸ ἐκτὸς τοῦ σφυροῦ, καὶ
νέμεται ἄνω παρὰ τοῦ ἀντικνημίου τὴν ἀντιβεβλημέ-
νην κερκίδα· παρὰ δὲ τὴν γαστροκνημίην οἷον σφεν-
δόνην πεποίηται· τὸ δ᾽ ἐντεῦθεν τέταται παρὰ τοῦ

distributed through the area of the back and the jugular along the spine interweaves the ribs with many small vessels. It continues into the vertebrae crosswise through the muscles, being well nourished and filled with blood. It also extends toward the buttock, through the muscle, and disappears under the femur. At the buttock near the acetabulum of the femur it sends off a branch through an opening on the femur's head, which supplies the femur with respiration. The vessel then comes out of the femur towards the bend of the knee, sending another vessel into the groin which has closely knit roots and is difficult to turn.[16] The branch that passes through the muscle (sc. of the ham) divides at the knee; along the bone at the front of the lower leg (i.e. the tibia) it sends off a vessel that acts as a conduit supplying the marrow with nourishment, and which emerges along the ankle at the articulation of the foot. A branch also passes by the knee-cap to the interior, deep inside the muscle of the calf, and then winds around the ankle on the inside, wide and filled with blood, and there vessels difficult to distinguish twine around the ankle and the Achilles tendon.

17. Another vessel runs down below the foot on the inferior surface of the flat of the foot, where a double blood-vessel forms a complex and attaches to the great toe. This vessel then bends its course under the skin upwards from the flat of the foot, and widening out makes its appearance at the outside of the ankle, whence it proceeds upward along the fibula opposite the tibia, forming something like a sling along the gastrocnemius muscle. From there it con-

---

[16] A difficult term; cf. Fuchs: "schwer aus ihrer Lage zu bringen."

γούνατος τὸ ἐντός· ἐπιβέβληκε δὲ καὶ τῇ ἐπιγουνατίδι
φλέβας, καὶ κατὰ τὸ ἐντὸς τῆς ἐπιγουνατίδος ἐπίκοι-
λον ἐμπέπλεχε φλέβα· ἤν τις ἦν[24] πονήσῃ, τάχιστα
συνάγει χολώδεα ἰχῶρα. διώρμηκε δὲ αὐτὴ κατὰ τὸ
ἐντὸς καὶ κοῖλον τοῦ γούνατος· ἀποκεκάρπωκε δὲ καὶ
εἰς τὰς ἰγνύας πολυπλόκους φλέβας, αἱ ἐντεῦθεν ἀνα-
τείνουσαι παρὰ τὰ ὑποκάτω νεῦρα τοῦ μηροῦ [καὶ][25]
κατερρίζωνται ἔς τε τοὺς ὄρχιας καὶ τὸν ἀρχόν, καὶ
περὶ τὸ ἱερὸν δὲ ὀστέον λελεπτυσμέναι ἠνωμέναι περι-
τέτανται.

18. Ἡ δὲ ἀφιγμένη παρὰ τὸ ἐντὸς τοῦ γούνατος
ἄνω παρὰ τοῦ μηροῦ τὸ ἐντὸς ἀνίεται[26] ἔς τε τὸν
βουβῶνα καὶ διὰ τοῦ ἰσχίου πέρην πρὸς τὴν ἄκανθαν
καὶ τὴν ψύαν ἐκτὸς ἔχουσα, παχεῖα καὶ πλατεῖα καὶ
ἔναιμος, ἄνω ὤρεκται πρὸς τὸ ἧπαρ· καὶ διακραίην
ἐκφύσασα ἔναιμον, κατέχει ἐς τὸν νεφρὸν <καὶ>[27] τὸν
δεξιὸν λοβὸν τὸν ἡπατιαῖον. αὕτη δὲ ὑποκάτω τὰ τοῦ
ἥπατος ὑπονησαμένη,[28] ἀπέσχισται ἐς φλέβα πα-
χέην· ἡ δ᾽ ὑποκλιθεῖσα[29] ἐσπέφυκεν ἐς τὸ παχὺ τοῦ
ἥπατος· καὶ τὸ μὲν αὐτῆς ἐπιπολάζον ἐπὶ τοῦ σπλάγ-
194 χνου πέφυκεν, ἐν ᾧπερ ἡ | χολή ἐστι, καὶ[30] πολύρριζος
διὰ τοῦ ἥπατος πεπλεκτανωμένη· τὸ δὲ διὰ τῶν ἐντὸς
αὐτοῦ ὠχέτευται. δύο δὲ ἐκπεπλώκασι φλέβες μεταξὺ
δύο λοβῶν τῶν πλατέων· καὶ μία μὲν διὰ τῶν κορυφῶν
καὶ τοῦ δέρματος διασχοῦσα ἐκ τοῦ ὀμφαλοῦ ἀνῆκται·
ἡ δ᾽ ἑτέρη πιέσασα ἐς τὴν ἄκανθαν καὶ ἐς τὸν νεφρὸν

---

[24] Ermerins: ἤν τις εἰ Littré: ἤν τις Μ.

tinues to the inside of the knee, sending off branches on to the knee-cap; along the inside of the knee-cap winds a hollow vessel which, if strained, quickly collects bilious serum. The vessel also passes through the hollow of the knee on the inside to send off into the ham very tangled vessels, which extend on from there along the deeper cords of the thigh to implant into the testicles and the anus, and thinning out in the region of the sacrum meet in an extended complex.

18. The vessel that arrives (sc. from below) at the knee on the inside continues upward along the inside of the thigh to the groin, passes by the hip towards the back to the psoas muscle on the outside—being at this point wide, flat and filled with blood—, and goes on to the liver above, where it forks to send off blood-vessels towards the (sc. right) kidney and the right lobe of the liver. As the latter vessel passes under the area of the liver, it sends off a branch into a wide vessel which turns aside into the wide part of the liver, and divides in turn sending one branch over the surface of the viscus to the region of the gall bladder, where it develops many roots that entwine through the liver, and another branch that forms a conduit to the inner parts of the liver. Two other vessels take form between the two flat lobes (sc. of the liver): one passes via the prominences and the skin to emerge out of the navel; the other one passes closely along the spine and the kidney to

---

25 Del. H.

26 Foes in note 68: ἀνιοῦνται M.   27 Littré.

28 Linden after Erotian and Galen: -νέμησ- M.

29 Duminil: ὑποκανθ. M.

30 Littré: τε M.

ἠγκυροβόληται ἐς τὴν κύστιν τε καὶ τὸ αἰδοῖον. ἐκ δὲ
τοῦ ἰσχίου ἀρχομένη ἀνιέναι ὑπὸ τὸ ἦτρον, πολλὰς
ἀπεπλάνησε φλέβας· καὶ τάς τε πλευρὰς καὶ τοὺς
σφονδύλους ἐνεκρίκωσε πρὸς τὴν ἄκανθαν, καὶ τάς τε
παραφυάδας ἐνεφλεβοτόμησε, καὶ τὰ ἔντερα καὶ τὴν
νηδὺν ἐνειλίξατο. καὶ αἱ μὲν ἀπὸ τοῦ ἤτρου ἔς τε τοὺς
μαζοὺς καὶ ὑπὲρ ἀνθερεῶνα καὶ τὰς ἀκρωμίας ἐπο-
ρεξάμεναι κατεπλάκησαν· ἡ δ' ἀφιγμένη παρὰ τὸ
παχὺ τοῦ ἥπατος καὶ ἀποσυριγγώσασα τὴν χολὴν
ἄνω ὑπὸ τὴν ἄκανθαν νέμεται διὰ τῶν φρενῶν ὁδὸν
ποιησαμένη. ἡ δ' ἐκ τῶν ἀριστερῶν φλὲψ τὰ μὲν ἄλλα
τὴν αὐτὴν φύσιν ἐρρίζωται τῇ ἐν τοῖσι δεξιοῖσι, ἐκ
τῶν ἀριστερῶν ἐς τὸ ἧπαρ ἀνιοῦσα οὐκ ἐκβάλλει, ἀλλ'
ἐς τὸν σπλῆνα ἐμπέφυκε κατὰ τὴν κεφαλὴν τὴν ἐν τῷ
πάχει αὐτοῦ· ἐντεῦθεν δὲ κατεδύσατο ἐς τὸ ἐντός,
<καὶ>[31] ἠραχνίωκε τοῦ σπληνὸς ἐναίμοισι φλεβίοις· ὁ
δὲ ὅλος ἐκ τοῦ ἐπιπλόου αἰωρεῖται τοῖσιν ἐξ ἑωυτοῦ
φλεβίοις ἐναιματώσας αὐτό. αἱ δ' ἀπὸ τῆς κεφαλῆς
τοῦ σπληνὸς πρὸς τὴν ἄκανθαν ἐγχρίμπτουσαι διὰ
τῶν φρενῶν διωρμήκασιν.

19. Ἐντεῦθεν δὲ κάτω καὶ ἡ δεξιὴ καὶ ἡ ἀριστερὴ
ὑπὸ τὸν πλεύμονα ἐλήλαται· αἱ δὲ ἐναίμονες ἐοῦσαι
ὑπὸ αὐτὸν [καὶ][32] ἐξοχετεύονται ἐς | αὐτόν. ὀλίγαιμοι
δὲ καὶ λεπταὶ αἱ ἀπὸ τοῦ πλεύμονος ἔσωθεν γενόμεναι
τῇ φύσει ἀραιοῦ ἐόντος, ἐς τὴν καρδίην, ἅτε ὑπ' αὐτοῦ
ἐξαθελγόμεναι, ἐγκεχαλίνωνται περὶ τὰ ὦτα αὐτῆς,
καὶ ἐς τὰ κοῖλα τὰ ἐντὸς διερρυήκασιν. ἐμβάλλουσι δὲ

anchor itself to the bladder and the penis. The latter then turns back from the hip and moves upward through the abdomen to distribute many branches: it encloses the ribs and the vertebrae in the form of a ring against the spine, it sends off many branching offshoots, and it winds around the intestines and the stomach. The offshoots from the abdomen extend to the breasts and over the neck and the tips of the shoulders to entwine them. The branch (sc. mentioned above) that arrives at the wide part of the liver also sends off a conduit to the gall bladder which leads upward close to the spine through the diaphragm, thereby providing a passage. The vessel out of the left parts implants in general according to the same scheme as the one on the right, except that in its passage upward on the left it does not send off a branch to the liver, instead attaching to the broad surface of the spleen at its head, whence it both sends branches down into the interior of the viscus and covers it on the surface with a venous network. The spleen as a whole is suspended from the omentum by vessels coming out of it, and it in turn provides the omentum with blood. Vessels leading from the head of the spleen converge at the spine and pass through the diaphragm.

19. From there both the right and the left vessels move beneath the lung, where, being filled with blood, they send down conduits into it. Vessels arising inside the lung, which become bloodless and narrow as they are drained off by the lung due to its naturally loose texture, pass to the heart where they take the form of a horse's bit around its auricles, before flowing through into its internal cavities. Both the preceding vessels and these later ones send

---

31 Froben.   32 Del. R.

καὶ αἱ πρότεραι καὶ αὗται ἐς αὐτήν· ἐν γὰρ στενο-
χωρίῃ τῆς διόδου ἐνίδρυται ὡς ἐκ παντὸς τοῦ σώματος
τὰς ἡνίας ἔχουσα· διὸ καὶ παντὸς τοῦ σώματος περὶ
τὸν θώρηκα μάλιστά ἐστιν ἡ αἴσθησις. καὶ τῶν χρωμάτων αἱ μεταβολαὶ[33] γίνονται, ταύτης ἀποσφιγγού-
σης τὰς φλέβας καὶ χαλώσης· χαλώσης μὲν οὖν,
ἐρυθρὰ τὰ χρώματα γίνεται καὶ εὔχροα καὶ διαφανέα·
συναγούσης δέ, χλωρὰ καὶ πελιδνά· τὰ τοιαῦτα δὲ
παραλλάσσει ἐκ τῶν παρεόντων ἑκάστῳ χρωμάτων.

33. HIR: μεταλλαὶ M.

branches into the heart, which is situated in a narrow space as if it were holding reins from every part of the body: for this reason sensations from the whole body are concentrated about the chest. Also changes of skin colour follow as the heart either constricts or dilates the vessels: when it dilates them, the complexion becomes rosy, fine and translucent; when it constricts them, it is pale and livid—these effects varying, of course, according to the individual's underlying colouring.

HEART

INTRODUCTION

Heart, which was apparently unknown to Erotian, has left three possible traces in antiquity. First, Galen describes an experiment to demonstrate that swallowed liquids pass partly to the lungs which is identical in substance, if not wording, to the one contained in chapter 2 of *Heart*:

ἀλλ' εἰ καὶ ζῷον ὅ τι ἂν ἐθελήσῃς διψῆσαι ποιήσεις ὡς κεχρωσμένον ὕδωρ ὑπομεῖναι πιεῖν, εἰ δοίης εἴτε κυανῷ χρώματι χρώσας εἴτε μίλτῳ, εἶτ' εὐθέως σφάξας ἀνατέμοις, εὑρήσεις κεχρωσμένον τὸν πνεύμονα. δῆλον οὖν ἐστιν ὅτι φέρεταί τι τοῦ πόματος εἰς αὐτόν.[1]

Second, Plutarch in a discussion of those who criticize Plato for saying that drink goes to the lung cites Hippocrates among the holders of the same view:[2] the only passage in the Hippocratic Collection which explicitly champions this belief is in *Heart* 2, although it may be implicit in other treatises: e.g. *Diseases II* 52 (Loeb vol. 5, 286); *Diseases III*

[1] P. De Lacy (ed.), *Galen on the Doctrines of Hippocrates and Plato*, Corpus Medicorum Graecorum V 4, 1, 2, Berlin, 1978–84, p. 538 = Galen vol. 5, 719.

[2] Plutarch, *Symposium*, 7, 1 (699c).

15 (Loeb vol. 6, 48). Hesychius includes four glosses on three words, some or all of which may refer to *Heart*:

ἀνακωχεῖν· ἀναχωρεῖν;[3] cf. *Heart* 11.
λάπτει· ἀναλαμβάνει, πίνει;[4] cf. *Heart* 1.
λαπτόμενος ἢ λάπτων· ἀναλίσκων. ἀπὸ τοῦ λά-
πτειν;[5] cf. *Heart* 1.
νηδύος· γαστρός, κοιλίας. καὶ νηδύοισι· ἐντέροις;[6]
cf. *Heart* 11.

The treatise *Heart* is a concise monograph on the human heart, arranged as follows:

  1: Shape of the heart and its position inside the pericardium.
2–3: Theory that drink passes mostly to the stomach but partly to the lung; experimental proof; the epiglottis; role of moisture and air in the chest.
4–6: The ventricles, their relationship to one another, structure and function of each individually.
  7: The orifices of the ventricles and the vessels attached to them.
  8: The auricles, their structure and function.
  9: Vessels bringing respiration to the ventricles, purpose.
10–12: Membranes of the heart, structure and function; localization of intelligence in the left ventricle.

---

[3] K. Latte (ed.), *Hesychii Alexandrini Lexicon*, Copenhagen, 1953–66, vol. 1, 153.

[4] Hesychius vol. 2, 572.

[5] Hesychius vol. 2, 573.

[6] Hesychius vol. 2, 709.

On many substantive points *Heart* differs significantly from the rest of the Hippocratic writings:

(i) Although a correlation between bodily structure and function is commonplace throughout the Collection, generally writers derive function from structure, which is taken as a given;[7] *Heart*, by contrast, posits a teleological relationship making function the purpose for which structures are created, and introduces a conscious "good handworker" (ch. 8).

(ii) Whereas the level of anatomical knowledge evident elsewhere in the Collection seems explainable as the product of clinical observations, animal investigations, and chance views into the human body, the intimate acquaintance with the interior of the human heart displayed in this treatise speaks strongly for the practice of human dissection.

(iii) Many clinical accounts in the Hippocratic writings implicitly assume the brain to be the centre of mental function (e.g. *Diseases III* 2–4), while others make this point directly (e.g. *Sacred Disease* 17): *Heart* is unique in localizing the γνώμη (understanding), as the ruling part of the soul, in the heart.[8]

(iv) In several points of anatomical nomenclature *Heart* deviates from general Hippocratic usage whereby φλέψ and φλέβιον are applied to any blood vessel,

[7] B. Gundert, "Parts and their Roles in Hippocratic Medicine," *ISIS* 83 (1992), 453–65, esp. p. 465.

[8] B. Gundert, "Soma and Psyche in Hippocratic Medicine," in J. P. Wright and P. Potter (edd.), *Psyche and Soma. Physicians and Metaphysicians on the Mind—Body Problem from Antiquity to Enlightenment*, Oxford, 2000, pp. 13–35, esp. 20–31.

on occasion even to other tubes such as the ureter: in *Heart* these terms are restricted to veins. ἀρτηρίη and ἀορτή seem in the few instances where they occur elsewhere in the Hippocratic writings to be interchangeable designations for the bronchial tubes: in *Heart* they are applied only to arteries—with no implication as to whether these contain air or blood. The gastro-intestinal tract is generally called the κοιλίη (cavity) in the Hippocratic Collection, with the two main divisions ἄνω κοιλίη (upper cavity) and κάτω κοιλίη (lower cavity): these names are displaced in *Heart* by terms which, while not unknown in the rest of the Collection, here take on a primary role that is exceptional, e.g. στόμαχος (oesophagus), γαστήρ (stomach), ἔντερον (intestine), νηδύς (gut, in general).

Whether these peculiarites in which the treatise *Heart* differs from the rest of the Hippocratic writings are the result of a different medical or intellectual tradition, of a later date, or of a different geographical or cultural milieu is unknown.[9]

Littré, in his introduction to the text (vol. 9, 76–78), correctly emphasizes the contrast between the author's accuracy in describing the heart's structure, and the rudimentary state of his physiological understanding. Despite an acquaintance with certainly two, but probably four, of the cardiac valves, *Heart's* account of their function makes them at most moveable gates which in some instances

[9] For a thorough and judicious account of the scholarly discussion on these questions see Duminil's introduction and notes.

close tightly, in others allow some flow of blood or air, or even of both simultaneously in opposite directions. Thus, although we may say that the author has seen and described the valves, we must hesitate to say that he has "discovered" them, if by that term we mean that he has grasped their functional significance. Any doubt in this matter can easily be set to rest by a comparison with Erasistratus' account of the four heart valves and their function as reported by Galen.[10] *Heart* belongs, in the final analysis, not to this Hellenistic world of mechanism,[11] but still, in spite of the differences discussed above, to a Hippocratic world view about materials and forces in which forces like the inherent attractiveness of a structure's being hollow, empty, or hot explain related bodily actions.

Besides finding a place in the collected Hippocratic editions and translations, and in two obscure Renaissance works by Jacobus Horstius (Frankfurt, 1563) and Joannes Nardius (Bologna, 1656), *Heart* has enjoyed considerable attention from twentieth century scholars: e.g.

> F. C. Unger, "Liber Hippocraticus Περὶ καρδίης," *Mnemosyne*, N. S. 51 (1923), 1–101.
>
> F. Kudlien, "Poseidonios und die Ärzteschule der Pneumatiker," *Hermes* 90 (1962), 419–29. (= Kudlien)
>
> C. R. S. Harris, *The Heart and the Vascular System in*

---

[10] De Lacy, p. 396 = Galen vol. 5, 548–50.

[11] For a description of Ctesibius' contemporary four-valved water pumping machine which may have influenced Erasistratus, see Vitruvius, *De Architectura*, 10, 7, and Hero Alexandrinus, *Pneumatica*, 1, 28.

Ancient Greek Medicine, Oxford, 1973, pp. 83–96. (=
 Harris)

I. M. Lonie, "The Paradoxical Text 'On the Heart,'"
 Medical History, 17 (1973), 1–15 and 136–53. (=
 Lonie)

P. Manuli and M. Vegetti, *Cuore, sangue e cervello*, Mi-
 lan, 1977, pp. 101–12 and 219–33.

It has appeared in two English translations made by F. R.
Hurlbutt Jr. ("PERI KARDIES. A Treatise on the Heart
from the Hippocratic Corpus: Introduction and Translation," *Bulletin of the History of Medicine* 7 (1939), 1104–
1113) and I. M. Lonie (in G. E. R. Lloyd, (ed.) *Hippocratic
Writings*, Harmondsworth, 1978, pp. 347–351), and recently in a Budé edition by M. P. Duminil (=Duminil).

The present edition is based on a collation from microfilm of the sole independent witness for the Greek text,
the manuscript V.

ΠΕΡΙ ΚΑΡΔΙΗΣ[1]

1. Καρδίη σχῆμα μὲν ὁκοίη πυραμίς, χροιὴν δὲ κατακορὴς φοινικέα. καὶ περιβέβλέαται χιτῶνα λεῖον· καὶ ἔστιν ἐν αὐτῷ ὑγρὸν σμικρὸν ὁποῖον οὖρον, ὥστε δόξεις ἐν κύστει τὴν καρδίην ἀναστρέφεσθαι. γεγένηται δὲ τούτου ἕνεκα, ὅκως θ[άλλεται]ρώσκῃ μὲν ὡς[2] ἐν φυλακῇ· ἔχει δὲ τὸ ὕγρασμα ὁκόσον μάλιστα καὶ πυρευμένη ἄκος. τοῦτο δὲ τὸ ὑγρὸν διουρέει ἡ καρδίη [πίνουσα, ἀναλαμβανομένη καὶ ἀναλίσκουσα][3] λάπτουσα τοῦ πνεύμονος τὸ ποτόν.

2. Πίνει γὰρ ὥνθρωπος τὸ μὲν πολλὸν ἐς νηδύν· ὁ γὰρ στόμαχος ὁκοῖον χόανος, καὶ ἐκδέχεται τὸ πλῆθος καὶ ἅσσα προσαιρόμεθα· πίνει δὲ καὶ ἐς φάρυγγα, τυτθὸν δὲ οἷον καὶ ὁκόσον ἂν λάθοι διὰ ῥύμης ἐσρυέν· πῶμα γὰρ ἀτρεκὲς ἡ ἐπιγλωσσίς, κἂν διήσει μεῖζον ποτοῦ οὐδέν. σημήϊον τοῦτο· ἢν γάρ τις κυάνῳ ἢ μίλτῳ φορύξας ὕδωρ δοίη δεδιψηκότι πάνυ πιεῖν, μάλιστα δὲ συί, τὸ γὰρ κτῆνος οὐκ ἔστιν ἐπιμελὲς οὐδὲ φιλόκαλον, ἔπειτα δὲ εἰ ἔτι πίνοντος ἀνατέμοις τὸν λαιμόν, εὕροις ἂν τοῦτον κεχρωσμένον τῷ ποτῷ.

---

[1] Holkhamensis Gr. 92 (XVI c.): κραδίης V.

58

HEART

1. The heart, in its shape, is like a pyramid, in colour, deep red. It is enclosed in a smooth tunic which contains a little urine-like liquid, so that you might imagine that the heart dwells in a bladder. This is so arranged in order that it may beat vigorously in safety, having a quantity of moisture just sufficient to protect it against being ignited. This liquid the heart passes through like urine after lapping up drink from the lung.

2. A person takes drink mostly into his gut, for the oesophagus, being shaped like a funnel, receives the greatest amount of what we consume; but he also takes some drink into his larynx, although just a little and only as much as escapes notice in flowing in through the narrow opening: for the epiglottis, being a close cover, will not let more of the drink pass through. Here is proof: if someone were to mix water with blue or red colouring and give it to a very thirsty animal to drink—especially a pig, as this animal is neither careful nor elegant—and then, while the animal was still drinking, you were to cut its throat, you would find this (i.e. the trachea) coloured by the drink. But this opera-

---

2 F. E. Kind in Kapferer / Sticker, part 16, p. XXX: θάλλεται ῥωσκημένως V.   3 Del. Kudlien, p. 425, n. 1; these three words are glosses taken over from Hesychius.

ἀλλ' οὐ παντὸς ἀνδρὸς ἡ χειρουργία· οὔκουν ἀπιστη-
82 τέον ἡμῖν περὶ τοῦ ποτοῦ, εἰ εὐτρε|πίζει τὴν σύριγγα
τῷ ἀνθρώπῳ. ἀλλὰ πῶς ὕδωρ ἀναιδὲς ἐνορουον[4] ὄχλον
καὶ βῆχα παρέχει πολλήν; οὕνεκα, φημί, ἀπάντικρυ
τῆς ἀναπνοῆς φέρεται. τὸ γὰρ διὰ τῆς ῥύμης[5] ἐσρέον,
ἅτε παρὰ τοῖχον ἰόν, οὐκ ἐνίσταται τῇ ἀναφορῇ τοῦ
ἠέρος, ἀλλά τινα καὶ λείην ὁδὸν οἱ παρέχει ἡ ἐπί-
τεγξις· τοῦτο δὲ τὸ ὑγρὸν ἀπάγει τοῦ πνεύμονος ἅμα
τῷ ἠέρι.

3. Τὸν μὲν οὖν ἠέρα χρή, γενόμενον θεραπείην,
ἀνάγκη ὀπίσω τὴν αὐτὴν ὁδὸν ἐκβάλλειν ἔνθεν ἤγα-
γε· τὸ δ' ὑγρόν, τὸ μὲν εἰς τὸν κουλεὸν αὐτῆς ἀποπτύει,
τὸ δ' αὖ ξὺν τῷ ἠέρι θύραζε χωρέειν ἐᾷ· ταύτῃ καὶ
διαίρει τὸν οὐραχόν,[6] ὁκόταν παλινδρομέῃ τὸ πνεῦμα.
παλινδρομέει δὲ κατὰ δίκην· οὐ γὰρ ἔστιν ἀνθρώπου
φύσιος τροφὴ ταῦτα· κῶς γὰρ ἀνθρώπου τροφὴ ἄνε-
μος καὶ ὕδωρ τὰ ὠμά; ἀλλὰ μᾶλλον τιμωρίη ξυγ-
γενέος πάθης.

4. Περὶ δὲ οὗ ὁ λόγος, ἡ καρδίη μῦς ἐστι κάρτα
ἰσχυρός, οὐ τῷ νεύρῳ, ἀλλὰ πιλήματι σαρκός. καὶ δύο
γαστέρας ἔχει διακεκριμένας ἐν ἑνὶ περιβόλῳ, τὴν μὲν
ἔνθα, τὴν δὲ ἔνθα· οὐδὲν δ' ἐοίκασιν ἀλλήλῃσιν. ἡ μὲν
γὰρ ἐν τοῖσι δεξιοῖσιν ἐπὶ στόμα κέεται ὁμιλέουσα τῇ
ἑτέρῃ φλεβί, ἡ δὲ δεξιή φημι τῶν ἐν λαιοῖς· ἡ γὰρ

---

[4] Linden after Foes' note 8: ἔνουρον V.   [5] Linden after
Aemilius Portus' emendation reported in Foes vol. 2, [1345]:
ὁρμῆς V.   [6] Parisinus Gr. 2255 (XIV c.): οὐρανὸν V.

tion might not be to every man's taste; nevertheless, our opinion concerning what is drunk is not to be dismissed, namely that in the human being it lubricates the windpipe. But how does it come that liquid which rushes in recklessly provokes such great trouble and coughing? Because, as I say, it collides with the breath coming out. What, on the other hand, flows in through the narrow opening, inasmuch as it passes along the wall, is not impeded by the air passing upward, but rather its moistening effect provides the air with a kind of smooth path; this moisture the person sends up from his lung along with the air.

3. Now whereas a person must of necessity expel the air, after it has fulfilled its office, back through the same passage by which he drew it in, the moisture he partly spits out into the sheath of the heart, and partly allows to go back with the air to the outside, the breath in this process raising the extremity (sc. of the epiglottis) as it flows back. It flows back according to the normal course of events, for such substances are not nourishing to a man's nature—indeed, how could air and water be human nutriments, crude as they are? Rather, they are the counterbalance to an inborn disposition.[1]

4. The subject of this discourse, the heart, is a muscle of particular strength, of flesh which is not cordlike, but compressed. It has two ventricles divided from each other in one covering, one on the one side, the other on the other. These ventricles do not resemble one another at all, for the one in the right parts—the right I mean of the parts on the

[1] I.e. to the heat of the heart; see chapter 5 below.

ΠΕΡΙ ΚΑΡΔΙΗΣ

πᾶσα καρδίη τούτοισι τὴν ἕδρην ἐμπεποίηται. ἀτὰρ
ἥδε καὶ πάμπαν εὐρυκοίλιος καὶ λαγαρωτέρη πολλῷ
τῆς ἑτέρης, οὐδὲ τῆς καρδίης νέμεται τὴν ἐσχατιήν,
ἀλλ᾽ ἐγκαταλείπει τὸν οὐραχὸν στερεόν, καί⁷ ἐστιν
84 ὥσπερ | ἔξωθεν προσερραμμένη.

5. Ἡ δὲ ἑτέρη κέεται ὑπένερθε μὲν μάλιστα, καὶ
κατ᾽ ἰθυωρίην μάλιστα μὲν μαζῷ ἀριστερῷ, ὅπῃ καὶ
διασημαίνει τὸ ἅλμα. περίβολον δὲ ἔχει παχύν, καὶ
βόθρον ἐμβεβόθρωται τὸ εἶδος εἴκελον ὅλμῳ. ἀλλὰ
γὰρ ἤδη καὶ τοῦ πνεύμονος ἐνδύεται μετὰ προσηνίης,
καὶ κολάζει τὴν ἀκρασίην τοῦ θερμοῦ περιβαλλομένη·
ὁ γὰρ πνεύμων φύσει ψυχρός· ἀτὰρ καὶ ψυχόμενος τῇ
εἰσπνοῇ.

6. Ἄμφω γε μὴν δασεῖαι τὰ ἔνδον καὶ ὥσπερ
ὑποδιαβεβρωμέναι, καὶ μᾶλλον τῆς δεξιῆς ἡ λαιή. τὸ
γὰρ ἔμφυτον πῦρ οὐκ ἐν τῇ δεξιῇ, ὥστε ⟨οὐ⟩⁸ θαῦμα
τρηχυτέρην γενέσθαι τὴν λαιὴν ἐμπλέην οὖσαν⁹
ἀκρήτου. ταύτῃ καὶ πάχετον ἐνδεδόμηται φυλακῆς
εἵνεκα τῆς ἰσχύος τοῦ θερμοῦ.

7. Στόματα δ᾽ αὐτῇσιν οὐκ ἀνεώγασιν, εἰ μή τις
ἀποκείρει τῶν οὐάτων τὴν κορυφὴν¹⁰ καὶ τῆς καρδίης
τὴν κεφαλήν· ἢν δ᾽ ἀποκείρῃ, φανήσεται καὶ δισσὰ
στόματα ἐπὶ δυσὶ γαστέροιν. ἡ γὰρ παχείη φλὲψ ἐκ
μιῆς ἀναθέουσα, πλανᾷ τὴν ὄψιν, ἢν ἀνατμηθῇ. αὗται
πηγαὶ φύσιος ἀνθρώπου, καὶ οἱ ποταμοὶ ἐνταῦθα ἀνὰ
τὸ σῶμα, τοῖσιν ἄρδεται τὸ σκῆνος, οὗτοι δὲ καὶ τὴν

---

⁷ καὶ στερεόν V.   ⁸ Add. Foes in note 20.

left, as the entire heart has its seat in these—lies up against an orifice, being in contact with one of the two veins (vena cava). This (sc. right) ventricle is altogether wide-chambered and much slacker than the other one, nor does it occupy the extremity of the heart, but rather it leaves the extremity solid, and is as if stitched on from the outside.

5. The other (i.e. left) ventricle lies beneath for the most part, and is oriented especially towards the left breast, where its beat is visible. It has a thick enclosing wall, and its interior is a pit which has the form of a mortar. This (sc. ventricle) is already clothed by the lung, for the sake of relief, and being thus covered counteracts the unmixed quality of its heat: for the lung is cold by nature, being cooled further by the inspired air.

6. Both ventricles are shaggy in their interior parts and, as it were, somewhat corroded, the left more so than the right. Now in the right ventricle there is no inborn fire, so that it is no wonder that the left ventricle is the rougher, being filled as it is with unmixed fire. Its construction is also thicker as a means of preserving the force of its heat.

7. The orifices into the ventricles are not open to view unless someone clips off the apex of the auricles and the top part of the heart; if he does clip them, double orifices on the two ventricles will be revealed. If, on the other hand, the wide vein running up from one of the ventricles (superior vena cava) is cut away, it spoils the view. These ventricles are the fountains of a person's being, and rivers pass from them through the body to water its frame; these

---

9 ἐμπλέην οὖσαν Duminil: ἐσπνέουσαν V
10 Littré: καρδίην V.

ζωὴν φέρουσι τῷ ἀνθρώπῳ, κἢν αὐανθέωσιν, ἀπ-
έθανεν ὤνθρωπος.

8. Ἀγχοῦ δὲ τῆς ἐκφύσιος τῶν φλεβῶν σώματα
τῇσι κοιλίῃσιν ἀμφιβεβήκασι, μαλθακὰ σηραγγώ-
δεα, ἃ κλῇσκεται μὲν οὔατα, τρήματα δὲ οὐκ ἔστιν
οὐάτων. ταῦτα γὰρ οὐκ ἐνακούουσιν ἰαχῆς· ἔστι δὲ
ὄργανα τοῖσιν ἡ φύσις ἁρπάζει τὸν ἠέρα. καί τε δοκέω
86 τὸ ποίημα | χειρώνακτος ἀγαθοῦ· κατασκεψάμενος
γὰρ σχῆμα στερεὸν ἐσόμενον τὸ σπλάγχνον διὰ τὸ
πιλητικὸν[11] τοῦ ἐγχύματος, ἔπειτα πᾶν ἐὸν ἑλκτικόν,
παρέθηκεν αὐτῷ φύσας, καθάπερ τοῖσι χοάνοισιν οἱ
χαλκέες, ὥστε διὰ τούτων χειροῦται τὴν πνοήν. τεκμή-
ριον δὲ τοῦ λόγου· τὴν μὲν γὰρ καρδίην ἴδοις ἂν
ῥιπταζομένην οὐλομελῆ, τὰ δὲ οὔατα κατ' ἰδίην ἀνα-
φυσώμενά τε καὶ ξυμπίπτοντα.

9. Διὰ τοῦτο δέ φημι καὶ φλεβία μὲν ἐργάζεται τὴν
ἀναπνοὴν ἐς τὴν ἀριστερὴν κοιλίην, ἀρτηρίη δ' ἐς τὴν
ἄλλην· τὸ γὰρ μαλακὸν ἑλκτικώτερον καὶ ἐπιδόσιας
ἔχον. ἔχρη δὲ ἡμῖν μᾶλλον τὰ ἐπικείμενα τῆς καρδίης
διαψύχεσθαι· βλαβερὸν[12] ἐστὶ γὰρ τὸ θερμὸν ἐν τοῖσι
δεξιοῖσι, ὥστε διὰ τὴν πάθην οὐκ ἔλαβεν εὐπετὲς
ὄργανον, ἵνα μὴ πάμπαν κρατηθῇ ὑπὸ τοῦ ἐσιόντος.

10. Λοιπός ἐστιν ὁ λόγος ὁ τῆς καρδίης ὑμένες
ἀφανέες, ἔργον ἀξιαπηγητότατον. ὑμένες γὰρ καὶ ἄλ-
λοι τινὲς ἐν τῇσι κοιλίῃσιν ὁκοῖον ἀράχναι διαπετέες
ζώσαντες πάντη τὰ στόματα, κτηδόνας ἐμβάλλουσιν

---

[11] Littré: πλατικὸν V.   [12] Potter: βέβλημα V.

(sc. rivers) carry life to a person, and if they dry up, the person dies.

8. Near the place where the veins grow out of the heart are bodies bestriding the cavities—soft, spongy things called auricles, although they do not have channels in them as real ears do. In fact, these auricles do not take in sound, but rather are the organs by which nature captures the air. And I think this is the creation of a good handworker, for when he recognized that the viscus was going to be of a solid frame on account of the thickness of its substance, and then highly attractive, he added bellows to it, just as bronze smiths do to their melting-pots, in order that through these it would be able to handle the respiration. Proof of this theory: the heart, as you can see, moves as a whole, but the auricles inflate and collapse individually.

9. For the same reason I also assert that certain small veins (pulmonary veins) bring about the respiration that enters the left ventricle, the artery (pulmonary artery) what enters the other one: what is soft is more attractive and can expand. It is more necessary in us for what lies over the heart[2] to be cooled, for heat is harmful to the right parts, so that through its disposition the organ there does not receive heat easily, in order not to be completely subdued by what comes into it.

10. There remains an explanation of the heart's hidden membranes, a work most worthy of the recounting. Now membranes and certain other structures in the cavities like spider-webs (cordae tendineae) spread out and completely encircle the orifices, and at the same time send off fibres

[2] Perhaps the right ventricle (Ermerins ad loc.): see chapter 4 above. This whole passage is very turbid.

ἐς τὴν στερεὴν καρδίην. οὗτοί μοι δοκέουσιν οἱ τόνοι
τοῦ σπλάγχνου καὶ τῶν ἀγγείων, ἀρχαὶ τῇσιν ἀορ-
88 τῇσιν. ἔστι δὲ αὐ|τῶν ζεῦγος καὶ θύρῃσι[13] μεμηχάνην-
ται τρεῖς ὑμένες ἑκάστῃ, περιφερέες ἐξ ἄκρου περ
ὁκόσον ἡμίτομα κύκλου, οἵ τε ξυνιόντες θαυμάσιον ὡς
κλείουσι τὰ στόματα, τῶν ἀορτέων πέρας. καὶ τὴν
καρδίην ἀποθανόντος ἤν τις ἐξεπιστάμενος τὸν ἀρ-
χαῖον κόσμον ἀφελὼν τῶνδε τὸν μὲν ἀποστηρίσῃ,[14]
τὸν δὲ ἐπανακλίνῃ, οὔτε ὕδωρ ἂν διέλθοι εἰς τὴν
καρδίην οὔτε φῦσα ἐμβαλλομένη· καὶ μᾶλλον τῶν τῆς
ἀριστερῆς· οὗτοι γὰρ[15] ἐμηχανήθησαν ἀτρεκέστερον
κατὰ δίκην. γνώμη γὰρ ἡ τοῦ ἀνθρώπου πέφυκεν ἐν τῇ
λαιῇ κοιλίῃ, καὶ ἄρχει τῆς ἄλλης ψυχῆς.

11. Τρέφεται δὲ οὔτε σιτίοισιν οὔτε ποτοῖσι τοῖσιν
ἀπὸ τῆς νηδύος, ἀλλὰ καθαρῇ καὶ φωτοειδεῖ περι-
ουσίῃ γεγονυίῃ ἐκ τῆς διακρίσιος τοῦ αἵματος. εὐπο-
90 ρέει δὲ τὴν τροφὴν ἐκ τῆς ἔγγιστα δεξα|μενῆς τοῦ
αἵματος, διαβάλλουσα τὰς ἀκτῖνας, καὶ νεμομένη
ὥσπερ ἐκ νηδύος [τῶν ἐντέρων][16] τὴν τροφήν, οὐκ ἐὸν
κατὰ φύσιν. † ὅκως δὲ μὴ ἀνακωχῇ τὸ σιτίον τὰ
ἐνεόντα ἐν τῇ ἀρτηρίῃ ἐν ζάλῃ ἐόν†,[17] ἀποκλείει τὴν
ἐπ᾽ αὐτὴν κέλευθον. ἡ γὰρ μεγάλη ἀρτηρίη βόσκεται

13 καὶ θύρῃσι Linden: αἱ θύρεσι V.
14 Duminil; ἀποστερήσει V.
15 οὗτοι γὰρ Ermerins: τοιγὰρ V.
16 Del. Kudlien, p. 425, n. 1.
17 "passage . . . très obscur" (Littré); see Lonie p. 14, n. 32 and
Duminil pp. 256f., n. 55.

into the solid heart (papillary muscles). These I believe to
be the bands of the viscus and of the chambers, the origins
to the aortae.[3] There is a pair of these, to each of which at
its gates three membranes are attached, rounded at their
margins and having the shape of semicircles, which in
coming together in some marvellous way close the orifices
and set the limit of the aortae. And if someone knowledge-
able of the ancient rite were to take out the heart of a man
who had died, and draw back one of these (sc. membranes)
and incline the other one,[4] neither water would be able to
go through into the heart nor air that was being forced—
and more so in the case of those on the left, for these are
constructed more tightly, as is fitting: for the intelligence
of man is established in the left cavity, and it rules over the
rest of his soul.

11. This intelligence is nourished not from the gut by
foods and drinks, but by a pure and luminous bath coming
from a distillate of the blood. It obtains its nutriment in
abundance from that which is most near, receiving it from
the blood, transmitting its rays, and feeding as if on nourishment out of the stomach and the intestines, but in a way
not according to normal nature. †In order that the contents of the artery do not send back food in a state of turbu-
lence†, it closes off the path to the ventricle.[5] For the large

---

[3] ἀορτή is little more than a variant of ἀρτηρίη, and in meaning both seem to occupy a middle ground between bronchus and
artery; cf. *Places in Man* 14, *Coan Prenotions* 394, and *Diseases II*
54.

[4] The mention of two rather than three valve cusps here suggests a knowledge of the mitral valve.

[5] Or "artery."

τὴν γαστέρα καὶ τὰ ἔντερα, καὶ γέμει τροφῆς οὐχ
ἡγεμονικῆς. ὅτι δὲ οὐ τρέφεται βλεπομένῳ αἵματι [ἡ
μεγάλη ἀρτηρίη][18] δῆλον ὧδε· ἀποπαγέντος τοῦ ζῴου,
σχασθείσης τῆς ἀριστερῆς κοιλίης, ἐρημίη φαίνεται
πᾶσα, πλὴν ἰχωρός τινος καὶ χολῆς ξανθῆς καὶ τῶν
ὑμένων, περὶ ὧν ἤδη μοι πέφανται· ἡ δὲ ἀρτηρίη οὐ
λειφαιμοῦσα, οὐδὲ ἡ δεξιὴ κοιλίη. τούτῳ μὲν οὖν τῷ
ἀγγείῳ κατ᾽ ἐμὸν νόον ἥδε ἡ πρόφασις τῶν ὑμένων.

12. Τὸ δ᾽ αὖ φερόμενον ἐκ τῆς δεξιῆς, ζυγοῦται μὲν
καὶ τοῦτο τῇ ξυμβολῇ τῶν ὑμένων, πλὴν οὐ κάρτα
ἔθρωσκεν ὑπὸ ἀσθενείης. ἀλλ᾽ ἀνοίγεται μὲν ἐς πνεύμονα, ὡς αἷμα παρασχεῖν αὐτῷ ἐς τὴν τροφήν, κλεί-
εται δὲ ἐς τὴν καρδίην οὐχ ἁρμῷ, ὅκως ἐσίη μὲν ὁ ἠήρ,
92   οὐ | πάνυ δὲ πολύς. ἀσθενὲς γὰρ ἐνταῦθα τὸ θερμόν,
δυναστευόμενον κρήματι ψυχροῦ· τὸ αἷμα γὰρ οὐκ
ἔστι τῇ φύσει θερμόν, οὐδὲ γὰρ ἄλλο τι ὕδωρ, ἀλλὰ
θερμαίνεται· δοκέει δὲ τοῖσι πολλοῖσι φύσει θερμόν.

Περὶ δὲ καρδίης τοιαῦτα εἰρήσθω.

18 Del. Littré.

artery feeds from the stomach and the intestines, and is full of nutriment which is not suitable for the ruling power. That it (i.e. the intelligence in the left ventricle) is not nourished by visible blood is made clear by the following: in an animal that has reached the state of *rigor mortis*, when the left cavity is cut open, it appears completely empty except for some serum and yellow bile, and the membranes mentioned above, but the artery has no shortage of blood, nor does the right cavity. Now to my mind, this is the reason for the membranes in this chamber.

12. The vessel (pulmonary artery) which passes out of the right (sc. ventricle) is also controlled by the meeting of membranes, except that it, on account of its weakness, is not well fitted with doors. It opens into the lung, in order to provide it with blood as nourishment, but is closed into the heart, although not by a completely tight joint, so that some air still goes in, but not very much. On the right the heat is weak, being dominated by an admixture of cold; indeed, blood is not warm by its nature any more than any other liquid is, but rather it becomes warm—it is only thought by most people to be warm by nature.

About the heart, let this much be said.

EIGHT MONTHS' CHILD

INTRODUCTION

Galen's approximately ten references to Hippocrates' or Polybus'—he seems undecided on authorship—*Eight Months' Child* contain occasional direct quotations, which confirm the identity of the text he is reading with the one transmitted in our medieval manuscripts.[1] However, whereas Galen and the Hippocratic manuscripts give the title in the singular, three other ancient sources offer the plural, *Eight Months' Children*: Clement of Alexandria,[2] Vindicianus,[3] and the Brussels *Yppocratis genus, vita, dogma*.[4] Erotian does not include any such title in his list of Hippocratic works, but the occurrence of one otherwise unattested expression in his *Glossary* makes it probable that he knew the treatise.[5]

Eight Months' Child is transmitted in different configurations in its independent manuscripts M and V. M includes ch. 1–9 under the title *Seven Months' Child*, followed by ch. 10–13 under the title *Eight Months' Child*: V presents the whole of ch. 10–13 and 1–9 in succession un-

[1]. Anastassiou / Irmer vol. II 1, 373–5; vol. II 2, 288–90.

[2]. *Stromateis* VI 16,6 (vol. 2, 502).

[3]. Ch. 5, p. 211.

[4]. J. Rubin Pinault, *Hippocratic Lives and Legends*, Leiden, 1992, p. 133.

[5]. Erotian K 20, p. 50; see Grensemann p. 66.

der the title *Eight Months' Child*, and then another short spurious text under the title *Seven Months' Child*. The claims of each of these arrangements to be the original have been much discussed in the scholarly literature, with no general concensus yet emerging. To avoid the unnecessary confusion a departure from Littré's chapter numbering would entail, I have kept his and M's order of the text, but adopted Joly's and V's title *Eight Months' Child* for the whole work.[6]

Eight Months' Child is an account of gestation and birth which attempts to explain common experience, including that of the pregnant woman herself, in terms of four special time periods: day, month, forty days, year. Three of these (day, month, year) have astronomical definitions, and were basic elements of the Greek calendar,[7] whereas the forty-day period possessed special significance in many areas of Greek thought.[8]

Ch. 9 explicates the author's theoretical position most generally, ch. 1 and 13b present specific calculations relating the different time periods to each other and to gestative events, and ch. 2–5 and 10a give an account of seven months' birth, and the stresses and dangers of the sixth

[6] See in particular Grensemann pp. 41–7; Joly pp. 149–55; J. Jouanna, "Tradition manuscrite et structure du traité hippocratique *Sur le foetus de huit mois*," in *Revue des études grecques* 86 (1976), 1–16; R. Joly, "La structure du *Foetus de huit mois*," in *L'Antiquité classique* 45 (1976), 173–80.

[7] See O. Wenskus, *Astronomische Zeitangaben von Homer bis Theophrast*, Stuttgart, 1990, pp. 93–6, 123.

[8] W. H. Roscher, *Die Tessarakontaden und Tessarakontadenlehren der Griechen und anderer Völker*, Leipzig, 1909, esp. pp. 85–101.

forty-day period. Ch. 6–8 and 10b-13a deal with the process of birth in the ninth, tenth, and eleventh month.

After finding a place in the collected editions and translations including Zwinger, *Eight Months' Child* was then edited separately twice in short succession:

H. Grensemann, *Hippokrates Über Achtmonatskinder* . . . , CMG I 2,1, Berlin, 1968 (= Grensemann).

R. Joly, *Hippocrate . . . Du Foetus de huit mois*, Budé XI, Paris, 1970 (= Joly).

The present edition is based on these studies.

ΠΕΡΙ ΟΚΤΑΜΗΝΟΥ

1. Οἱ δὲ ἑπτάμηνοι γίνονται ἐκ τῶν ἑκατὸν ἡμερέων καὶ ὀγδοήκοντα καὶ δύο καὶ προσεόντος μορίου. ἢν γὰρ τοῦ πρώτου λογίσῃ μηνὸς πεντεκαίδεκα ἡμέρας, τῶν δὲ πέντε μηνῶν ἑκατὸν καὶ τεσσαράκοντα καὶ ἑπτὰ καὶ ἥμισυ ἡμέρης—ἐν γὰρ ἑξήκοντα μιῆς δεού-σῃσιν ἡμέρῃσιν ἐγγύτατα δύο μῆνες ἐκτελεῦνται—, οὕτως οὖν τούτων ἐόντων ἐς τὸν ἕβδομον μῆνα περι-γίνονται ἡμέραι πλεῖον ἢ εἴκοσιν ἡμίσει τοῦ ἐνιαυτοῦ [καὶ]¹ τῆς ἡμέρης τοῦ μέρεος τῷ μέρει προσγινομέ-νου.² ὁκόταν οὖν ἐς τὴν ἀρχὴν τῆς τελειώσιος ἔλθῃ ταύτην, ἁδρυνομένου τοῦ ἐμβρύου καὶ τὴν ἰσχὺν πολὺ ἐπιδιδόντος ἐν τῇ τελειώσει μᾶλλον ἢ ἐν τοῖς ἄλλοις χρόνοις, οἱ ὑμένες, ἐν οἷσι τὴν ἀρχὴν ἐτράφη, ὥσπερ καὶ τῶν ἀσταχύων ἐξεχάλασαν πρόσθεν ἀναγκαζόμε-νοι ἢ τελείως ἐξαδρυνθῆναι τὸν καρπόν. τὰ οὖν ἰσχυ-

---

¹ Del. Littré after Galen.　　² Galen: ποστῳ γινομένῳ M, corr. to ποσγινομένου: τῷ πρὸς τῷ γιγνομένῳ V.

---

¹ The calculation is as follows:

Given (i) that a seven months' child is born exactly one half year after its conception;

(ii) that a year is (365 + a fraction) days long;

76

EIGHT MONTHS' CHILD

1. Seven months' children are born after 182 and a fraction days; indeed, if you reckon fifteen days for the first month, 147 ½ days for the next five months—since two months last very close to fifty-nine days—, then, this being so, there remain in the half year more than twenty days for the seventh month, since the fraction of a day is added to the other fraction.[1] Now as a fetus arrives at the onset of its final formation, it matures and gains much strength in the process, more than at any other time; the membranes in which it is nourished in the beginning become loose, just the way that ears of grain do when they are stretched before their fruit has reached its complete maturity. And so

(iii) that a child is conceived one half way through its mother's menstrual month;

(iv) that two months (lunar or menstrual) are 59 days long,

Then (i) a seven months' gestation will last: (365 + a fraction) / 2 = (182 + ½ + a fraction) days: "the fraction of a day is added to the other fraction";

(ii) month I of the gestation will last 15 days;

(iii) months II–VI of the gestation will last: (59 / 2) x 5 = 147½ days;

(iv) month VII of the gestation will last: (182½ + a fraction) – (15 + 147½) = 20 + a fraction days: "there remain in the half year more than twenty days for the seventh month."

ρότατα καὶ ἁδρότατα τῶν ἐμβρύων βιησάμενα καὶ
διαρρήξαντα τοὺς ὑμένας ἠνάγκασε τὸν τόκον γενέσθαι.

2. Καὶ τὰ μὲν πλεῖστα τούτων ἀπώλοντο, μικρὰ
438 γὰρ ἐόντα τῇ | μεταβολῇ μέζονι χρέωνται τῶν ἄλλων,
καὶ τὴν τεσσαρακονθήμερον κακοπαθείην ἀναγκάζον-
ται κακοπαθεῖν ἐξελθόντα ἐκ τῆς μήτρης, ⟨ἢ⟩[3] καὶ τῶν
δεκαμήνων πολλὰ ἀποκτείνει. ἔστι δὲ ἃ τούτων τῶν
ἑπταμήνων καὶ περιγίνονται, ἐκ πολλῶν ὀλίγα, ὅτι ὁ
λόγος καὶ ὁ χρόνος, ὅσον ἐτράφη ἐν τῇ μήτρῃ, κατ-
έστησεν ὥστε μετέχειν πάντων, ὧνπερ καὶ τὰ τελειότατα μετέχει καὶ μάλιστα περιγινόμενα, καὶ ἐξήλ-
λαξε τῆς μητρὸς πρόσθεν ἢ νοσῆσαι τὰ[4] ἐν τῷ ὀγδόῳ
μηνὶ νοσήματα.[5] τούτοισι γὰρ τοῖσι πόνοις ἢν ἐπιγίνηται ὥστ' ἐς τοὐμφανὲς ἐλθεῖν τὸ παιδίον, ἀδύνατον περιγενέσθαι διὰ τὰς πάθας τὰς προειρημένας, ἃς
ἐγώ φημι τὰ ὀκτάμηνα ἀποκτείνειν, πολλὰ[6] δὲ καὶ τῶν
δεκαμήνων.

3. Τὰ δὲ πολλὰ τῶν ἐμβρύων τῶν ἐν ταύτῃ τῇ
ἡλικίῃ τῇ ἑπταμήνῳ, ὅταν οἱ ὑμένες χαλάσωσι, μετ-
εχώρησαν ἐς τὸ ὑπεῖξαι καὶ ἐνταῦθα τὴν τροφὴν
ποιέεται, τὰς μὲν τεσσαράκοντα ἡμέρας τὰς πρώτας
πονέοντα τὰς μὲν μᾶλλον, τὰς δὲ ἧσσον διὰ τὴν
μεταβολήν, τὴν ἐκ τῶν χωρίων τῶν θρεψάντων μετ-
εβάλετο, καὶ ὅτι τὸν ὀμφαλὸν[7] ἔσπασε [καὶ μετεχώρη-

---

[3] Add. Ermerins.
[4] ἢ—τὰ Littré: ἢν ὅσης, αἵ τε Μ: νοσήσασαι τὰ V.

the most powerful and mature of these fetuses stretch out and break through their membranes, and thereby compel birth to occur.

2. But most of these die, being small and feeling the change more forcefully than other fetuses do, and being overcome by a forty-day period of distress after they have left the uterus, like the distress that also kills many ten months' children. Still, some seven months' children do survive, if only a few out of many, since the way and the time that such a child is nourished in the uterus provides it with a portion of everything that children who are the most mature and most likely to survive share, and since it is removed from its mother before it suffers the strains that occur in the eighth month. For if on top of these latter sufferings the child also undergoes the stress of coming into the light, it cannot survive, on account of the stresses I have mentioned before, which, as I say, kill eight months' children, and also many ten months' children.

3. The majority of fetuses at this seven month stage, when their membranes become loose, move in the direction that is giving way, and take their nourishment from there. They suffer during the following forty days—some days more, other days less—on account of their dislocation from the place where they were being nourished, due to tension on the umbilical cord, and as a result of

5 Grensemann: νενοσεύμενα M: νοσέοντα V.

6 Grensemann: τὰς (τοὺς V) ὀκταμήνους . . . πολλοὺς MV.

7 καὶ—ὀμφαλὸν Littré: καὶ ἀντὶ τοῦ ὀμφαλοῦ M: κάσι τοῦ τε ὀμφαλοῦ V.

σε]⁸ καὶ διὰ τῆς μητρὸς τοὺς πόνους. οἱ γὰρ ὑμένες
τεινόμενοι καὶ ὁ ὀμφαλὸς σπασθεὶς ὀδύνας ποιέει τῇ
μητρί, καὶ τὸ ἔμβρυον ἐκ τοῦ ἀρχαίου συνδέσμου
ἐκλυθὲν βαρύτερον γίνεται. πολλαὶ δὲ τῶν γυναικῶν
καὶ ἐπιπυρεταίνουσι τούτων γινομένων, αἱ δὲ καὶ
440 ἀπόλλυνται σὺν τοῖς ἐμβρύοισι. χρέωνται δὲ | πᾶσαι
ἑνὶ λόγῳ περὶ τούτου.⁹ φασὶ γὰρ τοὺς ὀγδόους τῶν
μηνῶν καὶ χαλεπώτατα φέρειν τὰς γαστέρας, ὀρθῶς
λέγουσαι. ἔστι δὲ¹⁰ οὐ μοῦνον ὁ χρόνος οὗτος, ἀλλὰ
καὶ ἡμέραι πρόσεισιν ἀπό τε τοῦ ἑβδόμου μηνὸς καὶ
ἀπὸ τοῦ ἐνάτου· ἀλλὰ τὰς ἡμέρας οὐχ ὁμοίως οὔτε
λέγουσιν οὔτε γινώσκουσιν αἱ γυναῖκες. πλανῶνται
γὰρ διὰ τὸ μὴ κατὰ τὸ αὐτὸ γίνεσθαι, ἀλλὰ τὸ μὲν
ἀπὸ τοῦ ἑβδόμου μηνὸς πλείονας ἡμέρας προσγενέ-
σθαι ἐς τὰς τεσσαράκοντα, τὸ δὲ ἀπὸ τοῦ ἐνάτου. ὧδε
γὰρ ἀνάγκη γίνεσθαι ὅπως ἂν τύχῃ ἡ γυνὴ ἐν γαστρὶ
λαβοῦσα τοῦ μηνός τε καὶ τοῦ χρόνου. ὁ δὲ δὴ ὄγδοος
ἀναμφισβήτητός ἐστι· κατὰ τοῦτον γὰρ κρίνεται ὥστε
εὐκρίτως ἔχειν, καὶ ἓν δὲ μέρος ἐν τοῖς δέκα¹¹ μησὶν ὁ
μής ἐστιν, ὥστε τοῦτον εὔμνηστον εἶναι.

4. Χρὴ δὲ οὐκ ἀπιστεῖν τῇσι γυναιξὶν ἀμφὶ τῶν
τόκων· λέγουσι γὰρ ταὐτὰ αἰεὶ καὶ λέγουσι ἅπερ ἂν
εἰδέωσι·¹² οὐ γὰρ ἂν πεισθείησαν οὔτ' ἔργῳ οὔτε λόγῳ
442 ἄλλο ἢ ὅ τι γνῶσιν¹³ ἐν τοῖσι σώμασιν αὐ|τῶν γινό-

---

8 Del. Grensemann.   9 ἑνὶ—τούτου Littré: ἐν ὀλίγῳ
περὶ τουτέου M: ὀλίγῳ πυρὶ τουτέῳ V.
10 Add. ὄγδοος μὴς M.

their mother's distress: for the loosening of the membranes and the tension on the umbilical cord cause pains in the mother, while the fetus, as it is released from its original supports, becomes more weighed down. Also, many women have fever when these things happen, and some even die along with their fetuses. All women have the same explanation for this: they say that in the eighth month it is most strenuous to carry their abdomens, and in this they are correct. But it is not only in the eighth month, for a number of days are also added out of the seventh month and out of the ninth month; but these extra days women do not report in a consistent way, nor are they clearly aware of them, erring because the process does not take place in the same manner in every case, sometimes more days being added from the seventh month to make up the forty, and at other times more from the ninth month. This follows necessarily in the individual case according to when the woman happens to have become pregnant in relationship to the time of the month. The eighth month, in any case, is agreed upon, for judgement is made in reference to this time, which is easy, since the unit month is a simple fraction in the ten months, so that it is easy to remember.

4. You should not distrust women about their giving birth, for they always say the same thing and they say what they know; they are not to be persuaded by either fact or argument to believe anything contrary to what they know is going on inside their own bodies. Although it is possible

---

11 V: ἔνδεκα M.

12 ταὐτὰ—εἰδέωσι Potter: πάντα καὶ αἰεὶ λέγουσι καὶ αἰεὶ ἐρέουσι M: ἄπερ καὶ εἰδέουσι V.

13 ἢ—γνῶσιν Potter: ὅτι γνῶναι καὶ M οὐ γίγνονται· ἢ τὸ V.

μενον. τοῖσι δὲ βουλομένοισιν ἄλλο τι λέγειν ἔξεστιν, αἱ δὲ κρίνουσαι καὶ τὰ νικητήρια διδοῦσαι περὶ τούτου τοῦ λόγου αἰεὶ ἐρέουσι καὶ φήσουσι τίκτειν καὶ ἑπτάμηνα καὶ ὀκτάμηνα καὶ ἐννάμηνα καὶ δεκάμηνα[14] καὶ ἑνδεκάμηνα, καὶ τούτων τὰ ὀκτάμηνα οὐ περιγίνεσθαι, τὰ δ' ἄλλα περιγίνεσθαι. φήσουσι δὲ καὶ τοὺς τρωσμοὺς πλείστους ἐν τῇ πρώτῃ τεσσαρακοντάδι γίνεσθαι, καὶ τὰ ἄλλα τὰ καταγεγραμμένα ἐν τῇσι τεσσαρακοντάσι καὶ ἐν τοῖσι μησὶν ἑκάστοισιν.

Ὅταν δὲ τῷ ἑβδόμῳ μηνὶ περιρραγέωσιν οἱ ὑμένες καὶ τὸ ἔμβρυον μεταχωρήσῃ, ὑπέλαβον οἱ πόνοι οἱ περὶ τὸν ὄγδοον μῆνα γεγενεαλογημένοι[15] καὶ περὶ τὴν ἕκτην τεσσαρακοντάδα. τούτου δὲ τοῦ χρόνου παρελθόντος ὅσῃσι μέλλει εὖ εἶναι αἱ φλεγμοναὶ ἐλύθησαν καὶ τοῦ ἐμβρύου καὶ τῆς μητρός, ἥ τε γαστὴρ ἐμαλθάχθη καὶ ὁ ὄγκος ὑποκατέβη ἀπὸ τῶν ὑποχονδρίων καὶ τῶν κενεώνων ἐς τὰ κάτω χωρία ἐς εὐτρεπείην τῆς ἐπὶ τοὺς τόκους στροφῆς.[16] καὶ τὴν ἑβδόμην τεσσαρακοντάδα ἐνταῦθ' ἐστὶ τὸ πλεῖστον[17] τοῦ χρόνου τὰ ἔμβρυα. τὰ γὰρ χωρία αὐτοῖς μαλθακά, καὶ αἱ μετακινήσιες αὐτοῖς εὐπετέστεραι γίνονται καὶ πυκνότεραι, καὶ διὰ ταῦτα κατέστη πρὸς τὸν τόκον εὐλυτώτερα. καὶ πάσης τῆς τεσσαρακοντάδος ταύτης[18] αἱ γυναῖκες φέρουσι τὰς τελευταίας ἡμέρας εὐπετέστερον τὰς γαστέρας, ἔστ' ἂν ὁρμήσῃ τὸ ἔμβρυον στρέφεσθαι. μετὰ δὲ ταῦτα αἵ τε ὠδῖνές εἰσι

14 καὶ δεκάμηνα om. M.

there may be some who wish to assert something different, in fact women who possess judgement and who furnish the most convincing arguments on this subject always say explicitly that they give birth in the seventh month, the eighth month, the ninth month, the tenth month, and the eleventh month, and that of these children, those born in the eighth month do not survive, whereas the others do. They also say that most miscarriages occur within the first forty-day period, and also what else is recorded in each forty-day period and month.[2]

When in the seventh month the membranes are torn through and the fetus shifts its position, pains immediately follow that are assigned to the eighth month and to the sixth forty-day period. When this time has passed, in women who are going to be alright, the febrile swellings of both the fetus and the mother go down; the belly becomes soft, and the mass descends from the hypochondrium and the flanks into the lower regions, in preparation for turning at the time of delivery. In the seventh forty-day period fetuses remain there for most of the time; for the region is soft for them, and their movements become freer and more frequent, so that they are more easily released at birth. Of this whole (sc. seventh) forty-day period women bear the final days more easily in their abdomen, until the fetus begins to turn. After that the pangs of childbirth and

[2] See ch. 9 below.

---

[15] Grensemann: γενεαλογημένοι M: ἐνγενεαλογημένον V.
[16] τῆς—στροφῆς Ermerins: τὴν . . . τροπήν M: τὴν ἀπὸ τοῦ στομάχου στρέφοι V.   [17] M: λοιπὸν V.
[18] Zwinger in marg.: ταύτας M: om. V.

καὶ οἱ πόνοι ἐπίκεινται, ἔστ᾽ ἂν ἐλευθερωθῇ τοῦ τε παιδίου καὶ τοῦ ὑστέρου. |

444   5. Ὅσαι δὲ τῶν γυναικῶν ἔτεκον πολλὰ παιδία καί τι αὐτῶν ἐξεγένετο χωλὸν ἢ τυφλὸν ἢ ἄλλο τι κακὸν ἔχον, φήσουσιν ἐπὶ τούτου τοῦ παιδίου τὸν ὄγδοον μῆνα χαλεπώτερον διαγαγεῖν ἢ ἐφ᾽ ὧν ἔτεκον οὐδὲν κακὸν ἐχόντων. τὸ γὰρ ἔμβρυον τὸ πηρωθὲν ἐν τῷ ὀγδόῳ μηνὶ ἰσχυρῶς ἐνόσησεν, ὥστε καὶ ἀπόστασιν ἐποίησεν ἡ νοῦσος, ὡς καὶ τοῖς ἀνδράσιν αἱ ἰσχυραὶ νοῦσοι ἐποίησαν. ὅσα δ᾽ ἂν τῶν ἐμβρύων ἐν ἄλλῳ[19] χρόνῳ ἰσχυρῶς νοσήσῃ, ἀπόλλυται πρόσθεν ἢ αὐτῷ ἀπόστασιν γενέσθαι. ὅσα δ᾽ ἂν τῶν [ὀκταμήνων][20] ἐμβρύων μὴ σφόδρα νοσήσῃ, ἀλλὰ κατὰ φύσιν ἐκ τῆς μεθόδου κακοπαθήσῃ, τὰς μὲν τεσσαράκοντα ἡμέρας διετέλεσεν ἀσθενέον⟨τα⟩[21] τὸ πλεῖστον ἐν τῇ μήτρῃ διὰ τὰς ἀνάγκας τὰς προειρημένας, ἐγένετο δὲ ὑγιαίνοντα. ὅ τι ἂν γένηται ἐν τῇσι τεσσαράκοντα ἡμέρῃσι ταύτῃσιν, ἀδύνατον περιγενέσθαι· νοσέοντι γὰρ αὐτῷ ἔτι ἐν τῇ μήτρῃ ἐπιγίνονται αἱ μεταβολαὶ καὶ αἱ κακοπάθειαι αἱ μετὰ τὸν τόκον.

6. Ὅ τι δ᾽ ἂν ἀπονοσῆσαν ἐν τῇ μήτρῃ ἐς τὸν ἔνατον μῆνα ἔλθῃ καὶ ἐν τούτῳ τῷ μηνὶ γένηται, περιγίνεται μὲν οὐκ ἔλασσον ἢ τὰ ἑπτάμηνα, ὀλίγα δὲ καὶ τούτων ἐκτρέφεται· οὔτε γὰρ τὴν παχύτητα ἴσχει ἥνπερ τὰ τελειότατα ἴσχει, οἵ τε[22] πόνοι οἱ ἐν τῇ μήτρῃ οὐ πάλαι πεπονημένοι εἰσὶν αὐτοῖς, ὥστε λεπτὸν γενέσθαι.

7. Σωθείη δ᾽ ἂν μάλιστα, εἰ τοῦ ἐνάτου μηνὸς

the pains impose themselves, until the woman is delivered of the child and the placenta.

5. Women who have borne many children, one of whom was lame at birth, or blind, or had some other defect, say that with this child they went through a more difficult eighth month than with their other children who were born without any defect. For a fetus that is maimed has been seriously ill in the eighth month, with the disease proceeding to an apostasis just as serious diseases in adults do. (Fetuses that have been seriously ill at another time perish before this apostasis can occur.) Fetuses that have not been especially ill, but have only suffered the distress of the process itself in the normal way, generally pass through this forty-day period in the uterus in a weakened state, because of the irresistible factors cited above, but then recover. Any fetus born within this forty-day period, however, cannot survive; for in addition to its own illness in the uterus, it suffers other disturbances and stresses after its birth.

6. A fetus that recovers from its illness in the uterus, and is born on reaching the ninth month, is no less likely to survive than seven months' children, but few of these, either, are brought through childhood. For they do not have the robustness that the most mature children have, and the pains they have recently suffered in the uterus make them thin.

7. Such children are saved most frequently if they are

---

19 M: ὀλίγῳ V.
20 Del. Grensemann.
21 Littré: ἐν ἀσθενίῃ M: ἀσθενέον V.
22 οἵ τε Cornarius in marg.: οὔτε MV.

446 γένοιτο ἐν ἐξόδῳ· ἰσχυρότερον γὰρ ἂν γένοιτο καὶ πλεῖστον ἀπέχον τῶν νούσων τῶν ἐν τῷ | ὀγδόῳ μηνὶ[23] γενομένων. καὶ γὰρ τὰ ἐν ἑπτὰ τεσσαρακοντάσι τικτόμενα, τὰ δεκάμηνα καλεόμενα, διὰ ταῦτα μάλιστα ἐκτρέφεται, ὅτι ἰσχυρότατά ἐστι καὶ πλεῖστον ἀπέχει τῶν γονίμων παιδίων τοῦ χρόνου ἐν ᾧ ἐκακοπάθησαν τὰς τεσσαράκοντα ἡμέρας τὰς νοσεομένας περὶ τὸν μῆνα τὸν ὄγδοον.

8. Σαφηνίζει δὲ περὶ τῶν παθημάτων τῶν τοῖς ὀκταμήνοισι γινομένων καὶ τὰ ἐννάμηνα λεπτὰ τικτόμενα κατὰ τὸ πλῆθος τοῦ χρόνου, ὃν γέγονε, καὶ κατὰ τὸ μέγεθος τοῦ σώματος, ἐκ νούσων καὶ κακοπαθείης ἀφιγμένα,[24] οὐχ ὥσπερ τὰ ἑπτάμηνα τίκτεται σεσαρκωμένα καὶ τὸ πάχος ἔχοντα εὐπρεπές, ὡς ἄνοσα διατετελεκότα τὸν χρόνον, ⟨ὅσον⟩[25] ἐν τῇ μήτρῃ διεφέρετο.

9. Τῇσι δὲ γυναιξὶν αἱ συλλήψιες τῶν ἐμβρύων καὶ οἱ τρωσμοί τε καὶ οἱ τόκοι ἐν τοῖσιν αὐτοῖσι χρόνοισι[26] κρίνονται, ἐν οἷσί περ αἵ τε νοῦσοι καὶ αἱ ὑγίειαι καὶ οἱ θάνατοι τοῖσι σύμπασιν ἀνθρώποισι· ταῦτα γὰρ πάντα τὰ μὲν καθ' ἡμέρας, τὰ δὲ κατὰ μῆνας ἐπισημαίνει, τὰ δὲ κατὰ τεσσαρακοντάδας ἡμερέων, τὰ δὲ κατ' ἐνιαυτόν. ἐν πᾶσι γὰρ τοῖς χρόνοισι τούτοις ἔνεστι πρὸς ἕκαστον πολλὰ μὲν συμφέροντα, πολλὰ δὲ πολέμια· ἐκ μὲν οὖν τῶν συμφερόντων αἵ τε ὑγίειαι γίνονται καὶ αἱ αὐξήσιες· ἐκ δὲ τῶν ἐναντίων αἵ τε νοῦσοι καὶ οἱ θάνατοι.

born at the end of the ninth month, for then they are born stronger, and are furthest removed from the diseases that occurred in the eighth month. For children born in the seventh forty-day period—what are called tenth months' children—are usually brought through childhood, inasmuch as they are the most robust of viable children, and furthest away from the time in which they suffered distress during the sickly forty days around the eighth month.

8. A clarification of the disorders that occur in eighth months' children is provided by the case of nine months' children, who are born thin in relation to the length of time after which they are born and to the length of their body: due to the diseases and the stress these children have suffered, they are not born fleshy like seven months' children, nor in possession of the pleasing robustness that results from the healthy time these have spent in the uterus.

9. In women, the conception of fetuses, their miscarriages, and their births are decided in the same periods of time in which diseases, convalescences, and deaths are decided in human beings in general. For all these appear either according to the number of days, or of months, or of forty-day periods, or the period of a year. For in all these time periods there are many factors favourable to each thing, and many others unfavourable; out of the useful come health and growth, and out of the contrary come diseases and death.

---

23 ἐν—μηνὶ Grensemann: τοῖς ὀκταμήνοις M: ἐν ὀκταμή-νοισιν V.   24 Linden: ἀφιγμένοισι(ν) MV.

25 Add. Grensemann.

26 τοῖσιν—χρόνοισι Grensemann: τουτέῳ τῷ χρόνῳ M: om. V.

Αἱ μὲν οὖν ἡμέραι ⟨αἱ⟩[27] ἐπισημόταταί εἰσιν ἐν τοῖσι πλείστοις αἵ τε πρῶται καὶ αἱ ἕβδομαι, πολλαὶ μὲν περὶ νούσους, πολλαὶ δὲ καὶ τοῖς | ἐμβρύοισι· τρωσμοί τε γὰρ γίνονται καὶ οἱ πλεῖστοι ταύτησι τῆσιν ἡμέρησιν—ὀνομάζονται δὲ τὰ τηλικαῦτα ἐκρύσιες, ἀλλ' οὐ τρωσμοί. αἱ δ' ἄλλαι ἡμέραι, ὅσαι ἐντὸς τῶν τεσσαράκοντα, ἐπίσημοι μὲν ἧσσον, πολλαὶ δὲ κρίνουσιν.

Ἐν δὲ τοῖσι μησὶ ταῦτά τε ἃ καὶ ἐν τῆσιν ἡμέρησι γινόμενα ἔνεστι κατὰ λόγον, καὶ τὰ καταμήνια τῆσι γυναιξὶ τῆσιν ὑγιαινούσῃσι φαίνεται καθ' ἕκαστον τῶν μηνῶν, ὡς ἔχοντος τοῦ μηνὸς ἰδίην δύναμιν ἐν τοῖσι σώμασιν. ἐξ ὧν δὴ καὶ οἱ ἕβδομοι μῆνες τῆσιν ἐν γαστρὶ ἐχούσῃσι τὰ ἔμβρυα ἐς τὴν ἀρχὴν καθιστᾶσι τῆς τελειώσιος. τοῖσι δὲ παιδίοις ἑπταμήνοισιν ἐοῦσι καὶ ἄλλα διαφέροντα γίνονται ἐν τοῖς σώμασι, καὶ οἱ ὀδόντες φαίνεσθαι ἄρχονται ἐν τούτῳ τῷ χρόνῳ.[28]

Ὁ δὲ αὐτὸς λόγος καὶ ἐπὶ κρισίμων, εἴ περ ἴσως καὶ τοῖσδέ τις συγχρέοιτο, ὁκοῖα ἔλεξα[29] ἱστορίας τε ἕνεκα εἰρήσθω. σκοπεῖν γὰρ χρὴ τὸν μέλλοντα ἰητρὸν ὀρθῶς στοχάζεσθαι τῆς τῶν καμνόντων σωτηρίας θεωροῦντα τὰς[30] μὲν περιττὰς πάσας, τῶν δὲ ἀρτίων τὴν τεσσαρεσκαιδεκάτην καὶ τὴν ὀγδόην εἰκοστὴν καὶ ⟨τὴν⟩[31] τεσσαρακοστὴν καὶ δευτέρην. οὗτος γὰρ ὁ ὅρος τίθεται τῷ τῆς ἁρμονίης λόγῳ πρός τινων καὶ ὁ ἀρτιφυής τε καὶ τέλειος ἀριθμός, δι' ἣν δὲ αἰτίην, μακρότερον ἂν εἴη ἐπὶ τοῦ παρόντος διελθεῖν. θεωρεῖν

Now the most significant days are generally the first and the seventh, and many such exist in the realm of diseases, and also of fetuses. Miscarriages occur in most instances on these days—in fact, what happen at that stage are actually called effluxions rather than miscarriages. The other days in the (sc. first) forty-day period are less significant, although many of them, too, are decisive.

In months, the same periods that exist in days are proportionately present, and the menses in a healthy woman appear monthly, since the month has its own particular power in their bodies. And indeed, for these same reasons the seventh month in pregnant women moves fetuses to the onset of their perfection; there are other changes, too, in children at seven months, e.g. the teeth begin to appear at that time.

The same logic applies to crises, as well; and if someone is taking up these questions, let what I have said be considered in the course of his investigation. For a person who intends correctly to assess the treatment of patients must make his investigation by attending to all the odd days, and of the even ones to the fourteenth, the twenty-eighth, and the forty-second. For this order is held by some people, on the basis of the principle of harmony, to be the true and perfect number system, for reasons it would be too long to go into on this occasion. Look at it this way, in terms

---

27 Add. Grensemann: ἡρημέναι V.
28 The remainder of this chapter is omitted from V.
29 Foes: ὁκοίας ἔλεξας M.
30 Grensemann: θεωροῦντας M.
31 Add. Linden.

δὲ χρὴ οὕτως· τριάσι τε καὶ τετράσι, ταῖς μὲν τριάσι
συνημμέναις ἁπάσαις, ταῖς δὲ τετράσι δύο μὲν παρὰ
δύο συνημμέναις, δύο δὲ παρὰ δύο διεζευγμέναις.[32]

Αἱ δὲ τεσσαρακοντάδες πρῶτον μὲν κρίνουσιν ἐπὶ
τῶν ἐμβρύων, ὅ τι δ' ἂν ὑπερβάλλῃ τὰς τεσσαράκοντα ἡμέρας τὰς πρώτας, ἐκφεύγει τοὺς τρωσμοὺς
450 ἐπὶ | παντὸς γινομένους. πλέονες δὲ γίνονται ἐν τῇ
πρώτῃ τεσσαρακοντάδι τρωσμοὶ ἢ ἐν ταῖς ἄλλαις
πάσαις.[33] τοῦδε τοῦ χρόνου παρελθόντος ἰσχυρότερά
ἐστι τὰ ἔμβρυα καὶ διακρίνεται καθ' ἕκαστα τῶν
μελέων τὸ σῶμα. καὶ τῶν μὲν ἀρσένων σφόδρα διάδη-
λα γίνεται πάντα, τὰ δὲ θήλεα ἐς τοῦτον τὸν χρόνον
σάρκες φαίνονται ἀποφύσιας μοῦνον ἔχουσαι· πλείονα γὰρ χρόνον τὰ ὅμοια ἐν τῷ ὁμοίῳ ὅμοιά[34] ἐστι καὶ
κρίνεται βραδύτερον διὰ τὴν συνηθείην τε καὶ φιλό-
τητα. καίτοι τά γ' ἄλλα, ὅταν χωρισθῶσι τῆς μητρός,
αἱ θυγατέρες τῶν κούρων θᾶσσον ἡβῶσι καὶ φρονέ-
ουσι καὶ γηράσκουσι διὰ τὴν ἀσθενείην τε τῶν σωμάτων καὶ τὴν δίαιταν.

Ἄλλη δὲ τεσσαρακοντάς, ἐν ᾗ περὶ τὸν ὄγδοον
μῆνα τὰ ἔμβρυα ἐν τῇ μήτρῃ νοσεῖ, περὶ ὧν ὅδε ὁ
λόγος σύμπας λέγεται. τρίτη δ', ἐν ᾗ τὰ παιδία, ὅταν
γένηται καὶ κακοπαθήσαντα [η][35] περιφύγῃ τὰς τεσσαράκοντα ἡμέρας, ἐφάνη ἰσχύοντά τε μᾶλλον καὶ

32 Grensemann: συνεζ. M.
33 Grensemann after Cornarius in marg.: ἐούσαις M.
34 Joly: ὅμοιόν M.
35 Del. Potter.

of triads and tetrads: triads are all conjunct;[3] but of tetrads alternate pairs are conjunct with one other, while at the same time being disjunct from their neighbouring pairs.[4]

Forty-day periods decide in fetuses first that any one which survives beyond the first forty days will escape the miscarriages which occur all that time, for more miscarriages occur in the first forty-day period than in all the others. When this term has passed, fetuses have become stronger, and the body is differentiated in all its limbs. In males everything becomes quite distinguishable, while in females at this stage their tissues seem only to have outgrowths, since like parts in like places remain similar for a longer time, and differentiate more slowly on account of their habituation and attraction. Inversely, after they have left their mother, daughters mature more quickly than boys, become sensible more quickly, and age more quickly, due to the weakness of their bodies and to their regimen.

Another forty-day period in which the fetus becomes ill in the uterus is around the eighth month—to this, my whole treatise is devoted. A third (sc. significant forty-day period) is when children, after being born and surviving for forty days in a sickly state, become visibly stronger and

3 Triads are three-day units of which the first and third days are critical: these triads are all conjunct with one another (e.g. 1–3, 3–5, 5 – 7), producing the series of critical days 1, 3, 5, 7, 9, 11, etc, i.e. all the odd days.

4 Tetrads are four-day units of which the first and fourth days are critical: succeeding pairs of these tetrads are conjunct internally with each other (e.g. 1–4, 4–7 and 8–11, 11–14), but disjunct externally with the preceding and following pairs of tetrads (e.g. 1–7, 8–14, 15–21): thus the special attention drawn above to the even-numbered critical days 14, 28, and 42.

φρονέοντα· καὶ γὰρ τὰς αὐγὰς ὁρᾷ σαφέστερον καὶ
τὸν ψόφον ἀκούει πρόσθεν οὐ δυνάμενα, ὡς ἐπίδοσιν
ἔχοντος τοῦ χρόνου τούτου κατὰ τὰ ἄλλα καὶ κατὰ τὴν
φρόνησιν τὴν διὰ τοῦ σώματος. τὸ μὲν γὰρ ἴδιον
φρόνημα δῆλόν ἐστιν ἐνεὸν ἐν τῷ σώματι τῇ πρώτῃ
ἡμέρῃ. ἔν τε γὰρ τοῖς ὕπνοισιν ἐνίοτε,[36] εὐθέως ἐπὴν
γένωνται, γελῶντα φαίνεται τὰ παιδία καὶ κλαίοντα,
καὶ ἐγρηγορότα γε αὐτόματα εὐθέως γελᾷ τε καὶ
κλαίει πρόσθεν ἢ τεσσαράκοντα ἡμέραι γενοίατο. οὐ
δὲ γελᾷ ψαυόμενά οὔτε κλαίει[37] ἐρεθιζόμενα πρόσθεν
ἢ αὐτὸς ὁ χρόνος οὗτος γένηται· ἀμβλύνονται γὰρ αἱ
δυνάμεις ἐν ταῖς † μίξεσι διὰ την μυρινην ὁ θανατος
ἔλαχεν,[38] † ὥστε παράδειγμα τοῖς πᾶσιν εἶναι ὅτι
πάντα φύσιν ἔχει ἐκ τῶν αὐτῶν ἐόντα μεταβολὰς
ἔχειν διὰ χρόνων τῶν ἱκνεομένων, σαφηνίζεται δὲ
ἄλλο τι ἐν ἑκάστοισι τῶν γινομένων καὶ ἀπογινο-
μένων.

Ἐν δὲ τῷ ἐνιαυτῷ τελεομένῳ πολλαὶ μὲν νοῦσαι
γίνονται, πολλαὶ δὲ ὑγίειαι κατὰ λόγον τοῦ χρόνου
452 πρὸς τοὺς μῆνάς τε καὶ τὰς ἡμέρας | ἑκάστας. τοῖσι
‹δ᾽›[39] ἑβδόμοις καὶ τὰ ἄλλα πολλὰ γίνεται διαφέροντα τοῖς σώμασι, τοῖσι δὲ παιδίοισιν οἱ ὀδόντες ἐκ-
πίπτουσι καὶ ἕτεροι φύονται. τὰ γὰρ ἐπὶ σώματα τὰ
γράψω.

10. [40]Περὶ δὲ ὀκταμήνου γενέσιός φημι δισσὰς
ἐφεξῆς κακοπαθείας γινομένας ἀδύνατον[41] εἶναι [ποιέ-

---

[36] H. Kühlewein in *Philologus* 42 (1884), p. 130: ἐοῦσι M.

more perceptive—the child sees rays of light more clearly, hears noises which before it was unable to—, experiencing a general improvement at this time both in the perceptive capacity present through their body, and in other ways. Now individual perception is clearly present in the body on the first day of life: thus sometimes when children are first born, in their sleep they are seen to laugh and to cry; and before forty days have gone by they spontaneously laugh and cry at once when they are awake. They do not, however, laugh or cry on being touched or provoked before just this time (i.e. forty days) has passed, since the powers are blunted. . . . Hence, this is an example of the principle that everything that consists of the same components has a nature that suffers changes through the particular periods of time that pass, and that they are otherwise explained through each of the things that come into existence or cease to be.

In the year, as it is passes, many diseases arise, but many of these remit at given times according to particular months and days. In seven years all kinds of other things happen to human bodies, e.g. in children the teeth fall out and others grow in. About bodies, this is what I shall write.

10. About the eight months' birth I contend that two distresses following immediately one upon the other are

---

37 Potter after Calvus' *nec . . . plorant*: τε καὶ M: οὔτε κλαίει ψαυόμενά τε καὶ Grensemann.

38 Grensemann: *locus nondum sanatus.*

39 Add. H. Diller in Grensemann

40 The text in V recommences at this point.

41 Ermerins: -άτους MV.

ειν]⁴² φέρειν τὰ παιδία, καὶ διὰ τοῦτο οὐ περιγίνεσθαι
τὰ ὀκτάμηνα. συγκυρεῖ γὰρ αὐτοῖς ἐφεξῆς κακοπαθεῖν τήν τε ἐν τῇ μήτρῃ γινομένην κακοπαθείην καὶ
τὴν ὅταν ὁ τόκος γένηται, καὶ διὰ τοῦτο τῶν ὀκτα-
μήνων οὐδὲν περιεγένετο. ἐπεὶ καὶ τὰ δεκάμηνα καλεό-
μενα, ⟨ἃ⟩⁴³ λέγω ἐν ἑπτὰ τεσσαράκονθ᾽ ἡμέραις μᾶλλον τίκτεσθαι—καὶ μάλιστα προσήκει ἐκτρέφεσθαι
καὶ τελειότατά ἐστιν ἐν τῆσι πρώτῃσι τεσσαράκοντα
ἡμέρῃσιν—ἐπὴν δὲ γένηται, πλείω ἀπόλλυται. ἀναγ-
κάζεται γὰρ πολλὰ μεταλαμβάνοντα ἐν ὀλίγῳ χρόνῳ
πολλὰ νοσεῖν, ἐξ ὧν οἱ θάνατοι γίνονται.

Ἄρχεται δὲ πονεῖν τὸ παιδίον τοῦ τόκου γινομένου
καὶ κινδυνεύει ἀπολέσθαι, ὅταν ἐν τῇ μήτρῃ στρέφηται. φύεται μὲν γὰρ πάντα ἄνω τὴν κεφαλὴν ἔχοντα,
454 τίκτεται δὲ τὰ πολλὰ ἐπὶ κεφαλήν, καὶ | ἀσφαλέ-
στερον ἀπαλλάσσει τῶν ἐπὶ πόδας τικτομένων. τὰ
γὰρ συγκαμπτόμενα τοῦ σώματος ἐπὶ κεφαλὴν οὐ
κωλύει ἰόντος τοῦ παιδίου, ἀλλὰ μᾶλλον, ὅταν ἐπὶ
πόδας ὁρμήσῃ, γίνεται τὰ ἐμφράγματα. αἱ γὰρ στρο-
φαὶ ἐν τῇ γαστρὶ καὶ ἄλλος κίνδυνος, καὶ οἱ ὀμφαλοὶ
πολλάκις ἤδη τῶν παιδίων ἐφάνησαν ἀμφὶ τὰς δει-
ράς. ἢν γὰρ κατὰ τὸ μόριον, ὁποτέρωθεν ἂν τύχῃ ὁ
ὀμφαλὸς τῇ μήτρῃ μᾶλλον παρατεταμένος, ταύτῃ τὸ
παιδίον τῆς κεφαλῆς τὴν περιαγωγὴν [τοῦ ὀμφαλοῦ]⁴⁴
στρεφόμενον ποιήσεται, ἢ περὶ τὸν αὐχένα ἢ κατὰ
τὸν ὦμον τὴν περιβολὴν τοῦ ὀμφαλοῦ ἀντιτείνει.
τούτου δὲ γινομένου καὶ τὴν μητέρα ἀνάγκη πονῆσαι
μᾶλλον καὶ τὸ παιδίον ἢ ἀπολέσθαι ἢ χαλεπώτερον

impossible for children to withstand, and that for this reason eight months' children do not survive. For it happens to these children in immediate succession that they suffer both the strain that occurs in the uterus and the one after birth has taken place, and therefore no eight months' child survives. Yet even so-called ten months' children, by which I mean those born in seven forty-day periods—and these are the most suitable to bring up and appear to be the most mature in the first forty-day period—, die in considerable numbers after they are born. For they must, when they experience many changes in a short period of time, suffer severe illnesses, which result in deaths.

A fetus begins to suffer strain as the birth process takes place, and is in danger of dying as it is turned in the uterus. For although all fetuses originally grow with their heads directed upward, many are born head first, and these are more safely delivered than those born feet first. For the folded parts of the body do not cause any impediment when the fetus moves out head first, whereas when it moves forward feet first impactions are more frequent. And in fact the turning that takes place in the abdomen represents another danger, as umbilical cords are often found at birth around fetuses' necks. For if at the place, on whichever side of the uterus the umbilical cord happens to be more distributed, the fetus becomes entangled when it rotates its head, it gets wound up by the twisting of the umbilical cord around its neck or over its shoulder. When this happens, both the mother must strain herself more, and the child must either die or be born with greater difficulty,

---

42 Del. Joly after Calvus. 43 H. Diller in Grensemann.
44 Del. Grensemann.

ἐξελθεῖν, ὥστε ἤδη πολλὰ καὶ ἔσωθεν[45] τὴν ἀρχὴν τῆς
νούσου τῶν παιδίων ἦλθεν ἔχοντα, ἐξ ἧς τὰ μὲν
ἀπώλετο, τὰ δὲ νοσήσαντα περιεγένετο.

11. Ὁκόσα δ᾽ ἂν εὐπορήσῃ καὶ ἀσφαλέως ἐς τοὐμφανὲς ἐξίῃ, ἀνεθέντα ἐξαίφνης ἐκ τῆς ἀνάγκης τῆς ἐν
τῇ γαστρὶ παχύτερα καὶ μέζω παραυτίκα ἢ κατὰ
λόγον ἐγένετο οὐκ αὐξήματος, ἀλλ᾽ οἰδή|ματος γενο-
μένου, ἐξ ὧν δὴ πολλὰ ἀπώλετο. ἢν γὰρ μὴ συνίζῃ τὸ
οἴδημα θᾶσσον ἢ τριταῖον ἢ ὀλίγῳ πολυχρονιώτερον,
αἱ νοῦσοι γίνονται ἀπ᾽ αὐτοῦ.

12. Αἵ τε τροφαὶ καὶ αἱ ἀναπνοαὶ σφαλεραὶ μεταλ-
λασσόμεναι. ἤν τι γὰρ νοσηλὸν ἐσάγωνται, κατὰ τὸ
στόμα καὶ κατὰ τὰς ῥῖνας ἐσάγονται. καὶ ἀντὶ τοῦ
τοσαῦτα εἶναι τὰ ἐσιόντα, ὅσα ἐξαρκεῖ, καὶ μὴ περιγίνεσθαι, πολλῷ πλείω ἐσέρχεται, ὥστε ἀναγκάζε-
σθαι ὑπὸ τοῦ πλήθεος τῶν ἐσιόντων καὶ ὑπὸ τῆς
διαθέσιος, ἧς διάκειται τὸ σῶμα τοῦ παιδίου ἤδη, τὰ
μὲν κατὰ τὸ στόμα τε καὶ τὰς ῥῖνας πάλιν ἐξιέναι, τὰ
δὲ κατὰ τὸ ἔντερον καὶ τὴν κύστιν κάτω περαιοῦσθαι,
πρόσθεν οὐδενὸς τούτων οὕτω γινομένου.

Καὶ ἀντὶ πνευμάτων τε καὶ χυμῶν οὕτω συγγενέων,
ὅκως αἰεὶ [δ᾽][46] ἀνάγκη ἐν τῇσι μήτρῃσι γίνεσθαι
συνηθείην τε ἔχοντα καὶ εὐμενείην, πᾶσι ξένοισι
χρῆται ὠμοτέροισί τε καὶ ξηροτέροισι καὶ ἧσσον
ἐξηνθρωπισμένοις, ἐξ ὧν[47] ἀνάγκη πόνους γενέσθαι
πολλούς, πολλοὺς[48] δὲ καὶ θανάτους. ἐπεὶ καὶ τοῖς ἀν-

45 Ermerins: ἐσώθη ἔνδον M: ἔνδον V.

so that in the past many of these children have carried forth from within the beginning of a disease, from which some have died and others have survived in an ailing state.

11. Children who are fortunate and come out safely into the light, but who, when suddenly released from the compression of the uterus, at once become thicker and larger than they should be, and this not through growth but through the onset of oedema, in many cases die. For if the oedema does not go down before the third day or a little later, it provokes diseases.

12. Nutriment and breath being dangerously altered: if children take in anything likely to cause disease, they take it in through their mouth or their nostrils. And if, instead of what is ingested being of an amount that is just adequate, and not forming residues, what enters is much more, it follows, on account of the fullness of what is being ingested and of the child's condition as it is already established, that one part of what goes in must necessarily come back out through the mouth and the nostrils, and that the rest must pass down through the intestines and the bladder, although nothing of this sort has taken place before.

Also, in place of the suitably congenial breath and humours in the uterus which must always produce familiarity and harmony, the new-born child has to deal with all kinds of foreign substances that are rawer, drier, and less humanized, and out of which many distresses necessarily result, and also many deaths. Likewise in adults, too, changes of

---

46 Del. Littré: οἷς δ' M: ὁκόσαι· εἰ δ' V.

47 Foes in note 11: ἔξω M: ἐξιὸν V.

48 Littré: πολλοῖσ(ι) MV.

δράσι πολλάκις αἱ μεταλλαγαὶ τῶν χωρίων τε καὶ τῶν
διαιτημάτων τὰς νούσους ἐμποιέουσιν. ὁ δ' αὐτὸς
λόγος ἐστὶ καὶ ἀμφὶ τῶν ἐσθημάτων· ἀντὶ γὰρ τοῦ
σαρκὶ καὶ χυμοῖς ἠμφιέσθαι χλιεροῖς τε καὶ ὑγροῖς
καὶ συγγενέσι, τοιαῦτα ἀμφιέννυνται τὰ παιδία οἷά
περ οἱ ἄνδρες.

Ὁ δ' ὀμφαλός, δι' οὗ αἱ ἔσοδοί εἰσι τοῖσι παιδίοισι
μοῦνον τοῦ σώματος, τῇ μήτρῃ προσέ[ρ]χεται,[49] καὶ
458 διὰ τούτου κοινωνεῖ τῶν | ἐσιόντων. τὰ δ' ἄλλα συμ-
μύει καὶ οὐκ ἀνεστομωμένα ἐστι πρόσθεν ἢ ἐν ἐξόδῳ ᾖ
τὸ παιδίον ἐκ τῆς γαστρός. ὁκόταν δ' ἐν ἐξόδῳ ᾖ, τὰ
μὲν ἄλλα ἀναστομοῦται, ὁ δ' ὀμφαλὸς λεπτύνεταί τε
καὶ συμμύει καὶ ἀποξηραίνεται. ὥσπερ δὲ τοῖς ἐκ τῆς
γῆς φυομένοις οἱ καρποὶ ἁδρυνόμενοι ἀποκρίνονταί τε
καὶ ἀποπίπτουσι κατὰ τὴν διάφυσιν, οὕτω καὶ τοῖσι
παιδίοισιν ἁδρυνομένοισί τε καὶ τελείοισι γινομένοις
ὁ μὲν ὀμφαλὸς συνέμυσε, τὰ δ' ἄλλα ἀνεστομώθη
ὥστε ἐσδέχεσθαί τε τὰ ἐσιόντα καὶ ἐξόδους ἔχειν
κατὰ φύσιν, οἷς ἀνάγκη τοὺς ζῶντας χρῆσθαι. ἕκαστα γὰρ χωρίζεται ῥέποντα κατὰ τὰς συλλοχίας.
[κρατιστεύει δὲ τὰ τῷ ἡλίῳ συντροφώτατα.][50]

13. Οἱ δὲ δεκάμηνοι τῶν τόκων καὶ ἑνδεκάμηνοι ἐκ
τῶν ἑπτὰ τεσσαρακοντάδων τὸν αὐτὸν τρόπον γίνονται καὶ ἐκ τοῦ ἡμίσεος τοῦ ἐνιαυτοῦ οἱ ἑπτάμηνοι· τῇσι
γὰρ πλείστῃσι τῶν γυναικῶν ἀναγκαῖόν ἐστιν ἐν
γαστρὶ λαμβάνειν μετὰ τὰ καταμήνια, ἢν ἴῃ ἡ λύσις.
δεῖ τοίνυν τῇ γυναικὶ χρόνον δοῦναι τοῦ μηνός, ἐν ᾧ
αὐτὴ ἡ κάθαρσις ἔσται, καὶ ὁ χρόνος οὗτος, ᾖσιν

location and regimen lead to diseases. This same principle applies to the infants' clothing, as well. In place of being surrounded by flesh and humours that are warm, moist, and congenial, new-born children are clothed with the same sorts of things that adults are.

The umbilical cord, through which pass the only channels out of the body into fetuses, is attached to the uterus, and through it the fetus has its share of the things that are ingested. All the other parts of the fetus' body are closed, and do not open up before the fetus is in the process of leaving the abdomen. When this is occurring, the other parts open up, whereas the umbilical cord becomes narrow, closes, and dries out. Just as in plants that grow out of the earth, at germination their fruits ripen, separate, and fall away, so too in infants that are ripening and becoming mature, the umbilical cord closes at the same time that the other parts open up in order to be able to receive what the body ingests, and structural exits form, which living beings must make use of. For each material is separated off, inclining to the place where they are severally collected.

13. Ten months' births and eleven months' births occur after the seventh forty-day period in the same way that seven months' births occur after half a year, since for most women, it must necessarily be after the menses that they become pregnant, after the evacuation passes. Accordingly, you should allow a woman the portion of the month in which the cleaning itself will be taking place, and this

---

49 Cornarius in marg.: -ἔρχεται M: -ἔρχονται V.
50 Del. Joly.

ἐλάχιστος γίνεται, τρεῖς ἡμέραι, τῇσι δὲ πλείστῃσι
καὶ πολὺ πλείονες· ἔστι δὲ καὶ ἄλλα πολλὰ κωλύματα
καὶ τῇσι γυναικὶ καὶ τοῖς ἀνδράσιν,[51] ἐξ ὧν βρα-
δύνεται ἡ σύλληψις. χρὴ δ' ἐν τοῖσι μάλιστα καὶ τόδε
λογίζεσθαι, ὅτι ἡ νεομηνίη μίη ἡμέρη ἐοῦσα ἐγγύ-
τατον τριηκοστημόριόν ἐστι τοῦ μηνός, αἱ δὲ δύο
460 ἡμέραι σχεδὸν πεντεκαιδεκατημόριον τοῦ μη|νός, αἱ
δὲ τρεῖς ἡμέραι δεκατημόριον τοῦ μηνός, καὶ τἆλλα
κατὰ λόγον τούτων, καὶ οὐχ οἷόν τε ἐν τοῖς ἐλάσσοσί
γε μορίοις μᾶλλον γίνεσθαι οὔτε τὴν λύσιν τῶν κατα-
μηνίων οὔτε τὴν σύλληψιν τῶν ἐμβρύων. ἐκ τούτων
οὖν ἁπάντων ἀναγκαῖόν ἐστι τῇσι πλείστῃσι τῶν
γυναικῶν περὶ διχομηνίην ἐν γαστρὶ λαμβάνειν καὶ
πορρωτέρω, ὥστε πολλάκις ἐπιλαμβάνειν τοῦ ἑνδεκάτου μηνὸς τὰς ὀγδοήκοντα καὶ διακοσίας ἡμέρας·
τοῦτο γάρ ἐστιν ἑπτὰ τεσσαρακοντάδες. ὅ τι γὰρ ἂν
ἔξω τῆς διχομηνίης συλλάβῃ ἡ γυνή, τοῦτο πᾶν
ἀναγκαῖόν ἐστι τοῦ ἑνδεκάτου μηνὸς ἐπιλαμβάνειν, ἤν
περ ἐς τὴν τελευταίην περίοδον καταστῇ.

---

[51] καὶ τῇσι—ἀνδράσιν Littré after Calvus' et foeminarum et
virorum: τοῖς ἀνδράσι M: τῇσι γυναικὶ καὶ τῇσιν V.

time, in women in whom it is least, is three days, but in most women it is many days more; there are also many other impediments in women and men that delay conception. You must also consider most especially in these calculations that when the new moon is one day old, this is very close to one thirtieth of the month, that two days are nearly a fifteenth of a month, that three days are a tenth of a month, and so forth after this fashion, and that it is not really possible for either the evacuation of the menses or the conception of fetuses to take place in a shorter period of time. Now, for all these reasons, most women must necessarily become pregnant around the middle of the month,[5] or beyond that, with the consequence that they often arrive at the 280th day only in the eleventh month—for this equals seven forty-day periods. In fact, whenever a woman conceives beyond the time around the middle of the month, all such fetuses must necessarily arrive at the eleventh month, if they remain in the uterus for their full term.[6]

[5] Here as elsewhere in the treatise "month" refers to a woman's menstrual month rather than to an astronomical lunar month.

[6] The author calculates that if conception occurs after the middle of the first month, so that month I has < 14 days, a full 280 day gestation will extend beyond months II–X ($59/2 \times 9 = 265\frac{1}{2}$ days) into month XI.

COAN PRENOTIONS

INTRODUCTION

Neither is *Coan Prenotions* included with *Prognostic*, *Prorrhetic I–II*, and *Humours* among the semeiotic works (Σεμειωτικά) in Erotian's census of Hippocratic writings, nor does any word definitely attributable to the treatise appear in his *Glossary*.[1] However Galen, writing a century later, clearly knows the work, citing the title six times in his *Commentary on Hippocrates, Epidemics III*.[2] He characterizes *Prorrhetic I* and *Coan Prenotions* as follows:

> For I have shown that all these (semeiotic) works have great value in regard to the sick, although if anyone were to follow too strictly everything said in *Prorrhetic* he would go completely wrong. I have also showed that much of what is contained in *Coan Prenotions* is similar in character, having some mixture of material originating from *Aphorisms* and *Prognostic* and some from the *Epidemics*, and that this part alone of what is in these books is true, whereas everything else in *Prorrhetic* and *Coan Prenotions* is unsound.[3]

[1] Erotian p. 9.

[2] E. Wenkebach, *Galeni In Hippocratis epidemiarum librum III*, Corpus Medicorum Graecorum V 10 2.1, Leipzig and Berlin, 1936, pp. 13, 59 (ter), 62 (bis). See also the abbreviated title ὡς ἐν Κῳακαῖς employed in Galen's *Glossary*, vol. 19, 81, s.v. ἄνθεα.

[3] Galen vol. 17A, 579 = CMG V 10 2.1, 62.

COAN PRENOTIONS

How *Coan Prenotions* came to share such a considerable part of its textual content, often verbatim, with *Prorrhetic I*, *Aphorisms*, and *Prognostic*, and what precisely the interdependencies among the writings are have proven to be difficult questions; the situation is summarized by W. H. S. Jones, Loeb *Hippocrates* vol. 2, xx–xxix.

The individual chapters of *Coan Prenotions* are arranged by topic:

These chapters consist mainly of general prognostic statements; in addition, one chapter (543) makes reference to a particular patient, and twenty-three pose apparently self-directed questions: e.g. 78 "Do these kinds of exacerbations also indicate phrenitis?"

With regard to its textual transmission, *Coan Prenotions* falls into two parts: ch. 1–274 as far as ἐν τῇ πρώτῃ περιόδῳ are preserved in two independent Greek manuscripts: A and I (derived from a now lost part of M); ch. 274 – 640 are transmitted only in A.[4] In the part of the treatise where both witnesses exist, each contains some chapters omitted from the other, and occasional variations of chapter order exist: A's chapter order in I's terms (which has been standard in editions since the Aldina) is: 1–2, 6–8, 3, 9–12, 14–20, 4–5, 21–274, omitting ch. 13, 41, 59, 151, and 241. In I, ch. 6 is located within ch. 11, and ch. 219, 221, and 275–640 are omitted.

Coan Prenotions, which is present in all the collected editions and translations of the Hippocratic Collection, received considerable special attention at the time when semeiotics played a central role in medical education and practice, appearing in four separate editions/translations with commentaries in the last quarter of the sixteenth century alone:[5]

> Jacobus Hollerius, *Magni Hippocratis Coaca praesagia . . . cum interpretatione et commentariis . . . nunc primum Desiderii Jacotii . . . opera editis . . .* , Lyon, 1576.

---

[4] See above pp. viiif. for manuscript details.
[5] See the bibliography at Littré vol 5, 586f.

COAN PRENOTIONS

Joannes Opsopoeus, *Hippocratis . . . Coaca praesagia
. . . Graecus et Latinus contextus accurate renovatus
. . .* , Frankfurt, 1587. (=Opsopoeus)

Ludovicus Duretus, *Hippocratis magni Coacae prae-
notiones, interprete et enaratore . . .* , Paris, 1588. (=
Duretus)

Illefonsus Lopes Pincianus, *Hippocratis prognosticum,
in quo omnes . . . tabellae . . . brevibus annotationibus
illustratae . . .* , Madrid, 1596.

Important modern scholarly works devoted to *Coan
Prenotions* include:

F. Z. Ermerins, *De Hippocratis doctrina a prognostice
oriunda*, Leiden, 1832.

O. Poeppel, *Die hippokratische Schrift* Κωακαὶ προ-
γνώσεις *und ihre Überlieferung*, Teil 1–2, Diss. Kiel,
1959.

An English translation appeared in:

J. Chadwick and W. N. Mann, *The Medical Works of
Hippocrates*, Oxford, 1950, pp. 219–78.

The present edition of *Coan Prenotions* is based on collations of the two independent manuscripts A and I, made
partly from microfilm and partly from autopsy.

ΚΩΑΚΑΙ ΠΡΟΓΝΩΣΕΙΣ

1. Οἱ ἐκ ῥίγεος περιψυχόμενοι, κεφαλαλγέες, τράχηλον ὀδυνώδεες, ἄφωνοι, ἐφιδροῦντες, ἐπανενέγκαντες θνήσκουσιν.

2. Αἱ μετὰ καταψύξιος δυσφορίαι κάκισται.

3.[1] Κατάψυξις μετὰ σκληρυσμοῦ, ὀλέθριον.

4.[2] Ἐκ καταψύξιος φόβος καὶ ἀθυμίη ἄλογος ἐς σπασμὸν ἀποτελευτᾷ.

5. Αἱ ἐκ καταψύξιος οὔρων ἀπολήψιες, κάκιστον.

6.[3] Μετὰ ῥίγους ἄπνοια[4] κακόν· κακὸν δὲ καὶ λήθη.

7. Τὰ κωματώδεα ῥίγεα ὑπολέθρια· καὶ τὸ φλογῶδες ἐν προσώπῳ μεθ᾽ ἱδρῶτος ἐν τούτοισι κακόηθες· ἐπὶ τούτοισι ψύξεις τῶν ὄπισθεν, σπασμὸν ἐπικαλέονται· καὶ ὅλως δὲ ψύξεις τῶν ὄπισθε, σπασμώδεες.

8. Αἱ ἐκ νώτου φρῖκαι πυκναί, καὶ ὀξέως μεταπίπτουσαι, δύσφοροι· οὔρου ἀπόληψιν ἐπώδυνον σημαίνουσιν· τὸ ἐφιδροῦν τούτοισι κάκιστον.[5]

---

[1] In ms. A ch. 6–8 are located before ch. 3.

[2] In ms. A ch. 4–5 are located after ch. 20.

[3] In ms. I ch. 6 is located within ch. 11 between δυσφορ. and ὅσα.   [4] A: ἄγνοια I.

[5] τὸ ἐφιδ.—κάκιστον om. A.

COAN PRENOTIONS

1. Persons who subsequent to a chill suffer a generalized cooling, headaches, pains in the neck, speechlessness, and sweating over the whole body recover consciousness, but then die.

2. Restlessness, if accompanied by a chill, is a very bad sign.

3. A chill together with constipation is a fatal sign.

4. A state of fear and groundless despondency arising subsequent to a chill ends with convulsions.

5. The stoppage of urine subsequent to a chill is a very bad sign.

6. Shortness of breath[1] in conjunction with a chill is a bad sign; bad also is forgetfulness.

7. Chills accompanied by coma are dangerous; in these patients, a fiery redness of the face together with sweating is also malignant, and chills of the back region announce a convulsion. In general, chills of the back region indicate a tendency to convulsions.

8. Frequent attacks of shivering starting from the back and moving rapidly through the body cause restlessness, and signal a painful blockage of the urine. For such patients to sweat over the whole body is a very bad sign.

---

[1] With I's text, "Loss of understanding."

9. Ῥῖγος ἐν συνεχείᾳ, τοῦ σώματος ἀσθενέοντος ἤδη, θανάσιμον.

10. Οἱ πυκνὰ ἐφιδροῦντες καὶ ἐπιρριγέοντες, ὀλέθριον· καὶ ἐπὶ τῇσι τελευτῇσιν ἀναφαίνονται ἐμπύημα ἔχοντες καὶ κοιλίας ταραχώδεας.

11. Τὰ ἐκ νώτου ῥίγεα δυσφορώτερα· ὅσα ἑπτακαιδεκάτῃ ἐπιρριγώσαντα, τετάρτῃ καὶ εἰκοστῇ ἐπιρριγεῖ, δύσκολα.

590   12. Οἱ φρικώδεες κεφαλαλγικοὶ ἐφιιδροῦντες, κακοήθεες.

13.[6] Οἱ φρικώδεες ἐφιδροῦντες πολλῷ, δύσκολοι.

14. Τὰ πολλὰ νωθρώδεα ῥίγεα, κακοήθεα.

15. Οἷσιν ἑκταίοισι ῥίγεα γίνεται, δύσκριτα.

16. Ὁκόσοισι φρῖκαι πυκναὶ ὑγιαίνουσιν,[7] οὗτοι ἐξ αἵματος ῥύσιος ἐμπυΐσκονται.

17. Τὸ φρικῶδες καὶ τὸ δύσπνοον ἐν τοῖσι πόνοισι, σημεῖα φθινώδεα.

18. Ἐξ ἐκπυήσιος πνεύμονος καὶ κατὰ κοιλίας ἐνίοτε ἀλγήματα καὶ κληῖδα, καὶ τὸ ὑπορέγχειν ἀσώδεα, σημαίνει πτυέλου πλῆθος ἐν τῷ πλεύμονι.

19. Οἱ φρικώδεες, ἀσώδεες, κοπιώδεες, ὀσφυαλγέες, κοιλίας καθυγραίνονται.

20. Τὰ δ' ἐπιρριγέοντα, καὶ ἐς νύκτα μᾶλλόν τι παροξυνόμενα, ἄγρυπνα, φλεδονώδεα, ἐν τοῖσιν ὕπνοισιν ἔστιν ὅτε οὖρον ὑπ' αὐτοὺς χαλῶντες, ἐς σπασμὸν ἀποτελευτᾷ.

[6] Ch. 13 om. A.   [7] Ὁκ.—ὑγιαίνουσιν om. A.

9. In a continuous fever, a chill occurring when the body is already weak is a deadly sign.

10. To suffer frequent sweats over the whole body and subsequent chills is a deadly sign. In the end such patients have an internal suppuration and disordered cavities.

11. Chills starting from the back cause great restlessness. If such chills re-occur on the seventeenth day and again on the twenty-fourth day, these conditions will be troublesome.

12. In headaches with shivering, those sweating over the whole body are in an evil way.

13. Shivering together with frequent sweating over the whole body is troublesome.

14. Frequent chills accompanied by torpor are malignant.

15. Patients in whom chills arise on the sixth day will have difficulty reaching their crisis.

16. Persons who have frequent attacks of shivering while they are healthy will develop internal suppuration after having a haemorrhage.

17. A tendency to shivering and difficult breathing on exertion are signs of consumption.

18. After suppuration in a lung, occasional pains in the areas of the cavity and the clavicle, together with slightly stertorous breathing and nausea, indicate a surfeit of sputum in the lung.

19. Conditions characterized by shivering, nausea, weariness, and pains in the loins lead to diarrhoea.

20. Chills occurring in diseases that have their exacerbations more at night, if accompanied by sleeplessness, loquaciousness, and occasional urinary incontinence in the sleep, end in convulsions.

21. Τὰ δὲ συνεχέα ῥίγεα ἐν ὀξέσι, πονηρόν.

22. Αἱ ἐκ ῥίγεος μετὰ κεφαλαλγίης ἐκλύσιες, ὀλέθριον· τὰ αἱματώδεα οὖρα ἐν τούτοισι, πονηρόν.

23. Ῥῖγος ὀπισθοτονῶδες κτείνει.

24. Τὰ φρικάσαντα καὶ ἀνιδρώσαντα κρισίμως, ἐς δὲ τὴν αὔριον φρίξαντα παραλόγως, ἀγρυπνεῦντα, μὴ πεπαινομένων, αἱμορραγήσειν οἴομαι.

25. Τὰ μετὰ ῥίγεος ἐπισχόμενα οὖρα, πονηρὰ καὶ σπασμώδεα, ἄλλως τε καὶ προκαρωθέντι· ἐλπὶς δὲ ἐπὶ 592 τούτοισι, καὶ τὰ παρ' | ὦτα.

26. Τὰ τριταιοφυέα ῥίγεα, τὴν ἐν μέσῳ παροξυνόμενα, πυρετῷ ἀτάκτῳ, πάνυ κακοήθεα· τἀναντία δὲ παροξυνόμενα . . . .

27. Τῶν σπώντων μετὰ ῥίγεος καὶ πυρετοῦ, ὀλέθριον.

28. Αἱ ἐκ ῥίγεος ἀφωνίαι τρόμῳ λύονται· καὶ τὰ ἐπιρριγεῦντα τρομώδεα γινόμενα κρίνει.

29. Οἱ ἐκ ῥίγεος μετὰ κεφαλαλγίης ἐκλυόμενοι, σφαλεροί· τὸ αἱματῶδες οὖρον τούτοισι κακόν.

30. Οἷσι ῥῖγος, οὔρου ἐπίστασις.[8]

31. Σπασμὸς ἐν πυρετῷ, χειρῶν καὶ ποδῶν πόνοι, κακόηθες· κακόηθες δὲ καὶ ἐκ μηροῦ ὁρμὴ ἀλγήματος· ἀλλ' οὐδὲ γουνάτων πόνος κρήγυον· ἀτὰρ καὶ γαστροκνημιῶν πόνοι, κακοήθεες, ποτὲ δὲ καὶ γνώμης παρά-

---

[8] I: ἀπόστασις A.

---

[2] With A's text, "an apostasis through their urine."

21. Continuous chills in acute diseases bode ill.

22. After a chill with headaches, faintness of the body is a fatal sign; bloody urines in such patients bode ill.

23. A chill occurring together with opisthotonus leads to death.

24. To shiver and reach a crisis with sweating, and on the following day to have unexplained shivering and sleeplessness in the absence of coction, indicates, I think, a haemorrhage.

25. For urines to be checked during a chill bodes ill and presages convulsions, especially if the person is already affected by drowsiness; in such cases swelling beside the ears is also to be expected.

26. In a person with an irregular fever, if chills occurring in the pattern of a tertian fever have their exacerbations on the middle day, it is a very malignant sign; but when the exacerbations are on the other days, it is a good sign.

27. Convulsive disorders in conjunction with chills and fever are fatal.

28. Speechlessness developing out of a chill is resolved by trembling; when trembling is added to chills, a crisis occurs.

29. If, after a chill with headache, there is a faintness of the body, such patients are in a precarious state; bloody urines in these patients are a bad sign.

30. Persons with a chill will have a stoppage of their urine.[2]

31. A convulsion during a fever, along with pains of the hands and feet, is a malignant sign; malignant also is the onset of a pain from a thigh; nor is pain of the knees a good symptom. Pains of the calves are also malignant, and

φοροι, ἄλλως τε καὶ ἢν οὖρον ἐναιωρηθῇ.

32. Οἱ ἐξ ὑποχονδρίων ἀλγήματος πυρετοί, κακο-
594 ήθεες· τὸ | καρῶδες ἐπὶ τούτοισι, κακόν.

33. Οἱ μὴ διαλείποντες, ἐφιδροῦντες πυκνά, μετὰ
ὑποχονδρίου ἐντάσιος, ἐπὶ πολὺ κακοήθεες· καὶ τὰ ἐς
ἀκρώμιον καὶ κληῖδας ἐνστηρίζοντα ἀλγήματα ἐν
τούτοισι πονηρά.

34. Οἱ τριταιοφυέες ἀσώδεες πυρετοί, κακοήθεες.

35. Αἱ ἐν πυρετῷ ἀναυδίαι, κακόν.

36. Κοπιώδεες, ἀχλυώδεες, ἄγρυπνοι, κωματώδεες,
ἐφιδροῦντες, ἀναθερμαινόμενοι, κακόν.

37. Οἱ κοπιώδεες, μετὰ φρίκης, ἐφιδρώσαντες κρισίμως, ἀναθερμανθέντες, ἐν ὀξεῖ, κακόν, ἄλλως τε καὶ
ἢν ἐπιστάζῃ· περὶ ταῦτα ἰκτεριώδεες κατακορέες θνή-
σκουσι· λευκὸν διαχώρημα τούτοισι προσδιέρχεται.

38. Οἱ τριταιοφυέες πλανώδεες, ἐς ἀρτίας μετα-
πεσόντες, δύσκολοι.

39. Οἱ ἐν κρισίμοισιν ἀλυσμοὶ ἀνιδρωτὶ περιψυχόμενοι, καὶ πάντες δὲ οἱ ἄνευ ἱδρῶτος καὶ ἀκρίτως,
κακόν· καὶ οἱ ἐπιρριγώσαντες ἐκ τούτων, ἐμέσαντες
ἄκρητα χολώδεα, ἀσώδεες, τρομώδεες, ἐν πυρετῷ, κα-
κόν· καὶ φωνὴ δὲ ὡς ἐκ ῥίγεος.

40. Τὰ δ' ἐκ ῥινῶν σμικροῖσιν ἱδρῶσι περιψύχοντα,
κακόν.

41. Οἱ ἐφιδροῦντες, ἄγρυπνοι, ἀναθερμαινόμενοι,
κακόν.

sometimes cause derangement of the mind, especially if the urine contains suspended material.

32. Fevers beginning from a pain in the hypochondria are a malignant sign; drowsiness in these cases is bad.

33. Frequent perspiration over the whole body in fevers that do not intermit, if accompanied by tension of the hypochondrium, is generally a malignant sign; pains that become fixed in the tips of the shoulders and the clavicles in such patients also bode ill.

34. Fevers of a tertian kind when present together with nausea are a malignant sign.

35. Loss of speech during a fever is a bad sign.

36. If patients with weariness, dimness of vision, sleeplessness, coma, and perspiring over their whole body become warm, it is a bad sign.

37. In an acute disease, if patients who have weariness together with shivering perspire over their whole bodies during a crisis and become warm, it is a bad sign, especially if they have a nose-bleed; about that time they die with a very intense jaundice; they also pass white stools.

38. Irregular fevers of the tertian type that change so that they falls on even days indicate trouble.

39. Patients who are restless at their crisis and cool down without a sweat, and also those with restlessness who neither sweat nor have a crisis, are in a bad way; also those who in their fever have a subsequent chill, who vomit unmixed bilious material, who are nauseated, and who tremble. A voice that sounds as if it has originated from a chill is also a bad sign.

40. For patients to be cooled down by mild sweating after a nose-bleed is a bad sign.

41. For sleepless persons who perspire over the whole body to become warm is a bad sign.

42. Οἱ ἐφιδρῶντες ἐν πυρετῷ, κακοήθεες.

43. Οἷσι, χολώδεος διαχωρήσιος ἐούσης, περὶ στῆθος δῆξις καὶ πικρότης, κακόν.

596   44. Ἐν πυρετῷ, κοιλίης | ἐμφυσωμένης, πνεῦμα μὴ διεκπῖπτον, κακόν.

45. Κοπιώδεες, λυγγώδεες, κάτοχοι, κακοί.

46. Ἐκ νώτου πυκινῇσι καὶ λεπτῇσι φρίκῃσιν ἐφιδροῦντες, δύσφοροι· οὔρου ἀπόληψιν ἐπώδυνον σημαίνει· τὸ ἐφιδροῦν τούτοισι, κακόν.

47. Τὸ παρὰ τὸ ἔθος ποιέειν τι, οἷον προθυμέεσθαι προσδέχεσθαί τι πρότερον μὴ εἰθισμένον, ἢ τοὐναντίον, πονηρὸν καὶ πλησίον παρακοπῆς.

48. Τὰ ἐν πονηροῖσι σημείοισι κουφίζοντα, καὶ τὰ ἐν χρηστοῖσι μὴ ἐνδιδόντα, δύσκολα.

49. Οἱ ἐφιδροῦντες καὶ μάλιστα κεφαλὴν ἐν ὀξέσιν, ὑποδύσφοροι, κακόν, ἄλλως τε καὶ ἐν οὔροισι μέλασι· καὶ τὸ θολερὸν ἐν τούτοισι πνεῦμα, κακόν.

50. Ἄκρεα ταχὺ ἐπ᾽ ἀμφότερα μεταπίπτοντα, καὶ δίψα δὲ τοιαύτη, πονηρόν.

51. Ἐκ κοσμίου θρασεῖα ἀπόκρισις, φωνὴ ὀξεῖα, κακόν· ὑποχόνδρια τούτοισιν εἴσω εἴρυαται.

42. To perspire over the whole body during a fever is a malignant sign.

43. While bilious evacuations are taking place, an irritation around the chest with discomfort is a bad sign.

44. In patients who have a collection of wind in their cavity during a fever, for the air not to escape is a bad indication.

45. Patients with weariness, hiccups, and catalepsy are in a bad way.

46. In patients with frequent mild attacks of shivering originating from the back, to perspire over the whole body causes restlessness: this announces a painful blockage of the urine. For such patients to perspire over the whole body is a bad sign.

47. To do something outside one's custom, as for example to desire to have something that was not one's habit before, or just the opposite, bodes ill and indicates that delirium is near.

48. Both conditions that are relieved in the presence of bad signs, and those that do not relent in the presence of favourable signs will be difficult.

49. For persons who perspire over the whole body—but especially their head—in acute diseases to become somewhat restless is a bad sign, particularly in association with the passage of dark urines; laboured breathing in these patients is also a bad sign.

50. Rapid changes in the extremities in both directions (sc. between cold and hot), and the same with regard to thirst, bode ill.

51. An insolent reply in a raised voice from a polite person is a bad sign; in such persons the hypochondrium is drawn tight inside.

52. Τὰ ἐκ καταψύξιος ἱδρώδεος ταχὺ ἀναθερμαινόμενα, κακόν.

53. Οἱ ἐν ὀξέσιν ἐφιδροῦντες, ὑποδύσφοροι, κακόν.

54. Οἱ παραλόγως, κενεαγγείης μὴ ἐούσης, ἀδύνατοι, κακόν.

55. Ἐν πυρετῷ ἕλξις οἷον ἀπὸ ἐμέτου εἰς ἀνάχρεμψιν ἀποτελευτῶσα, κακόν.

56. Νάρκαι ἐς ἀμφότερα ταχὺ μεταπίπτουσαι, κακόν.

57. Στάξιες αἱ ἐλάχισται, κάκισται.

58. Κακὸν δὲ πάντως ἐν ὀξεῖ δίψα παραλόγως λυθεῖσα.

59. Οἱ πρὸς χεῖρα ἀναΐσσοντες, κακοί.

598   60. Οἷσιν ἅμα πυρετῷ καυ|σώδει οἰδήματα ὑπνώδεα νενωθρευμένα, ἐς πλευρὸν ὀδύνη ἐσελθοῦσα, παραπληκτικῶς κτείνει.

61. Πνιγμὸς ἐν ὀξέσιν, ἰσχνοῖσιν, ὀλέθριον.

62. Ἐπὶ τοῖσιν ἤδη ὀλεθρίοισι τὰ σμικρὰ τρομώδεα, καὶ ἰώδης ἔμετος, ἐν τοῖσι ποτοῖσιν ὑποψοφέοντες καὶ ὑποβορβορύζοντες ξηροῖσι, καὶ οἱ χαλεπῶς καταβροχθίζοντες πνεύματι βηχώδει, ὀλέθριοι.

63. Ἐν ὀξέσι ὑποκατεψυγμένοισι ἐν χερσὶ καὶ ποσὶν ἐρυθήματα, ὀλέθρια.

64. Οἱ ἐκφυσῶντες καὶ ἀνακεκλασμένοι ἐν τοῖς ὕπνοισιν ὑποβλέποντες, ἰκτερώδεες κατακορέες θνήσκουσιν· λευκὸν διαχώρημα τούτοισι προδιέρχεται.

52. To become warm rapidly after having been cooled by a sweat is a bad sign.

53. For patients who perspire over their whole body in acute diseases to become somewhat restless is a bad sign.

54. For patients to become weak for no reason at a time when they are not undergoing evacuant treatment is a bad sign.

55. In a fever, for a retching as if to vomit to terminate in expectoration is a bad sign.

56. Numbness that moves quickly between the two sides (sc. of the body) is a bad sign.

57. Very minor nose-bleeds are very bad signs.

58. In an acute disease, it is always bad for thirst to go away for no reason.

59. Persons who start up when touched are in a bad way.

60. In patients who together with ardent fever have swelling, drowsiness, and torpor, a pain invading the side leads to death by apoplexy.

61. In acute diseases, suffocation with no swelling (sc. in the throat) is a fatal sign.

62. In patients with mild trembling and greenish vomitus who are already in a fatal condition, to have faint bowel sounds when they drink and faint rumblings when they are dry, and to have difficulty gulping anything down because their breath is interrupted by coughing, indicates death.

63. In patients with acute diseases that have had a slight cooling, for redness to appear on the hands and feet is a fatal sign.

64. Persons who snore and are bent backward in their sleep with their eyes slightly opened die of intense jaundice: they pass white stools.

65. Αἱ ἐν πυρετοῖσιν ἐκστάσιες σιγῶσαι μὴ ἀφώνῳ, ὀλέθριαι.

66. Τὰ πελιδνὰ γινόμενα ἐν πυρετῷ σύντομον θάνατον σημαίνει.

67. Οἷσιν ἐν πυρετῷ, ἀλγήματος πλευροῦ ἐγγενομένου, κοιλίης ὑδατόχολα πολλὰ διαδιδούσης, ῥηΐζει, ἀσιτίαι δὲ παρακολουθοῦσι κακαί·[9] ἱδρῶτες μετὰ προσώπου εὐχροίης, καὶ κοιλίης ὑγρῆς, καί τι καὶ καρδιαλγίης, οὗτοι μακροτέρως νοσήσαντες περιπνευμονικῶς τελευτῶσιν.

68. Πυρέσσοντι ἐν ἀρχῇ μέλαινα χολὴ κάτω ἢ ἄνω διελθοῦσα, θανάσιμον.

69. Αἱ μετὰ καταψυξίων οὐκ ἀπύρων ἐφιδροῦντι ἄνω δυσφορίαι,[10] φρενιτικαί τε καὶ ὀλέθριοι.

600   70. Ἐν ὀξεῖ | τὰ ἐπ' ὀλίγον ὀξέα ἀλγήματα ἐς κληῖδας καὶ τὰ νῶτα ἐμπίπτοντα, ὀλέθρια.

71. Ἐν μακροῖσιν ὀλεθρίοισιν, ἕδρης ἄλγημα, θανάσιμον.

72. Τοῖσιν ἀσθενῶς ἤδη διακειμένοισι, τὸ μὴ βλέπειν, ἢ μὴ ἀκούειν, ἢ διαστρέφεσθαι χεῖλος ἢ ὀφθαλμὸν ἢ ῥῖνα, θανάσιμον.

73. Ἐν πυρετοῖσι βουβῶνος ἄλγημα νοῦσον χρονίην σημαίνει.

74. Αἱ ἐν πυρετοῖσιν ἀκρισίαι χρόνους μὲν ποιέουσιν, ἀτὰρ οὐχὶ ὀλέθριοι.

75. Οἱ ἐξ ἀλγημάτων ἰσχυρῶν πυρετοί, πολυχρόνιοι.[11]

120

65. Silent trances during fevers, in a patient who has not lost his speech, are a fatal sign.

66. Becoming livid in a fever signals a rapid death.

67. In fever patients with pain in the side, for the cavity to pass copious, watery, bilious material brings relief, but a malignant loss of appetite follows closely upon this. In these, sweats accompanied by a good colour of the face, diarrhoea, and pain in the cardia lead—as the disease continues for a longer time—to death by pneumonia.

68. In a person with a fever, for dark bile to pass at the beginning either downwards or upwards is a mortal sign.

69. Restlessness together with a general cooling that does not end the fever, in a person who is perspiring over the upper part of their body, presages phrenitis and death.

70. In an acute disease, for sharp pains to set in briefly in the clavicles and the back is a fatal sign.

71. In chronic, terminal conditions pain of the seat indicates death.

72. In patients already in a state of weakness, the loss of sight or hearing, or the distortion of a lip, an eye, or a nostril is a sign of death.

73. In fevers a pain in the inguinal region signals a long disease.

74. If no crisis occurs in fevers, this makes them long, but is not a fatal sign.

75. Fevers arising from severe pains will be of long duration.

---

9 κακαί A: καὶ I.   10 I: -φορίη A.

11 ἰσχυρ.—πολυχρόνιοι Aldina: πυρετοὶ πολλοὶ ὀλέθριοι ἐς χρόνιοι (-νον above the line) A: ἰσχυρῶν πολυχρόνιοι I.

76. Αἱ τρομώδεες, παρακρούσιες ψηλαφώδεες, φρενιτικαί· καὶ οἱ κατὰ γαστροκνημίην πόνοι ἐν τούτοισι, γνώμης παράφοροι.

77. Ὅσοι ἐν ξυνεχεῖ ἄφωνοι κείμενοι, μύοντες καρδαμύσσουσιν, ἤν, αἵματος ῥυέντος ἐκ ῥινῶν [ῥυῇ],[12] ἐμέσαντες φθέγξωνται, καὶ παρ' αὐτοῖσι γένωνται, σῴζονται· μὴ γενομένων δὲ τούτων, δύσπνοοι γενόμενοι θνήσκουσι ξυντόμως.

78. Οἱ λαβόντες, ἐς τὴν αὔριον παροξυνθέντες, τρίτην ἐπισχόντες, τετάρτην παροξυνθέντες, κακόν· ἦρα καὶ φρενιτικοὶ οἱ τοιοῦτοι παροξυσμοί;

79. Οἷσιν ἐκλείπουσιν πυρετοὶ μὴ κατὰ κρισίμους, ὑποτροπικόν.

80. Οἱ ἐν ἀρχῇ λεπτοὶ μετὰ κεφαλῆς σφυγμοῦ καὶ οὔρου λεπτοῦ,[13] πρὸς κρίσιν παροξύνονται· θαῦμα δὲ οὐδέν, εἰ καὶ παρακοπὴ καὶ ἀγρυπνίη γένοιτο.

81. Ἐν ὀξέσι κίνησις, ῥιπτασμός, ὕπνος ταραχώδης, σπασμὸν ἐνίοισι σημαίνει.

82. Αἱ ταραχώ|δεες θρασύτητι ἐγέρσιες παράφοροι, πονηρὸν καὶ σπασμώδεες,[14] ἄλλως τε καὶ μεθ' ἱδρῶτος· σπασμώδεες δὲ καὶ τραχήλου καὶ μεταφρένου δοκέουσι ψύξιες, καὶ ὅλου δὲ τοῦ σώματος, ἐν τούτοισι καὶ ὑμενώδεες οὐρήσιες.

83. Αἱ ἐν καύμασι παρακρούσιες, καὶ σπασμώδεες.

602

---

12 Del. Froben.
13 μετὰ κεφαλῆς—λεπτοῦ om. A.
14 παράφοροι—σπασμώδεες I: πονηρόν A.

76. Patients with trembling, delirium, and groping with the hands are suffering from phrenitis; pains in their calves lead to a disturbance of their mind.

77. If patients in a continuous fever who lie speechless, closing their eyes and blinking, recover their voice after haemorrhaging from the nostrils and vomiting, and return to their senses, they are saved. If these things do not occur, their breathing becomes difficult and they rapidly succumb.

78. Fevers that set in on one day, have an exacerbation on the next day, remit on the third day, and have an exacerbation on the fourth day, are a bad indication: do these kinds of exacerbations also indicate phrenitis?

79. Patients in whom fevers remit at times other than their crises are subject to relapses.

80. Fevers that are mild at the beginning and accompanied by throbbing in the head and thin urine have an exacerbation towards their crisis; it would be no wonder in such cases if delirium and sleeplessness also set in.

81. In acute diseases, movement, tossing about, and disturbed sleep sometimes announce a convulsion.

82. Disturbed awakenings with over-boldness and derangement of the mind are a bad sign, and announce convulsions—especially if they occur together with a sweat. Chills of the neck and the back also seem to indicate convulsions, as do those of the whole body; such patients will also have urines containing membranous fibres.

83. In ardent fevers, attacks of delirium also point to convulsions.

84. Αἱ ἐπ᾽ ὀλίγον θρασέες παρακρούσιες, θηριώδεες, καὶ σπασμοὺς προσημαίνει.

85. Ἐν τοῖσι μακροῖσι κοιλίης ἄλογοι[15] ἐπάρσιες, σπασμώδεες.

86. Τὰ εὐθὺ ταραχώδεα, ἄγρυπνα, ἐπιστάζοντα ἐκ ῥινῶν ἑκταῖα, κουφισθέντα νύκτα, πονήσαντα δ᾽ ἐς αὔριον, ἐφιδρώσαντα, κατενεχθέντα, παρακρούσαντα, αἱμορροεῖ λάβρως, καὶ λύεται τὰ πάθεα· τὸ ὑδατῶδες οὖρον τοιαῦτα σημαίνει, εἰ μετὰ τῶν εἰρημένων.

87. Τῶν ἐξισταμένων μελαγχολικῶς, οἱ τρομώδεες γενόμενοι, κακοήθεες.

88. Παραφροσύνη ἐν πνεύματι καὶ ἱδρῶτι, θανατώδης· θανατώδης δὲ καὶ ἐν πνεύματι καὶ λυγμῷ.

89. Ἐνύπνια τὰ ἐν φρενίτιδι, ἐναργῆ.

90. Ἐν φρενίτιδι λευκαὶ διαχωρήσιες, καὶ νωθρότης, κακόν· ῥῖγος τούτοισι κάκιστον.

91. Ἐν τοῖσι φρενιτικοῖσιν ἐν ἀρχῇσι τὰ ἐπιεικῶς ἔχοντα, πυκνά τε μεταπίπτοντα, κακόν.[16]

92. Τῶν ἐξισταμένων μελαγχολικῶς, οἷς τρόμοι ἐπιγίνονται, κακόν.

93. Οἱ ἐξιστάμενοι μελαγχολικῶς, τρομώδεες γινόμενοι καὶ πτυαλίζοντες, ἦρα φρενιτικοί;

94. Οἱ ἐξαναστάντες ὀξέως ἐπιπυρέξαντες, φρενιτικοὶ γίνονται.

95. Οἱ φρενιτικοὶ βραχυπόται, ψόφου καθαπτόμενοι, | τρομώδεες σπασμώδεες.

604

---

[15] ἄλογοι om. A.   [16] I: Ἐν τοῖσι δὲ μεταπίπτοντα κακόν· καὶ πτυελισμὸς κακόν. A.

84. Derangement of the mind characterized by over-boldness lasting for a short time is a malignant sign, and foretells convulsions.

85. In long diseases, swellings of the cavity for no reason point to a convulsion.

86. Persons who immediately become disturbed and sleepless, who bleed from the nostrils on the sixth day, who experience relief during that night but suffer again on the next day, who perspire over their whole body, and who are semicomatose and delirious, will have a violent haemorrhage which resolves their sufferings. Aqueous urine indicates the same outcome, if it occurs together with the things mentioned.

87. In patients who become deranged with melancholy, the occurrence of trembling is a malignant sign.

88. Delirium in conjunction with difficult breathing and perspiration is a mortal sign; it is also a mortal sign in association with difficult breathing and hiccups.

89. Dreams that occur in phrenitis are vivid.

90. In phrenitis white evacuations together with torpor are a bad sign; a chill in such patients is a very bad sign.

91. In cases of phrenitis, if the signs are reasonable at the beginning, but frequently vary, this bodes ill.

92. Among patients who become deranged with melancholy, those in whom trembling occurs are in a bad way.

93. Patients who become deranged with melancholy, tremble, and salivate: are they given to phrenitis?

94. In persons out of their wits, to be attacked suddenly by an acute fever brings on phrenitis.

95. In patients with phrenitis, to drink little and to be over-sensitive to noise indicates the onset of trembling and a convulsion.

96. Τὰ ἐν φρενιτικοῖσι νεανικῶς τρομώδεα, θανάσιμα.

97. Αἱ περὶ ἀναγκαῖα παραφροσύναι, κάκισται, οἱ ἐκ τούτων παροξυνόμενοι, ὀλέθριοι.

98. Αἱ παρακρούσιες, φωνῇ κλαγγώδεες, γλώσσῃ σπασμώδεες, καὶ αὐτοὶ τρομώδεες γινόμενοι, ἐξίστανται· σκληρυσμὸς τούτοισιν ὀλέθριος.

99. Αἱ προεξαδυνατησάντων παραφροσύναι, κάκισται.

100. Τὰ ἐν φρενιτικοῖσι πυκνὰ μεταπίπτοντα, σπασμώδεα, πονηρά.

101. Οἱ ἐν πυρετοῖσι[17] μετὰ καταψύξιος πτυελίζοντες, μέλανα ἔμετον δηλοῦσιν.

102. Τοῖς ποικίλως διανοσέουσι καὶ παρακρούουσι, πυκινὰ κωματώδεσι, προσδέχεσθαι μέλανα ἔμετον.

103. Τὰ παροξυνόμενα τρόπον σπασμώδεα, κάτοχα.

104. Τὰ παρ' οὖς ἐπάρματα ἐν μακροῖσι, σμικρά, αἱμορρώδεα καὶ σκοτώδεα ἐπιφαινόμενα, ὀλέθρια.

105. Οἱ λυγγώδεες πυρετοὶ ἄνευ εἰλέων καὶ μετ' εἰλέων, ὀλέθριοι.

106. Τοῖσι πνευματίῃσιν ἐοῦσιν πυρετὸς ὕστερος ὀξὺς λύει·[18] μετὰ ὑποχονδρίου συντόνου παροξυνθεῖσι[19] παρ' οὖς μέγα ἔπαρμα.

---

[17] πυρ. A: φρενιτικοῖσι I.
[18] ὕστερος ὀξ. λύει A: ὕστερον, ὀξύς I; cf. ch. 166 below.
[19] παρ. A: καταψυχθεῖσι I.

96. Cases of phrenitis in which there is violent trembling are fatal.

97. Derangement of the mind concerning necessities is a very bad sign, and exacerbations that follow this are fatal.

98. Patients with delirium, shrillness of the voice, and spasms of the tongue will lose their senses, if they begin to tremble. Constipation in such cases is a fatal sign.

99. In persons already greatly debilitated, derangement of the mind is a very bad sign.

100. Frequent changes in patients with phrenitis announce convulsions, and bode ill.

101. Patients with fevers,[3] who experience a general cooling and salivate, will exhibit dark vomitus.

102. In patients with changing symptoms, delirium, and frequent attacks of coma, expect dark vomitus.

103. Exacerbations of a convulsive kind presage catalepsy.

104. In long diseases, small swellings beside the ear occurring in combination with haemorrhages and dizziness are a fatal sign.

105. Fevers accompanied by hiccups—whether with or without an intestinal obstruction—are a fatal sign.

106. In patients with difficult breathing, a subsequent acute fever brings relief; in those who have an exacerbation together with tension in the hypochondrium, there will be a great swelling beside the ear.

[3] With I's reading, "with phrenitis."

107. Οἷσιν ἂν ἐν πυρετῷ ὀδύναι γενόμεναι περὶ ὀσφὺν καὶ τὰ κάτω χωρία, φρενῶν ἅπτονται, ἐκλείπουσαι τὰ κάτω, ὀλέθριαι, ἄλλως τε καὶ ἢν ἄλλο τι σημεῖον προσγένηται πονηρόν· ἢν δὲ τὰ ἄλλα σημεῖα μὴ γένηται πονηρά, ἔμπυον γενήσεσθαι ἐλπίς.

108. Παιδίοισιν ὀξὺς πυρετὸς καὶ κοιλίης ἐπίστασις μετὰ ἀγρυπνίης, καὶ τὸ ἐκλακτίζειν, καὶ τὸ χρῶμα μεταβάλλειν, καὶ ἴσχειν ἔρευθος, σπασμώδεες.

109. Τὰ εὐθὺ ταραχώδεα, ἄγρυπνα, μέλανα δὲ τὰ σύνθετα, αἱμορροεῖ ἔνια.

110. Τὰ ἀγρυπνήσαντα ἐξαίφνης ἀλυσμῷ, αἱμορροεῖ, ἄλλως τε καὶ ἤν τι προερρυήκῃ· ἦρά γε καὶ μεταφρίξαντες;

111. Οἱ ἐπ᾽ ὀλίγον περιψύχοντες, περὶ δὲ τοὺς παροξυσμοὺς βήσσοντες, καὶ ἐφιδροῦντες σμικρόν, κακοήθεες· ἐς πλευρὸν ὀδύνης καὶ πνιγμοῦ προσγενομένου, οὗτοι ἐκπυοῦνται.

112. Οἷσιν ἐν συνεχέσι φλυζάκια κατὰ πᾶν τὸ σῶμα ἐκθύει, θανάσιμον, μὴ ἐπιγινομένου πυώδεος ἀποστήματος· μάλιστα δὲ τούτοισιν εἴθισται γίνεσθαι παρ᾽ οὖς.

113. Ἐν ὀξεῖ τὰ μὲν ἔξωθεν περιψύχεσθαι, τὰ δὲ εἴσωθεν καίεσθαι, καὶ διψῆν,[20] κακόν.

114. Οἱ δὲ συνεχέες διὰ τρίτης ἐπιτείνοντες, ἐπικίνδυνοι· οἷσιν ἄν[21] ποτε πυρετὸς διαλίπῃ, ἀκίνδυνον.

---

[20] καὶ διψῆν om. A.
[21] ἐπικίν.—ἄν I: ἀκίνδυνοι ὡς ἤν A.

107. In patients with a fever, for pains arising in the loins and the lower parts to seize the diaphragm at the same time they go away in the lower parts is a fatal sign, especially if some other ominous sign is also present. If the other signs do not take an evil turn, expect the patient to suppurate internally.

108. In children, an acute fever and stoppage of the cavity in association with sleeplessness, kicking, a change of colour, and flushing announce a coming convulsion.

109. Of patients who immediately become disturbed, sleepless, and who have solid, dark stools, some haemorrhage.

110. Patients who are sleepless and who suffer a sudden restlessness haemorrhage, especially if something has run off before: does this also happen after an attack of shivering?

111. Patients that have a general cooling for a short time, cough around the time of their exacerbations, and perspire weakly over their whole body are in a malignant state: when pain in the side and suffocation follow, such cases suppurate.

112. Patients who in continuous fevers have an outbreak of blisters over their whole body are doomed, unless an abscession of pus occurs: this is most likely to happen beside the ear in these patients.

113. In an acute disease for the outward parts to have a general cooling but the inward parts to blaze, and for thirst to be present, bode ill.

114. Continuous fevers that increase over three days are dangerous, but such patients in whom the fever once remits are out of danger.

115. Ἐν μακροῖσι πυρετοῖσιν ἢ φύματα ἢ ἐς ἄρθρα πόνοι γίνονται, κἂν γένωνται, οὐκ ἄχρηστοι.

608   116. Κεφαλαλγίη ἐν | ὀξεῖ πυρετῷ,[22] μὴ ῥυέντος αἵματος ἐκ ῥινῶν, ἐς φρενιτικὸν περιίσταται.

117. Τὰ λειπυρικά, μὴ χολέρας ἐπιγενομένης, οὐ λύεται.

118. Ἴκτερος πρὸ μὲν τῆς ἑβδόμης ἐπιγενόμενος, κακόν· ἑβδόμῃ δέ, καὶ ἐνάτῃ, καὶ ἑνδεκάτῃ, καὶ τεσσαρεσκαιδεκάτῃ, χρήσιμον,[23] μὴ σκληρύνων ὑποχόνδριον, ἢ ἐνδοιαστῶς.[24]

119. Αἱ πυκναὶ διὰ τῶν αὐτῶν ὑποστροφαί, περὶ κρίσιν ἐμετώδεες, μελάνων ἔμετον ποιέουσιν· γίνονται δὲ καὶ τρομώδεες.

120. Τὰ ἐν τριταίοισιν ἅμα πυρετοῖσιν ἀλγήματα παροξυνόμενα τριταιογενῆ, ποιέεται θρομβώδεα αἵματα διαχωρέειν.

121. Ἐν πυρετοῖσι κατὰ φλέβα τὴν ἐν τῷ τραχήλῳ σφυγμὸς καὶ πόνος ἐς δυσεντερίην ἀποτελευτᾷ.

122. Τὸ μεταβάλλειν πολλάκις χρῶμα καὶ θερμασίην, χρήσιμον.

123. Τοῖσι δὲ χολώδεσι πνεῦμα μέγα, καὶ πυρετὸς ὀξὺς μετὰ ὑποχονδρίου ἐπάρσιος,[25] τὰ παρ' οὖς ἀνίστησιν.

124. Οἱ ἐκ μακρῶν ἀναλαμβάνοντες, εὔσιτοι, μηδὲν ἐπιδιδόντες, ὑποστρέφουσι κακοήθεες.

---

[22] πυρετῷ Α: ὑποχόνδριον ἀνεσπασμένον Ι.
[23] χρήσ. Α: κρίσιμον Ι.

115. In long fevers, either growths or pains in the joints may arise, and if they do, they are not unfavourable.

116. Headache in an acute fever, unless there is a haemorrhage from the nostrils, develops into phrenitis.

117. Remittent fevers are not resolved unless cholera comes on.

118. Jaundice coming on before the seventh day (sc. of a disease) is a bad sign; if on the seventh, ninth, eleventh, or fourteenth day, it is a favourable sign, unless it causes the hypochondrium to become hard, or ambiguous.

119. Around the crisis, frequent relapses of the same kind accompanied by vomiting produce dark material; they also provoke trembling.

120. Pains in tertian fevers, that have their exacerbations on the third days, like the fever, provoke the passage of clotted, bloody stools.

121. In fevers, throbbing and pain in the vessel of the neck end with the arrival of dysentery.

122. For the skin colour and temperature to change often is a favourable sign.

123. In bilious diseases, deep breathing and an acute fever in conjunction with a swelling[4] of the hypochondrium lead to swelling beside the ear.

124. Patients who recover their appetite after long diseases, but do not put on weight, will have a malignant relapse.

---

[4] According to I, "tension."

---

24 ἢ ἐνδοιαστῶς A: ἢν δὲ μή, ἐνδοιαστόν I.
25 ἐπάρσιος A: ἐνστάσιος I.

125. Οἷσιν ἐν πυρετοῖσι φλέβες αἱ ἐν κροτάφοισι σφυγματώδεες, καὶ πρόσωπον ἐρρωμένον, καὶ ὑποχόνδριον μὴ λαπαρόν, χρόνιον· καὶ οὐ παύονται χωρὶς αἵματος ῥύσιος ἐκ ῥινῶν πολλῆς, ἢ λυγγός,[26] ἢ σπασμοῦ, ἢ ὀδύνης ἰσχίων.

126. Ἐν καύσῳ κοιλίη καταρραγεῖσα, θανάσιμον.

127. Ἐκ κοιλίης ἀλγήματος ἐπιπόνου πυρετὸς καυσώδης, ὀλέθριον.

128. Ἐν τοῖσι καυσώδεσιν, ἤχων προσγενομένων μετὰ ἀμβλυωγμοῦ καὶ κατὰ ῥῖνας βάρους, ἐξίστανται μελαγχολικῶς, μὴ αἱμορραγήσαντες.[27]

129. Τοὺς ἐν καύσοις τρόμους παρακοπὴ λύει.

130. Ἐν καύσῳ ῥύσις ἐκ μυκτῆρος τριταίῳ,[28] κακόν, ἢν μή τι ἄλλο ἀγαθὸν συμπέσῃ· πεμπταίῳ δέ, ἧσσον κινδυνῶδες.

131. Ἐν τοῖσι καυσώδεσιν ὑποπεριψύχουσι, διαχωρήμασιν ὑδατοχόλοισι συχνοῖσιν, ὀφθαλμῶν ἴλλωσις[29] κακόν, ἄλλως τε κἢν κάτοχοι γένωνται.

132. Καῦσος, ῥίγεος ἐπιγενομένου, λύεται.

133. Καῦσοι ὑποτροπιάζειν εἰώθασι, ἡμέραις ε[30] ἐπισημήναντες, εἶτα ἐξιδροῦσιν· εἰ δὲ μή, τῇ ἑβδόμῃ.

134. Τοὺς καυσώδεας διακρίνουσιν αἱ τεσσαρεσκαίδεκα ἡμέραι, κουφίζουσαι ἢ ἀναιροῦσαι.

---

[26] ἢ λυγγός om. A.

[27] μὴ αἱμ. om. A.

[28] τριτ. A: τεταρταίῳ I.

[29] ὀφ. ἴλλ. L. Servin in Foes' Variae Lectiones: ὀφθαλμοῖσι codd.   [30] ἡμέραις ε´ A: καὶ ἡμέρας τέσσαρας I.

125. In patients with fevers, throbbing in the vessels of the temples, healthy colour of the face, and the hypochondrium not being soft indicate chronicity. These conditions do not cease without a copious haemorrhage from the nostrils, or hiccups, or a convulsion, or pain in the hips.

126. In an ardent fever, for the cavity to have a violent discharge is a fatal sign.

127. An ardent fever coming from a troublesome pain of the cavity is a fatal sign.

128. Patients with ardent fever in whom there are ringing in the ears, dullness of vision, and a sensation of heaviness in the nose, become deranged with melancholy, unless they have haemorrhages.

129. Tremors in ardent fevers are relieved by delirium.

130. In an ardent fever, a haemorrhage from the nostril on the third[5] day is a bad sign, unless some other good sign coincides with it; but on the fifth day it is less dangerous.

131. In ardent fevers with slight general cooling and recurring watery, bilious evacuations, for the eyes to look awry is a bad sign, especially if catalepsy occurs.

132. Ardent fever is relieved if a chill comes on.

133. Ardent fevers often relapse, appearing for five days, after which the patient perspires over his whole body; if not then, then on the seventh day.

134. Patients with ardent fevers have their crisis in fourteen days, which relieve them or carry them off.

[5] With I's reading, "fourth."

135. Ἐκ καύσου, μὴ γενομένου παρ' οὓς ἀποστήματος πυώδεος,[31] οὐ πάνυ σῴζονται.

136. Οἱ ληθαργικοὶ τρομώδεες ἀπὸ χειρῶν, ὑπνώδεες, δύσχρωτες, οἰδηματώδεες, σφυγμοῖσι νωθροῖσι,[32] καὶ μετάρσια τὰ ὑποφθάλμια, καὶ ἱδρῶτες ἐπιγίνονται, καὶ κοιλίας[33] χολώδεας καὶ ἀκρατέας ἢ
καταξήρους ἴσχουσιν,[34] οὖρα καὶ διαχωρήματα προϊ-
όντα λαθραίως, τὸ οὖρον <οἷον> ὑποζυγίου,[35] πιεῖν τε
οὐκ αἰτέουσιν, οὐδὲ θάτερον οὐδέν· ἔμφρονες δὲ γενόμενοι, τράχηλον ἐπώδυνόν φασιν ἔχειν, καὶ διὰ τῶν
612 ὤτων | ἤχους διαΐσσειν· ὁπόσοι δὲ σῴζονται τῶν
ληθαργικῶν, ἔμπυοι ὡς ἐπὶ τὸ πολὺ γίνονται.

137. Οἷσιν ἐν πυρετοῖσιν ἀκρίτως τὰ τρομώδεα
παύεται, τούτοισι χρόνῳ ἐς ἄρθρα ἀποστάσιες ὀδυνώ-
δεις ἐκπυοῦσαι, καὶ κύστις ἐπώδυνος.

138. Τῶν πυρεσσόντων οἷσιν ἐρυθήματα ἐπὶ προσώπου καὶ πόνος κεφαλῆς ἰσχυρός, καὶ σφυγμὸς φλεβῶν, αἵματος ῥύσις τὰ πολλὰ γίνεται· οἷσι δὲ ἆσαι,
καὶ καρδιωγμοί, καὶ πτυαλισμοί, ἔμετος. οἷσι δὲ
ἐρευγμοί, φῦσαι, ψόφοι κοιλίης, καὶ ἐπάρσις καὶ ἐκτά-
ραξις κοιλίης.

139. Τοῖσι χρονίζουσιν ἀσφαλῶς ἐν πυρετῷ ξυνεχεῖ, χωρὶς πόνου, ἢ φλεγμονῆς, ἢ ἄλλης προφάσιος,
ἀπόστασιν προσδέχεσθαι μετὰ πόνου καὶ οἰδήματος,
καὶ μᾶλλον ἐς τὰ κάτω χωρία· προσδέχεσθαι δὲ δεῖ

31 Α: πυώδους π. οὓς ἀποστ. Ι.
32 δύσχρω.—νωθροῖσι om. Α.

134

135. Few patients are saved from ardent fever unless purulent abscesses arise beside the ear.

136. Patients with lethargy have trembling of the hands, sleepiness, a bad colour of the skin, and oedema with torporous throbbing; the areas below the eyes are puffed up; sweating comes on; the cavities have bilious, involuntary evacuations or are very dry; the urines and stools pass without their notice; the urine is like that of cattle. Such patients do not ask to drink, nor for anything else; when they regain their senses, they say their neck is painful, and that sounds are rushing through their ears. Patients with lethargy who survive usually suppurate internally.

137. In patients with fever, in whom the shivering ceases without a crisis, painful abscessions will occupy the joints after a time and produce pus, and the bladder will have pain.

138. In fever patients who have redness of the face, a strong pain in the head, and throbbing of the vessels, a haemorrhage often occurs; if they have nausea, heartburn, and salivation, there will be vomiting; if they have eructations, winds, noises of the cavity, and swelling, diarrhoea also supervenes.

139. In patients who continue in a safe state for a long time during a continuous fever, without a pain, inflammation, or any other obvious cause, expect an abscession with pain and swelling, and most likely to the lower parts. You

---

33 Add. καὶ ὑποιδέουσι καὶ A.

34 ἴσχ. om. A.

35 Potter, after Cornarius' *velut*: τὸ οὖρ. ὑποζ. om. A.

τὰς ἀποστάσιας τοῖσιν εἰς τριάκοντα ἔτεα μᾶλλον·
ὑποσκέπτεσθαι δὲ τούτοισι τὰς ἀποστάσιας, ἢν
τὰς εἴκοσιν ἡμέρας ὁ πυρετὸς ὑπερβάλλῃ· τοῖσι δὲ
πρεσβυτέροισιν ἧσσον γίνονται, καὶ πολυχρονιώτε-
ραι πολλῶν γινομένων τῶν πυρετῶν· οἱ δὲ διαλείποντες καὶ λαμβάνοντες πεπλανημένως, φθινοπώρου
μάλιστα ἐς τεταρταῖον ἐπιεικῶς ἐφίστανται, καὶ
μάλισθ᾽ οἷσιν ἐπὶ τριήκοντα ἔτεα γεγονόσιν· αἱ δὲ
ἀποστάσιες τοῦ χειμῶνος γίνονταί τε μᾶλλον,[36] καὶ
παύονται βραδύτερον, καὶ ἧσσον[37] παλινδρομέουσι.

614   140. Τοῖσι δὲ πολλάκις ὑπο|τροπιασθεῖσιν, ἢν ἑξά-
μηνον ὑπερβάλλωσιν, ἰσχιαδικὴ φθίσις ἐπιεικῶς ἐπιγίνεται.

141. Ὅσα πυρετῷ ἀντιδίδοται, καὶ μὴ ἀποστη-
ματώδεα σημεῖα, κακοήθεα.

142. Τῶν πυρετῶν οἱ μήτε ἐν ἡμέρῃσι κρισίμῃσι,
μήτε μετὰ σημείων λυτηρίων ἀφιέντες, ὑποτροπιάζου-
σιν.

143. Τὰ ὀξέα τῶν νοσημάτων ἐν ἡμέρῃσι κρίνεται
τεσσαρεσκαίδεκα.

144. Τριταῖος ἀκριβὴς ἐν πέντε, ἢ[38] ἐν ζ´ περιόδοισιν, ἢ τὸ μακρότατον ἐν θ´ κρίνεται.

145. Οἷσιν ἀρχομένοισι πυρέσσειν, αἵματος στά-
ζοντος ἐκ ῥινῶν, ἢ πταρμοῦ γενομένου, λευκὴν ὑπόστασιν οὖρον ἴσχει περὶ τῇ τετάρτῃ, λύσιν ἐν τῇ
ἑβδόμῃ σημαίνει.

146. Τὰ ὀξέα λύεται, αἵματος ῥυέντος ἐκ ῥινέων ἐν
κρισίμῳ, καὶ ἱδρῶτος πολλοῦ γενομένου, καὶ οὔρου

must expect such abscessions more in persons up to thirty years; be on the look-out for these abscessions if the fever goes beyond twenty days. In older persons the abscessions occur less often, but are more chronic if the fevers are long. In fall, fevers that remit and attack in an erratic way are most likely to turn into quartans, and most often in persons over thirty years. In winter, abscessions occur more often, cease more slowly, and relapse less often.

140. If, in patients who have frequent relapses, these (sc. abscessions) go beyond six months, a consumption of the hips is likely to occur.

141. Everything that takes the place of fever, without being a sign of abscession, is malignant.

142. Any fevers that go away neither on critical days nor with signs that indicate resolution, relapse.

143. Acute diseases have their crisis in fourteen days.

144. A precise tertian fever has its crisis in five or seven periods, at the longest in nine.

145. When patients who at the beginning of a fever bleed from the nostrils or sneeze have a white sediment in their urine around the fourth day, this indicates that a resolution will occur on the seventh day.

146. Acute diseases are resolved at the crisis by a haemorrhage from the nostrils, by copious perspiration, by the

---

36 αἱ δὲ ἀποσ.—μᾶλλον om. A.

37 ἧσσον om. A.

38 ἀκριβ.—ἣ om. A.

πυώδεος καὶ ὑλώδεος γενομένου, ὑπόστασιν χρηστὴν
ἔχοντος, καὶ ἀθρόου γενομένου, καὶ ἀποστήματος ἀξι-
ολόγου, καὶ κοιλίης μυξώδεος καὶ αἱματώδεος, καὶ
ἐξαπίνης καταρραγείσης, καὶ ἐμέτων οὐ μοχθηρῶν
κατὰ κρίσιν.

147. Ὕπνοι βαθεῖς, μὴ ταραχώδεες, βεβαίαν κρίσιν σημαίνουσιν· οἱ δὲ ταραχώδεες μετ' ἀλγήματος
σώματος, ἀβέβαιοι.

148. Ἑβδομαίοισιν, ἢ ἐναταίοισιν, ἢ τεσσαρεσ-
καιδεκαταίοισι ῥύσιες ἐκ ῥινέων λύουσιν ὡς ἐπὶ τὸ
πολὺ τοὺς πυρετούς· ὁμοίως δὲ καὶ κοιλίης ῥύσις
616 χολώδης ἢ | δυσεντεριώδης, καὶ πόνος γουνάτων ἢ
ἰσχίων, καὶ οὖρον πεπανθὲν πρὸς τὴν κρίσιν, γυναιξὶ
δὲ καὶ ἐπιμηνίων ῥύσις.

149. Οἱ ἐν πυρετοῖσιν αἱμορραγήσαντες ἱκανῶς
ὁποθενοῦν, ἐν ταῖς ἀναλήψεσι κοιλίας καθυγραίνονται.

150. Οἱ ἐν πυρετοῖσιν ἐφιδροῦντες, κεφαλαλγέες,
κοιλίην ἀπολελαμμένοι, σπασμώδεες.

151.[39] Αἱ ἐπ' ὀλίγον θρασέες παρακρούσιες, καὶ
θηριώδη καὶ σπασμὸν σημαίνουσιν.

152. Σπασμὸς ἐν πυρετῷ γενόμενος, παύει τὸν
πυρετὸν αὐθημερόν, ἢ τὴν αὔριον, ἢ τριταίῳ.

153. Σπασμὸς ἐν πυρετῷ αὐθήμερος παυόμενος,
ἀγαθόν· ὑπερβάλλων δὲ τὴν ὥρην ἐν ᾗ ἤρξατο, καὶ μὴ
διαπαυόμενος, κακόν.

154. Οἱ διαλείποντες, ἀνωμάλως δὲ χλιαινόμενοι,
κοιλίης ἐμφυσωμένης, σμικρὰ διαδιδούσης, ὀσφυαλ-

urine being purulent and turbid with a favourable sediment and occurring in great amounts, by a considerable abscession, by the cavity suddenly discharging mucous, bloody stools, and by vomiting that is not strenuous.

147. Sleeping deeply and not being disturbed indicates a safe crisis; disturbed sleep with bodily pain is unreliable.

148. Occurring on the seventh, ninth, or fourteenth day, haemorrhages from the nostrils generally resolve fevers; similarly also a bilious or dysenteric flux of the cavity, pain in the knees or hips, the urine becoming concocted toward the crises, and in women a flux of the menses.

149. Patients who in fevers have sufficient haemorrhages from some part of their body or other, will have diarrhoea during their recovery.

150. Patients with fevers, who perspire over their whole body and have headaches and blockages of the cavity, are subject to convulsions.

151. Derangements of an over-bold kind for a short time indicate the arrival of wildness and convulsions.

152. A convulsion occurring in a fever stops the fever on the same day, the next day, or the third day.

153. In a fever, for a convulsion to arise and cease on the same day is a good sign, but for it to go beyond the hour at which it began and not to cease is a bad sign.

154. Patients whose fevers intermit, but who have irregular warming, flatulence of the cavity, scanty excre-

---

39 Ch. 151 om. A.

γέες μετὰ κρίσιν, τούτοισι κοιλίαι καταρρήγνυνται· οἱ
δὲ περικαέες πρὸς χεῖρα, νωθροί, διψώδεες, ἀσώδεες,
κοιλίης ἀπειλημμένης, βαρυνόμενοι, ἐκχλοιοῦνται·
ἔστι δ' ὅτε καὶ τὰ ἐξέρυθρα ἐν ποσὶ καύματα τὰ αὐτὰ
σημαίνει.

155. Οἱ χειμερινοὶ τεταρταῖοι πυρετοὶ ἐπιεικῶς
μεθίστανται ἐς τὰς ὀξείας νούσους.

156. Κεφαλῆς πόνος σύντονος μετ' ὀξέος πυρετοῦ
καὶ ἄλλου σημείου τῶν δυσκόλων, θανάσιμον· ἄνευ δὲ
σημείου φαύλου, ὑπερβάλλων τὰς εἴκοσιν ἡμέρας,
618 αἵματος ῥύσιν ἢ πύεος ἐκ ῥινῶν, | ἢ ἀποστάσιας ἐς τὰ
κάτω σημαίνει· μᾶλλον δὲ τοῖσι νεωτέροισι τῶν τριήκοντα πέντε τὰς ῥύσιας τοῦ αἵματος, τοῖσι δὲ
πρεσβυτέροισι τὰς ἀποστάσιας προσδέχεσθαι, περὶ
μέτωπον δὲ καὶ κροτάφους ὄντος τοῦ πόνου, τὰς
ῥύσιας.

157. Οἷσι δὲ κεφαλαλγίαι καὶ ἦχοι ἀπυρέτοισι, καὶ
σκοτοδινίη, καὶ φωνῆς βραδυτής, καὶ νάρκαι χειρῶν,
ἀποπλήκτους ἢ ἐπιληπτικοὺς προσδέχου τούτους ἔσεσθαι, καὶ ἐπιλήσμονας.

158. Οἱ κεφαλαλγέες, κατόχως παρακρούοντες, κοιλίης ἀποληφθείσης, ὄμμα θρασυνθέντες, ἀνθηροί,
ὀπισθοτονώδεις γίνονται.

159. Τὰ ὑποσείοντα κεφαλάς, ὄμματα ἐξέρυθρα,
παρακρούοντα σαφῶς, ὀλέθρια· οὐ συναποθνῄσκει τὸ
τοιοῦτον, ἀλλὰ παρ' οὓς οἴδημα ποιέει.

160. Κεφαλαλγίη μεθ' ἕδρης καὶ αἰδοίων ἀλγήματος, νωθρότητα καὶ ἀκρασίην παρέχει, καὶ φωνὴν

tions, and pain in the loins after their crisis, are subject to violent discharges of the cavity. Those who are burning hot to the touch and have torpor, thirst, nausea, a stoppage of the cavity, and a sensation of heaviness, grow sallow. Sometimes burning heat and redness of the feet announce the same things.

155. Quartan fevers in winter are likely to change to acute diseases.

156. Intense pain in the head, together with an acute fever and any other troublesome sign, indicates death. If the pain is without an indifferent sign, and goes beyond twenty days, it points to a flux of blood or pus from the nostrils or apostases to the lower parts. Expect the flow of blood more in people younger than thirty-five, and the apostastes in older ones: when the pain is in the region of the face and the temples, expect the fluxes.

157. Persons without a fever who have headache and ringing in their ears, and vertigo, slowness of speech, and numbness of the arms, you should expect to suffer from apoplexy or epilepsy, and also forgetfulness.

158. Persons with headaches, cataleptic derangement of the mind, stoppage of the cavity, protrusion of the eyes, and a florid complexion will be attacked by opisthotonus.

159. Shaking of the head, intense redness of the eyes, and obvious delirium are fatal signs. Such symptoms do not accompany the patient to his death, but bring about swelling beside the ear.

160. Headache, in association with pain in the seat and the genital parts, provokes torpor and weakness, and paral-

παραλύει· ταῦτα οὐ χαλεπά· ὑπνώδεες δὲ καὶ λυγ-
γώδεις γίνονται. ἐνάτῳ μηνὶ ἐκ τοιούτων, φωνῆς λυ-
θείσης, ἐς ταὐτὸ καθίστανται, ἀσκαριδώδεες γινόμενοι.

161. Ἐν κεφαλαλγίῃ, κώφωσις καὶ κῶμα παρακο-
λουθοῦντα, τὰ παρ' οὓς ἐπαίρει.

162. Οἱ κεφαλαλγέες, κατόχως ὀδυνώδεες, ὄμμα
ἐξέρυθροι, αἱμορραγικοί.

163. Τὰ σείοντα κεφαλήν, ἠχώδεα, αἱμορροεῖ,[40]
γυναικεῖα καταβιβάζει, ἄλλως τε κἢν κατὰ ῥάχιν
620   καῦμα παρακολουθῇ· ἴσως δὲ καὶ δυσ|εντερικά.

164. Οἱ καρηβαρικοί, κατὰ βρέγμα ὀδυνώδεες,
ἄγρυπνοι, αἱμορροοῦσιν, ἄλλως τε κἢν τι ἐς τράχηλον
συντείνῃ.

165. Τὰ ἐν κεφαλαλγίῃσιν ἰώδεα ἐμέσματα μετὰ
κωφώσιος, ἀγρύπνοισι, ταχὺ ἐκμαίνει.

166. Οἷσι κεφαλῆς καὶ τραχήλου πόνος, καὶ ὅλου
δέ τις ἀκράτεια τρομώδης, αἱμορραγίη λύει· ἀτὰρ καὶ
οὕτω χρόνῳ λύονται· ἆρα τούτοις κύστιες[41] ἀπολαμ-
βάνονται;

167. Ἐν τῇσιν ὀξείῃσι κεφαλαλγίῃσι, καὶ ἐν τῇσι
ναρκώδεσι μετὰ βάρεος, φιλεῖ[42] σπασμώδεα γίνεσθαι.

168. Κεφαλαλγίην λύει πῦον διὰ ῥινῶν, ἢ πτύαλα
παχέα καὶ ἄνοσμα· λύει δὲ καὶ ἑλκέων ἔκθυσις, ποτὲ
δὲ καὶ ὕπνος, καὶ κοιλίης ῥύσις.

40 ἢ γυναικὶ τὰ add. I.
41 ἆρα—κύστ. A: αἱ δὲ κύστιες ἐν τούτῳ I.

ysis of the voice. These things are not hard to bear: such patients develop drowsiness and hiccups. In the ninth month after this happens, the voice is recovered, but these patients return to the same state and suffer from worms.

161. If in headache deafness and coma follow closely, the area beside the ear swells up.

162. Patients with headaches who also suffer from a painful catalepsy and intense redness of the eyes tend to have haemorrhages.

163. Shaking of the head, in association with ringing in the ears, provokes haemorrhages or brings down the menses, especially if followed closely by a burning heat along the spine: there is also some probability of dysentery.

164. Patients with heaviness in the head, pain in the bregma, and sleeplessness have haemorrhages, especially if there is any tension in the neck.

165. During headaches, for sleeplessness persons with deafness to vomit rust-coloured material indicates a rapidly approaching mania.

166. In persons with pain of the head and neck, and a degree of tremulous disability of the whole body, haemorrhage brings resolution; but even thus, the relief comes only with time: do the bladders of these patients become blocked?

167. In cases of acute headache and in persons who have numbness accompanied by heaviness, it is likely that convulsions will occur.

168. Headache is relieved by a passage of pus through the nostrils, or by a thick, odourless sputum. Also an outbreak of ulcers on the skin, sometimes sleep, or a flux of the cavity can relieve it.

---

42 φιλ. A: ἐθέλει I.

169. Κεφαλῆς ἄλγημα μέτριον μετὰ δίψης, μὴ ἰδίουσιν,[43] ἢ μεθ᾽ ἱδρῶτος μὴ λύοντος τὸν πυρετόν, ἀπαστάσιας ἐν οὔλοισιν ἢ παρ᾽ οὓς σημαίνει, μὴ κοιλίης ἐκταραχθείσης.

170. Κεφαλαλγίη καρώδης μετὰ βάρεος ποιέει τι σπασμῶδες.

171. Οἱ κεφαλαλγικοί, διψώδεες, ὑπάγρυπνοι, ἀσαφέες, ἀδύνατοι, ἐπὶ κοιλίῃ ὑγρῇ κοπιώδεες, ἆρά γε ἐξίστανται;

172. Κεφαλαλγέες, ὑπόκωφοι, χεῖρας τρομώδεες, τράχηλον ὀδυνώδεες, οὐρέοντες μέλανα δεδασυσμένα, ἐμέοντες μέλανα, ὀλέθριοι.

173. Οἱ κεφαλαλγέες, ἐφιδροῦντες, κοιλίην ἀπειλημμένοι, σπασμώδεες.

174. Τὸ καρῶδες πανταχοῦ κακόν.

622　175. Οἱ κωματώδεες ἐν ἀρχῇσι[44] | μετὰ κεφαλῆς, ὀσφύος, ὑποχονδρίου, τραχήλου ὀδύνης, ἀγρυπνέοντες, ἦρα φρενιτικοί; μυκτὴρ ἐν τούτοισιν ἀποστάζων, ὀλέθριον, ἄλλως τε καὶ τεταρταίοισιν ἐοῦσιν, ἢ ἀρχομένοισιν· κακὸν δὲ καὶ κοιλίης περίπλυσις ἐξέρυθρος.

176. Οἱ [κωματώδεες][45] ἐξ ἀρχῆς ἐφιδρώσαντες, οὔροισι πέποσι, καυστικοί, ἀκρίτως περιψύχοντες, διὰ ταχέων περικαέες, νωθροί, κωματώδεες, σπασμώδεες, ὀλέθριοι.

---

[43] μὴ ἰδ. Opsopoeus in Foes' note: νηδιούσης A: μὴ ἰδείουσιν I.

[44] Add γενόμενοι I.

[45] Del. Littré.

169. A moderate pain in the head together with thirst—as long as the patients are not sweating, or if they do have a sweat it does not resolve the fever—announces abscessions in the gums or beside the ear, unless there is a discharge of the cavity.

170. Headache with drowsiness and a feeling of heaviness provokes something of a convulsion.

171. Do patients with headache, thirst, mild sleeplessness, confusion, and debility, who after a bout of diarrhoea suffer weariness, become deranged in their minds?

172. If patients with headache, partial loss of hearing, trembling of the hands, and pain in the neck pass dark cloudy urines and dark vomitus, they are doomed.

173. Patients with headaches who sweat over their whole body and have a blockage of the cavity are marked for convulsions.

174. Drowsiness is a totally bad sign.

175. Do patients who are comatose at the beginning of their fever, and lie awake with pains in the head, loins, hypochondrium and neck develop phrenitis? For a nostril to pass drops of blood in these is a fatal sign, especially if it is on the fourth day or at the beginning. A very red discharge from the cavity is also bad.

176. Patients who right from the beginning sweat over their whole body and have concocted urines, who have burning heat, are cooled without a crisis and then quickly become burning hot again, and who suffer torpor, coma, and convulsions, are doomed.

177. Οἱ κωματώδεες ὕπνοι, καὶ αἱ καταψύξιες, ὀλέθριον.

178. Κωματώδεας, κοπιώδεας, κεκωφωμένους, κοιλίης κατερρωγυίης, ἐρυθρὰ διελθόντα περὶ κρίσιν ὠφελέει.

179. Κωματώδεες, ἀσώδεες, ὑποχόνδριον ὀδυνώδεες, μικρὰ ἐμετώδεες, τὰ παρ᾽ οὖς ἴσχουσι, πρόσθεν δὲ περὶ τὸ πρόσωπον ἐπάρματα.

180. Τὰ μετὰ κώματος, ἐξαίφνης παρακρούσαντα ἀλυσμῷ, αἱμορροϊκά.

181. Τὰ κωματώδεα, ἀσώδεα, ὀδυνώδεα ὑποχόνδρια, θαμινὰ[46] μικρὰ πτύοντα, τὰ παρ᾽ οὖς ἐπαίρει· τὸ κωματῶδες ἆρα ἔχει τι σπασμῶδες;

182. Κωματώδεα, μεμωρωμένα, κάτοχα, ποικίλλοντα ὑποχόνδρια καὶ κοιλίην ἐπηρμένοι, ἄσιτοι, ἀπολελαμμένοι, ἐφιδροῦντες· ἦρα τούτοισι τὸ θολερὸν πνεῦμα καὶ τὸ γονοειδὲς ἐλθὸν λύγγα σημαίνει; κοιλίης[47] δὲ ἆρα χολῶδες διέρχεται; τὸ λαμπῶδες ἐν τούτοισιν οὐρηθὲν ὠφελέει, καὶ κοιλίαι δὲ τούτοισιν ἐπιταράσσονται. |

624 183. Ἐγκεφάλου σφακελίσαντος, οἱ μὲν ἐν τῇσι τρισὶν ἡμέρῃσιν, οἱ δὲ ἐν τῇσιν ἑπτὰ τελευτῶσι, ταύτας δὲ διαφυγόντες, σῴζονται· οἷσι δ᾽ ἂν τμηθεῖσι τῶν τοιούτων διεστηκὸς εὑρεθῇ τὸ ὀστέον, ἀπόλλυνται.

184. Τοῖσι κεφαλαλγικοῖσιν ὀστέα ῥαγεῖσιν ἐκ τῶν ὄπισθεν, ῥύσις ἐκ μυκτῆρος λάβρως, παχεῖα, κακόν· ὀφθαλμὸν προαλγήσαντες οὗτοι ῥιγέουσιν·

177. Comatous sleep with generalized cooling is a fatal sign.

178. In patients with coma, weariness, and deafness, a downward discharge of red material from the cavity around the time of their crisis is of benefit.

179. Patients with coma, nausea, and pain in the hypochondrium who vomit a small amount, will swell up beside the ear, but before that have swellings about the face.

180. In conjunction with coma, a sudden delirium with restlessness is an indication of haemorrhage.

181. Patients with coma, nausea, and pain in the hypochondrium who frequently produce a small amount of sputum will swell up beside the ear: does being comatose have something in common with convulsions?

182. Patients with coma, stupor, catalepsy, variation in the state of the hypochondrium, swelling of the cavity, loss of appetite, constipation, and perspiration over their whole body: does laboured breathing and the passage of urine resembling seed in these indicate the onset of hiccups? Does bilious material pass through their cavity? For these patients to pass urine with a scum on it is an advantageous sign; their cavities will also be disturbed.

183. Sphacelus of the brain: some patients die in three days and others in seven, but if they survive beyond that, they recover. If, in the ones who are incised, the bone is found to be separated, they die.

184. In patients with headache whose skull-bones are fractured from behind, a thick, violent haemorrhage from a nostril is a bad sign; if such patients have had a pain in the

---

46 ἀσῶδ.—θαμινὰ I: ὑποχόνδρια A.
47 -ίης Potter: -ίη codd.

ἆρα αἱ κατὰ κρόταφον ὀστέων διαρραγαὶ σπασμώδεες;

185. Ὠτὸς πόνος σύντονος, μετὰ πυρετοῦ ὀξέος, καὶ ἄλλου του σημείου τῶν ὑποδυσκόλων,[48] τοὺς μὲν νέους ἑβδομαίους κτείνει ἢ πρόσθεν,[49] παραφρονήσαντας, μὴ ῥυέντος πολλοῦ πύου ἐκ τοῦ ὠτός, ἢ ἐκ τῶν ῥινῶν αἵματος, μηδ' ἄλλου του σημείου χρηστοῦ γενομένου· τοὺς δὲ πρεσβυτέρους βραδύτερον καὶ ἧσσον ἀναιρεῖ· τά τε γὰρ ὦτα φθάνει ἐκπυοῦντα, καὶ παραφρονέουσιν ἧσσον· ὑποστρέφουσι δ' οἱ πολλοὶ τούτων, καὶ οὕτως ἀπόλλυνται.

186. Κώφωσις ἐν ὀξέσι καὶ ταραχώδεσι παρακολουθοῦσα, κακόν· κακὸν δὲ καὶ ἐν τοῖσι μακροῖσιν· ἄγει δ' ἐν τούτοισι καὶ ἐς ἰσχία πόνος.

187. Ἐν πυρετῷ κώφωσις κοιλίην ἐφίστησιν.

188. Ὦτα ψυχρὰ καὶ διαφανέα καὶ συνεσταλμένα, ὀλέθριον.

189. Βόμβος δὲ[50] καὶ ἦχος ἐν ὠσί, θανάσιμον.

190. Ἦχοι μετ' ἀμ|βλυωσμοῦ, καὶ κατὰ ῥῖνας βάρεος, παρακρουστικόν, καὶ αἱμορροεῖ.

191. Οἷσι κώφωσις μετὰ καρηβαρίης, καὶ ὑποχονδρίου ἐντάσιος, καὶ πρὸς αὐγὰς ἐνοχλεῖσθαι,[51] αἱμορροεῖ.

192. Ἐν ὀξεῖ πυρετῷ ὦτα κωφοῦσθαι, μανικόν.

193. Οἱ δύσκωφοι, ἐν τῷ λαμβάνειν τρομώδεες, γλῶσσαν παραλελυμένοι, νωθροί, κακόν.

---

[48] καὶ ἄλλου—ὑποδυσ. om. A.
[49] ἢ πρόσ. A: καὶ συντομώτερον I.

626

148

eye before, they will have chills. Do fractures of the bones in the temple presage convulsions?

185. Intense earache with an acute fever, in conjunction with some other rather troubling sign, kills young persons on the seventh day or before, in a state of delirium, unless there is a copious flux of pus from the ear or of blood from the nostrils, or some other favourable sign appears. It carries off older persons more slowly and less surely, for their ears have time to suppurate, and they suffer less delirium. Many such patients have relapses and die in the manner described.

186. For deafness to follow closely in acute, disruptive diseases is a bad sign: it is also bad in long diseases. In the latter, it also brings pains in the hips.

187. In a fever, deafness stops the cavity.

188. For the ears to be cold, transparent, and contracted is a fatal sign.

189. To have buzzing and ringing in the ears is a fatal sign.

190. Ringing (sc. in the ears) together with dullness of vision and heaviness in the nose presages delirium and haemorrhage.

191. Deafness together with heaviness of the head, tension of the hypochondrium, and irritation by the light, indicates a coming haemorrhage.

192. For the ears to become deaf in an acute fever presages mania.

193. Persons who are hard of hearing, have trembling when they reach for something, are paralysed in their tongue, and have torpor are in a bad way.

---

50 δὲ A: ἐν ὀξέσι I.
51 Littré in *app. crit.*, Ermerins: ὀχλεῖ A: ἐνοχλεῖν I.

194. Προηκούσης ἀρρωστίης, κώφωσις, καὶ οὖρον ὑπέρυθρον, ἀκατάστατον, ἐναιωρεύμενον, παρακρουστικόν· τὸ ἰκτεροῦσθαι ἐν τούτοισι κακόν[52] κακὸν δὲ καὶ ἐπὶ ἰκτέρῳ μώρωσις· τούτους ἀφώνους, αἰσθανομένους δέ, συμβαίνει γίνεσθαι· τάχα δὲ καὶ κοιλίη πονηρεύεται τούτοισι.

195. Τὰ ὀδυνηρῶς παρ' οὖς ἀνιστάμενα, ὀλέθρια.

196. Τὰ παρ' οὖς ἐκ τοῦ ἔμπροσθεν ἀλγήματος ἐρυθήματα ἐν πυρετῷ γινόμενα, σημεῖον μὲν ἐρυσιπέλατος ἐπὶ προσώπου ἐσομένου· ἀτὰρ καὶ σπασμοὶ ἐκ τῶν τοιούτων γίνονται μετὰ ἀφωνίης καὶ ἐκλύσιος.

197. Τὰ παρ' οὖς, ἐπὶ πᾶσι τοῖσι λυώδεσι,[53] πυρετῷ ὀξεῖ, ὑποχονδρίῳ συντόνῳ χρονιωτέρως, ἀρθέντα, κτείνει.

198. Τὰ παρ' οὖς, φλαῦρα τοῖσι παραπληκτικοῖσιν.

199. Τὰ παρ' οὖς ἐν μακροῖσι, μὴ ἐκπυοῦντα, θανάσιμον· κοιλίαι δὲ τούτοισι καταφέρονται.

200. Ἆρά γέ εἰσι τὰ παρ' ὦτα κεφαλαλγικοί; ἆρά γε ἐφιδροῦσι τὰ ἄνω ἢ καὶ ἐπιρριγέουσιν; ἆρά γε | καὶ κοιλίαι καταρρήγνυνται; καί τι καὶ κωματώδεες; ἆρα καὶ ὑδατῶδες οὖρον, ἐναιωρεύμενον λευκοῖσι, καὶ τὰ ποικίλα, ἔκλευκα, δυσώδεα;

201. Τὰ παρ' οὖς λαπάσσει καὶ βηχία μετὰ πτυαλισμῶν ἰόντα.

202. Οὖρα τοῖσι παρ' ὦτα ταχὺ καὶ ἐπ' ὀλίγον

---

[52] Προηκούσης—κακόν om. A.
[53] Potter, cf. Galen, Glossary s. v.: λ*ώδεσι A: δυσώδεσι I.

194. When an illness is advanced, deafness in conjunction with reddish urine that sets down no deposit, but which contains suspended material, presages delirium: for these patients to become jaundiced is a bad sign; also a bad sign is stupor coming after jaundice. These patients lose their speech but at the same time retain their mental faculties. In such cases, it is also likely that the cavity will be in a bad state.

195. Painful swellings beside the ear are a fatal sign.

196. Redness beside the ear arising during a fever from an earlier pain is a sign that erysipelas will involve the face; but convulsions also arise in such cases, together with loss of speech and faintness.

197. In all cases of delirium with acute fever and tension of the hypochondrium for a longer time, swellings that occur beside the ear kill the patient.

198. Swelling beside the ear is an indifferent sign when paralysis is present.

199. Swelling beside the ear in long diseases, unless suppuration occurs, is a fatal sign. The cavities in such patients evacuate downwards.

200. Are patients with swelling beside the ear subject to headaches? Do they perspire weakly over the upper part of their body or also have chills? Do the cavities also have violent discharges? Are they also somewhat comatose? Is there also watery urine with white suspended material? And variegated, very white, and foul-smelling urines?

201. Mild coughs that produce sputa also bring down swellings beside the ear.

202. In patients with swellings beside the ears, urines that are passed early and that quickly become concocted

πεπαινόμενα, φλαῦρα· καὶ τὸ καταψύχεσθαι ὧδε, πονηρόν.

203. Τὰ παρ' οὖς ἐν τοῖσι χρονίοισιν ἐκπνεύμενα
μὴ λευκῷ σφόδρα καὶ ἀνόσμῳ, κτείνει, καὶ μάλιστα
γυναῖκας.

204. Τὰ παρ' οὖς μάλιστα τῶν ὀξέων ἐν τοῖσι
καυσώδεσι γίνεται· κἢν μὴ κρίσιν ποιήσῃ, ἢ ἐκπε-
παίνηται, ἢ ἐκ ῥινῶν αἷμα ῥυῇ, ἢ οὖρα ὑπόστασιν
παχεῖαν λάβῃ, ἀπόλλυνται· τὰ δὲ πολλὰ τῶν τοιούτων
οἰδημάτων ἀποκαθίσταται· προσεπιθεωρεῖν δὲ καὶ
τοὺς πυρετούς, ἤν τε ἐπιτείνωσιν, ἤν τε ἀνιῶσι, καὶ
οὕτως ἀποφαίνεσθαι.

205. Ἐπὶ κωφώσει καὶ νωθρίῃ ἐκ ῥινῶν ἀποστά-
ζειν, ἔχει τι δύσκολον· ἔμετος τούτοισιν ἁρμόζει καὶ
κοιλίης ταραχή.

206. Ἐκ κωφώσιος ἐπιεικῶς τὰ παρ' ὦτα, ἄλλως τε
καὶ ἢν ἀσῶδές τι γίνηται· καὶ τοῖς κωματώδεσιν ἐπὶ
τούτοισι μᾶλλόν τι τὰ παρ' ὦτα.

207. Κώφωσιν ἐν πυρετῷ ῥύσις ἐκ ῥινῶν λύει καὶ
κοιλίης ταραχή.

208. Πρόσωπον ἐκ μετεώρου ταπεινούμενον, καὶ
φωνὴ λειοτέρη καὶ ἀσθενεστέρη γινομένη, καὶ πνεῦμα
μανότερον καὶ λεπτότερον, ἄνεσιν ἐς τὴν ἐπιοῦσαν
σημαίνει.

209. Προσώπου διαφθορή, θανάσιμον· ἧσσον δὲ
ἢν δι' ἀγρυπνίην, ἢ λιμόν, ἢ κοιλίης ἐκτάραξιν |
630 γένηται· καθίσταται δ' ἐν ἡμέρῃ καὶ νυκτὶ τὸ διὰ
ταῦτα διαφθαρέν· γένοιτο δ' ἂν τοιοῦτον, ὀφθαλμοὶ

are an indifferent sign. To have a chill in this condition also bodes ill.

203. In chronic diseases, swellings beside the ears that suppurate with pus that is not very white and odourless bring death, and especially to women.

204. Swellings beside the ear occur most often, among acute diseases, in ardent fevers; and unless they have a crisis and come to coction, or there is a haemorrhage from the nostrils, or the urines acquire a thick sediment, such patients die. Most of this kind of swellings subside; you must also consider the fevers—whether they are increasing in intensity or diminishing—and on this basis make your decision.

205. In the presence of deafness and torpor, to pass drops of blood from the nostrils indicates some degree of trouble. Vomiting is favourable in such cases, and also an evacuation of the cavity.

206. After deafness, swelling beside the ears is likely to occur, especially if some degree of nausea is present; among these patients any with coma have an even greater tendency to swelling beside the ears.

207. Deafness in a fever is resolved by a haemorrhage from the nostrils, or by an evacuation of the cavity.

208. For the face to come down after having been swollen, and the voice to become soft and weak and the breathing less often and less deep, indicate a remission on the following day.

209. Falling in of the face is a sign of death; but less so if it is the result of sleeplessness, fasting, or an evacuation of the cavity: it recovers in a day and a night if it fell in because of these. The signs of this condition would be as

κοῖλοι, ῥὶς ὀξεῖα, κρόταφοι συμπεπτωκότες, ὦτα ψυχρὰ καὶ συνεσταλμένα, δέρμα σκληρόν, χρῶμα χλωρὸν μελανοῦν·[54] πελιαινόμενον δὲ ἐπὶ τούτοισι βλέφαρον, ἢ χεῖλος, ἢ ῥίς, συντόμως θανάσιμον.

210. Προσώπου εὔχροια καὶ σκυθρωπότης ἐν ὀξεῖ,
κακόν· μετώπου συναγωγὴ ἐπὶ τούτοισι, φρενιτικόν.

211. Περὶ πρόσωπον εὔχροια καὶ ἱδρῶτες ἀπυρέτοισι, κόπρανα παλαιὰ ὑπεόντα σημαίνει, ἢ διαίτης
ἀταξίην.

212. Τὰ κατὰ ῥῖνας ἐρυθήματα, κοιλίης ὑγραι-
νομένης σημεῖα· τοῖσι κατὰ ὑποχόνδρια ἢ πλεύμονα
πόνοισι ἐκπυημάτων κακῶν.[55]

213. Ὀφθαλμῶν καθαρότης καὶ τὰ λευκὰ αὐτῶν ἐκ
μελάνων ἢ πελίων καθαρὰ γίνεσθαι, χρήσιμον· ταχέως μὲν οὖν καθαιρομένων,[56] ταχεῖαν σημαίνει κρίσιν, βραδέως δὲ βραδυτέρην.

214. Τὸ ἀχλυῶδες ὀφθαλμῶν, ἢ τὸ λευκὸν ἐρυ-
θραινόμενον ἢ πελιαινόμενον, ἢ φλεβίων μελάνων
πληρούμενον, οὐκ ἀστεῖον· φλαῦρον δὲ καὶ τὴν αὐγὴν
φεύγειν, ἢ δακρύειν, ἢ διαστρέφεσθαι, καὶ τὸν ἕτερον
ἐλάσσω γενέσθαι· πονηρὸν δὲ καὶ τὰς ὄψιας πυκνὰ
632 διαρρίπτειν, ἢ λημία σμικρὰ περὶ αὐτάς, ἢ | αἰγίδα
λεπτὴν ἴσχειν, ἢ τὸ λευκὸν μέζον γίνεσθαι, τὸ δὲ
μέλαν ἐλάσσω, ἢ κρύπτεσθαι τὸ μέλαν ὑπὸ τὸ ἄνω
βλέφαρον· πονηρὸν δὲ καὶ κοιλότης ὀμμάτων, καὶ
ἔκθλιψις ἔξω σφοδρή, καὶ λαμπηδόνος ἔκλαμψις,[57]

---

[54] χλωρ. μελ. Α: ὦχρον ἢ μέλαν I.

154

follows: eyes hollow, nose pointed, temples emaciated, the ears cold and contracted, the skin hard and darkish green in colour; if in addition the eyelid, lip, or nose becomes livid, it is a sign of rapidly approaching death.

210. A good colour of the face in association with sullenness in an acute disease is a bad sign; a contraction of the forehead besides indicates phrenitis.

211. A good colour of the face together with sweating in persons without fever indicates that old fecal material lies hidden, or that there is an irregularity of the regimen.

212. Redness in the area of the nose is a sign of approaching diarrhoea; together with pains in the region of the hypochondria or a lung, it indicates evil suppurations.

213. Clearness of the eyes, or their whites becoming clear after they have been dark or livid, is a favourable sign. Now if they become clear quickly, it indicates a rapid crisis, if slowly, then a slower crisis.

214. Dimness of the eyes, or the whites becoming red or livid, or for them to become filled with dark vessels is not good; also indifferent are for the eyes to turn away from the light, to pass tears, or to look awry, and for one eye to become smaller than the other; also bad is for the eyes to move around frequently in an indiscriminate manner, or for there to be small eye-sores around them, or for them to have a thin speck, or for the white to become larger, or the black smaller, or for the black to be hidden under the upper eyelid. Bad signs are also a hollowness of the eyes, an excessive protrusion, a brightness of the lustre so

---

55 ἔκπυη. κακῶν A: ἐμπυομένοισι κακόν I.

56 ταχέως—καθαιρομένων om. I.   57 A: ἔκθλιψις I.

ὥστε μὴ δύνασθαι τὴν κόρην ἐκτείνεσθαι, καὶ βλεφαρίδων καμπυλότης καὶ πῆξις ὀμμάτων, συνεχῶς τε μύειν, καὶ χρῶμα μεταβάλλειν· καὶ βλέφαρα μὴ συμβάλλειν ἐν τῷ καθεύδειν, ὀλέθριον· κακὸν δὲ καὶ ἰλλαίνων ὀφθαλμός.

215. Ὀφθαλμῶν ἔρευθος ἐν πυρετῷ γινόμενον, κοιλίης πονηρίην χρόνιον σημαίνει.

216. Αἱ παρ' ὀφθαλμὸν ἀναστάσιες ἐν τῇσιν ἀνακομιδῇσι, κοιλίην καταρρηγνύουσιν.

217. Ἐπὶ ὀμμάτων διαστροφῇ, κοπιώδει, πυρετώδει, ῥῖγος, ὀλέθριον· καὶ οἱ κωματώδεες ἐν τούτοισι, κακόν.

218. Ὀφθαλμιῶντι,[58] πυρετοῦ ἐπιγενομένου, λύσις· εἰ δὲ μή, κίνδυνος τυφλωθῆναι, ἢ ἀπολέσθαι, ἢ ἀμφότερα.

219. Οἷσιν ὀφθαλμιῶσι κεφαλαλγίη προσγίνεται, καὶ παρακολουθεῖ χρόνον πολύν, κίνδυνος τυφλωθῆναι.

220. Ὀφθαλμιῶντι διάρροια ἀπὸ ταυτομάτου,[59] χρήσιμον.

221. Ὀμμάτων ἀμαύρωσις, καὶ τὸ πεπηγός, ἀχλυῶδες, κακόν.[60]

222. Ὀμμάτων ἀμαύρωσις ἅμα ἀψυχίῃ, σπασμῶδες συντόμως.

223. Ὀμμάτων ὀρθότης ἐν ὀξεῖ, ἢ κίνησις ὀξεῖα, καὶ ὕπνος ταραχώδης, ἢ ἀγρυπνίη, ποτὲ δὲ καὶ στάξις

that the pupil cannot dilate, an inversion of the eyelashes (trichiasis), fixation of the eyes, continual blinking, and for their colours to change. For the eyelids not to come together in sleep is a fatal sign. It is also a bad sign for one eye to look awry.

215. For redness of the eyes to develop in a fever indicates there will be a chronic disturbance of the cavity.

216. Abscesses beside an eye during the process of recovery announce a violent discharge of the cavity.

217. In a person with strabismus, weariness, and fever, a chill is a fatal sign; patients who in addition fall into a coma are in an evil way.

218. For a person with ophthalmia, the arrival of fever signals resolution; if this does not happen, there is a danger that he will lose his sight, or his life, or both.

219. Patients with ophthalmia in whom headache comes on and follows them for a long time are in danger of losing their sight.

220. For a person with ophthalmia to have a spontaneous bout of diarrhoea is a favourable sign.

221. For the eyes to lose their power of vision in conjunction with fixation and cloudiness is a bad sign.

222. For the eyes to lose their power of vision together with fainting indicates an impending convulsion.

223. Fixation of the eyes in an acute disease, or a sharp movement of the eyes together with disturbed sleep or sleeplessness, sometimes also provokes a haemorrhage

---

58 Add ἀνδρὶ I.

59 ἀπὸ ταυτομάτου om. A.

60 (218) ἢ ἀπολέσθαι—(219) τυφλωθῆναι and ch. 221 om. I.

ἐκ ῥινῶν· πρὸς τὴν ἀφὴν μὴ περικαέες, φρενιτικοὶ
γίνονται, καὶ μᾶλλον ἢν αἷμα ῥυῇ. |

634   224. Γλῶσσα κατ᾽ ἀρχὰς μὲν πεφρικυῖα, τῷ δὲ
χρώματι διαμένουσα, προϊόντος δὲ τοῦ χρόνου τρηχυνομένη, καὶ πελιαινομένη, καὶ ῥηγνυμένη, θανάσι-
μον· σφόδρα δὲ μελαινομένη, τῇ τεσσαρεσκαιδεκάτῃ
κρίσιν σημαίνει[61] χαλεπωτάτη δέ ἡ μέλαινα καὶ
χλωρή.

225. Γλώσσης παρὰ τὸ δικροῦν ὥσπερ σιάλῳ λευκῷ καταλείφεσθαι, σημεῖον ἀνέσεως πυρετοῦ· παχέος
μὲν ἐόντος τοῦ ἐπιγεννήματος, αὐθημερόν· λεπτοτέρου
δέ, ἐς τὴν ὑστεραίην· ἔτι δὲ λεπτοτέρου, τριταίην· τὰ
δ᾽ αὐτὰ σημαίνει καὶ ἐπ᾽ ἄκρην τὴν γλῶσσαν γινόμενα, ἧσσον δέ.

226. Γλῶσσα τρομώδης, μετ᾽ ἐρυθήματος κατὰ
ῥῖνας καὶ κοιλίης ὑγρῆς, τὰ δ᾽ ἄλλα[62] ἀσήμως ἔχοντα
κατὰ πλεύμονα, πονηρὰς καὶ ὀξείας καθάρσιας ὀλε-
θρίους σημαίνει.

227. Γλῶσσα παρὰ λόγον ἀπαλυνομένη ἀσώδεσι,[63]
μεθ᾽ ἱδρῶτος ψυχροῦ, ἐπὶ κοιλίῃ ὑγρῇ, μελάνων ἐμέτων σημεῖον· τὸ κοπιῶδες ἐν τούτοισι κακόν.

228. Αἱ τρομώδεες γλῶσσαί τισι καὶ κοιλίην καθυ-
γραίνουσιν·[64] μελανθεῖσαι δ᾽ ἐν τούτοισι, ταχὺν θάνατον σημαίνουσιν· ἆρα τρομώδης γλῶσσα σημεῖον
οὐχ ἱδρυμένης γνώμης;

229. Αἱ δασεῖαι, κατάξηροι, φρενιτικαί.

[61] σημ. Α: γενέσθαι δηλοῖ I.

from the nostrils. Such patients that are not burning hot to the touch develop phrenitis, especially if a haemorrhage occurs.

224. For the tongue to tremble at the beginning of a disease, although it retains its normal colour, but then with the passage of time to become rough, livid, and broken, is a mortal sign. If it becomes very dark, this indicates that the crisis will take place on the fourteenth day. Worst is a dark and green tongue.

225. For the bifurcation of the tongue to be smeared with a kind of white saliva signals a remission of fever—if the coating is thick, on the same day, if it is thinner, on the next day, if it is thinner still, then on the third day. If this occurs on the tip of the tongue, it has the same significance, only less so.

226. A trembling of the tongue, in association with redness in the region of the nose and diarrhoea—if the lung is otherwise without any sign—, is a bad sign and indicates that acute cleanings will have a fatal outcome.

227. For the tongue to become soft for no reason, in patients with nausea, cold sweating, and diarrhoea, is a sign that dark material will be vomited up. Weariness in such patients is a bad sign.

228. Trembling of the tongue may also presage diarrhoea in some patients; if the tongue is also dark in such cases, it foretells an early death. Does trembling of the tongue indicate that the mind is unsettled?

229. Rough, very dry tongues indicate phrenitis.

---

62 Add οὐκ A.
63 -δεσι Potter: -δες εἰ A: -δει I.
64 καθυγρ. A: ὑγρήν ποτε ποιέουσι I.

230. Ὀδόντας συνερείδειν ἢ πρίειν, ᾧ μὴ σύνηθες ἐκ παιδίου, μανικὸν καὶ θανάσιμον· ἢν δὲ παραφρονέων ποιέῃ τοῦτο, παντελῶς ὀλέθριον· ὀλέθριον δὲ καὶ ξηραίνεσθαι τοὺς ὀδόντας.

231. Ὀδόντος σφακελισμὸς ἀπόστημα παρὰ οὖλον γενόμενον λύει.

232. Ἐπὶ ὀδόντος σφακελισμῷ πυρετὸς ἐπιγενόμενος σφοδρός, καὶ παραφροσύνη, | θανάσιμον· ἢν δὲ σῴζωνται, καὶ ἕλκεα ἐκπυήσῃ, ὀστέα ἀφίσταται.

233. Οἷσι περὶ τὴν ὑπερῴην ὑγροῦ σύστασις γίνεται, ὡς τὰ πολλὰ πυοῦται.

234. Τὰ περὶ γένυν ἀλγήματα σφοδρὰ κίνδυνος εἰς ὀστέου ἀνάπλευσιν ἐλθεῖν.

235. Χεῖλος συσπώμενον σημαίνει κοιλίης χολώδεος κατάρρηξιν.

236. Τὰ ἀπὸ οὖλων αἵματα ἐπὶ κοιλίῃ ὑγρῇ, ὀλέθρια.

237. Πτυάλου ἀναχρέμψιες ἐν πυρετῷ πελιαί, μέλαιναι, χολώδεες, ἐπιστᾶσαι μέν, κακόν· ἀποχωρέουσαι δὲ κατὰ λόγον, χρήσιμον.

238. Οἷσιν ἁλμώδεα πτύαλα καὶ βὴξ προσίσταται, ἐν τούτοισι χρὼς ἐρυθραίνεται, οἷον ἐξανθήμασι, πρὸ δὲ τῆς τελευτῆς τρηχύνεται.

239. Ἀνάχρεμψις πυκνή, ἢν δή τι καὶ ἄλλο σημεῖον προσῇ, φρενιτικόν.

240. Αἱ μετ' ἐκλύσιος ἀφωνίαι, κάκιστον.

241.[65] Αἱ ἐπ' ὀλίγον θρασέες παρακρούσιες, πονηρὸν καὶ θηριῶδες.

230. To grind and saw the teeth, for a person who has not had this habit since his childhood, is a sign of mania and death. If a person who is deranged does this, it is a very ill omen. It is also a fatal sign for the teeth to become dry.

231. Sphacelus of a tooth resolves an abscession along the gum.

232. After the sphacelus of a tooth, the addition of a powerful fever and of derangement of the mind foretells death. If such patients survive, and their lesions produce pus, their bones separate.

233. Patients who form a collection of fluid in their palate usually suppurate.

234. When there are severe pains in the area of the jaw, there is a danger that the case will come to a separation of the bone.

235. A contracted lip indicates a violent downward discharge of bilious material.

236. Bleeding from the gums in diarrhoea is a fatal sign.

237. In a fever, the expectoration of livid sputa that are dark and bilious is a bad sign if they stop; but if they proceed as they should, it is a favourable sign.

238. In patients who begin to cough up salty sputa, the skin becomes red, resembling an efflorescence, and before the end it becomes rough.

239. Frequent expectoration, if some other sign is present as well, indicates phrenitis.

240. Loss of speech together with faintness is a very bad sign.

241. Short periods of delirium characterized by overboldness are a bad sign and presage wildness.

---

65 Ch. 241 om. A.

242. Οἷσι φωνὴ ἅμα πυρετῷ ἐκλείπει μετὰ ἀκρισίης, τρομώδεις θνήσκουσιν.

243. Αἱ ἐν πυρετῷ ἀφωνίαι σπασμώδεα τρόπον, ἐκστᾶσαι σιγῇ, ὀλέθριον.

244. Αἱ ἐκ πόνου ἀφωνίαι, δυσθάνατοι.

245. Αἱ μετ᾽ ἐκλύσιος κατόχως ἀφωνίαι, ὀλέθριοι.

246. Αἱ κατακλώμεναι φωναὶ μετὰ φαρμακείην, ἆρα πονηρόν; τούτων οἱ πλεῖστοι ἐφιδροῦσι, καὶ κοιλίας καθυγραίνονται.

638   247. Ἐν ἀφωνίῃ πνεῦμα οἷον | τοῖσι πνιγομένοισι πρόχειρον, πονηρόν· ἆρά γε καὶ παρακρουστικόν;

248. Αἱ ἐκ κεφαλαλγίης ἀφωνίαι ἅμα ἱδρῶτι πυρετώδεες· χαλῶντα ὑπ᾽ αὐτούς, ἐπανιόντα, χρονιώτερα· ἐπιρριγοῦν τούτοισιν, οὐ πονηρόν.

249. Αἱ μετὰ ἀφωνίης ἐκστάσιες, ὀλέθριοι.

250. Αἱ τοῖσιν ἐπιρριγοῦσιν ἀφωνίαι, θανάσιμον· εἰσὶ δὲ καὶ κεφαλαλγέες οἱ τοιοῦτοι ἐπιεικῶς.

251. Αἱ μετ᾽ ἐκλύσιος ἀφωνίαι ἐν πυρετῷ ὀξεῖ ἀνιδρωτί εἰσι μὲν θανάσιμοι, ἧττον δὲ τῷ ἐφιδροῦντι, χρόνον δὲ σημαίνει· ἴσως δὲ οἱ ἐξ ὑποστροφῆς παθόντες τι τοιοῦτον, ἀσφαλέστατοι, ὀλεθριώτατοι δὲ τῶν τοιούτων,[66] οἷσι τὰ ἐκ ῥινῶν, καὶ οἷσι κοιλίαι καθυγραίνονται.

252. Ὀξυφωνίη κλαυθμώδης, καὶ ὀμμάτων ἀμαύρωσις, σπασμῶδες· οἱ ἐς τὰ κάτω πόνοι τούτοισιν εὔφοροι.

[66] ἀσφαλ.—τοιούτων om. A.

242. Patients who lose their speech during a fever that does not reach a crisis die with tremors.

243. In fevers, loss of speech with convulsions that lead to a silent delirium is a fatal sign.

244. Loss of speech arising from exertion brings a hard death.

245. Loss of speech in conjunction with a fit of cataleptic fainting is a fatal sign.

246. A broken, feeble voice after the administration of a purging medication: is this an evil sign? Most of these patients perspire over their whole body and have diarrhoea.

247. In persons who have lost their speech, perceptible breathing like that heard in suffocation bodes ill: do these patients also become delirious?

248. Loss of speech subsequent to a headache, if accompanied by sweating, indicates fever; spontaneous evacuations or remissions indicate that the disease will be quite long. For such patients to have chills afterwards is not an evil sign.

249. Derangements of the mind in association with loss of speech are a fatal sign.

250. Loss of speech in patients with chills is a sign of death. Such patients are also likely to have headaches.

251. In an acute fever, loss of speech together with faintness but unaccompanied by sweating is a mortal sign; but less so if sweating comes on, although this does indicate chronicity. Patients suffering something of this sort as a relapse are perhaps in less danger, whereas it is more likely fatal in those who have an epistaxis and those with diarrhoea.

252. A high-pitched, broken voice and dimness of the eyes indicate convulsions. Pains invading the lower parts in such patients are easy to bear.

253. Ἅμα φωνῇ τρομώδει, λύσις κοιλίης παράλογος, ἐν τοῖσι διεστηκόσι χρονίοισιν ὀλέθριον.

254. Αἱ πυκναὶ ὑποκαρώδεις ἀφωνίαι σύστασιν
φθινώδεα προσημαίνουσιν.

255. Πνεῦμα πυκνὸν μὲν ἐὸν καὶ μικρόν, φλεγμονὴν
καὶ πόνον[67] ἐν τοῖσιν ὑπὲρ τῶν φρενῶν[68] τόποισι
σημαίνει· μέγα δὲ καὶ διὰ πολλοῦ χρόνου, παραφροσύνην ἢ σπασμόν· ψυχρὸν δέ, θανάσιμον· θανάσιμον δὲ καὶ τὸ πυρετῶδες καὶ λιγνυῶδες πνεῦμα,
ἧσσον δὲ τοῦ ψυχροῦ· καὶ τὸ μέγα ἔξω πνεόμενον,
640   σμικρὸν δὲ εἴσω, καὶ τὸ σμικρὸν ἔξω, | μέγα δὲ
εἴσω,[69] κάκιστον δὲ καὶ πλησίον θανάτου· καὶ τὸ
ἐκτεῖνον καὶ κατεπεῖγον, ἀμαυρόν, καὶ διπλῇ εἴσω
ἐπανάκλησις, ὁκοῖον ἐπεισπνέουσιν· εὔπνοια δὲ ἐν
πᾶσιν, ὁπόσα σὺν πυρετῷ ὀξεῖ, κἢν ἐν τεσσαράκοντα
ἡμέρῃσι κρίνηται, μεγάλην ἔχει ῥοπὴν ἐς σωτηρίην.

256. Τράχηλος σκληρὸς καὶ ἐπώδυνος, καὶ γενύων
σύνδεσις, καὶ φλεβῶν σφαγιτίδων παλμὸς ἰσχυρός,
καὶ τενόντων σύντασις, ὀλέθριον.

257. Τὰ ἐν φάρυγγι ἰσχνῇ ἀλγήματα πνιγώδεα,
ἀπὸ κεφαλῆς ἀλγηδόνος ὁρμώμενα, σπασμώδεα.

258. Αἱ τραχήλου καὶ μεταφρένου ψύξιες, δοκέου-
σαι καὶ ὅλου δὲ τοῦ σώματος, σπασμώδεες· ἐν τούτοισι κριμνώδεις οὐρήσιες.

259. Οἷσι κατὰ φάρυγγα ἐρεθισμοί, ἐπιεικῶς τὰ
παρ' οὖς ἐπάρματα.

[67] καὶ π. om. A.

253. In conjunction with a tremulous voice, a relaxation of the cavity for no reason in irregular chronic diseases is a fatal sign.

254. Frequent losses of speech in conjunction with a mild stupor signal in advance a consumptive disturbance.

255. Breathing that is frequent and shallow indicates an inflammation and pain in the parts above the diaphragm. If the breaths are deep and at long intervals, they indicate a disordering of the mind or convulsions. If they are cold, they signal death; it also signals death if the breath is febrile and sooty, but less so than if it is cold. Large expirations with small inspirations, and small expirations with large expirations are the worst sign and occur near death; also bad are an extended respiration, a hurried, obscure respiration, and a double inspiration, as if patients are breathing in again. In all cases with acute fever, even if they only have a crisis on the fortieth day, healthy breathing makes a great contribution towards salvation.

256. A stiff and painful neck, contraction of the jaws, a powerful throbbing of the jugular vessels, and contraction of the tendons are a fatal sign.

257. To have suffocating pains in the throat, that take their origin from a headache, in the absence of any swelling, gives an indication of convulsions.

258. Chills of the neck and back that also seem to occupy the whole body indicate convulsions. In such cases the urines have a thick sediment like meal.

259. Patients with irritations in the area of the throat will probably also have swellings beside the ear.

---

68 ὑπὲρ τ. φρενῶν A: καιρίοισι I.
69 καὶ τὸ σμικρὸν ἔξω, μέγα δὲ εἴσω om. A.

260. Φάρυγξ ἐπώδυνος, ἰσχνή, μετὰ δυσφορίης, ὀξέως ὀλέθριον.

261. Οἷσι πνεῦμα ἀνέλκεται, καὶ πνιγμώδης φωνή, σφόνδυλος ἐγκάθηται τούτοισιν ἐπὶ τῇσι τελευτῇσιν ὡς συσπῶντός τινος τὸ πνεῦμα γίνεται.

262. Φάρυγξ τρηχυνθεῖσα ἐπ᾽ ὀλίγον, καὶ κοιλίη κενεῇσιν ἀναστάσεσι, μετώπου ἄλγημα, ψηλαφώδεες, ὀδυνώδεες· τὰ ἐκ τούτων αὐξανόμενα, δύσκολα.

263. Τὰ κατὰ φάρυγγα ἰσχυρὰ ἀλγήματα παρ᾽ οὓς ἔπαρμα καὶ σπασμοὺς ἐργάζεται.

264. Καὶ τραχήλου καὶ νώτου ἄλγημα, μετὰ πυρετοῦ ὀξέος, | σπασμῷ, ὀλέθριον.

265. Τραχήλου καὶ πήχεων ἄλγημα, σπασμῶδες· ἀπὸ προσώπου δὲ ταῦτα· οἱ κατὰ φάρυγγα ὄχλοι, ἰσχνοί, πτυαλίζοντες, ἐν τούτοισιν, οἱ ἐν ὕπνοις ἱδρῶτες ἀγαθόν· ἆρά γε καὶ τῷ ἱδρῶτι κουφίζεσθαι, τοῖσι πλείστοισιν οὐ πονηρόν; οἱ ἐς τὰ κάτω πόνοι τούτοισιν, εὔφοροι.

266. Ἐν ἀλγήματι νώτου καὶ στήθεος αἱματώδης οὔρησις ἐπιστᾶσα, ὀλέθριος ἐπιπόνως.

267. Τραχήλου πόνος, κακὸν μὲν ἐν παντὶ πυρετῷ, κάκιστον δὲ καὶ ἐν οἷσι ἐκμανῆναι ἐλπίς.

268. Ἐπὶ στήθεος ἀλγήματι πυρετώδει κοιλίη ταραχώδης, ναρκώδης, σημεῖον μελαινῶν ὑποχωρησίων.

269. Τὰ ἐν ὀξέσι κατὰ φάρυγγα μικρὰ ὀδυνώδεα, ὅτε χάνοι, μὴ ῥηιδίως συνάγοντι, ἰσχνῷ, παρακρουστικά· ἐκ τούτων φρενιτικοί, ὀλέθριον.

260. A painful throat with no swelling, if accompanied by restlessness, is an acutely fatal sign.

261. Patients in whom the breath is drawn short, the voice choked, and the spine depressed, breathe in the end stages like a person having a spasm.

262. The throat becoming rough over a short period, the cavity having empty contractions, pain of the forehead; the patient groping with his hands and having pains: the sequelae of these things are troublesome.

263. Severe pains in the area of the throat cause swellings beside the ear and convulsions.

264. Pain of the neck and the back, in conjunction with an acute fever, indicates death from a convulsion.

265. Pain of the neck and forearm is an indication of convulsions; these originate from the face. For patients who have obstructions without swelling in the throat, which produce saliva, to sweat during sleep is a good sign. Is it not injurious in most cases to be relieved by the sweat? Pains moving to the lower parts in such patients are easy to bear.

266. With pains of the back and chest, a bloody urine that stops portends suffering and death.

267. A pain in the neck is a bad sign in every fever, but worst in patients in whom there is reason to expect delirium.

268. During a pain of the chest accompanied by fever, disturbance of the cavity and numbness indicate the passage of dark stools.

269. In acute diseases, slight pains in the throat felt by a person who on opening his mouth cannot easily close it, and who has no swelling, announce delirium. Any of these who have phrenitis are doomed.

270. Φάρυγξ ἑλκουμένη ἐν πυρετῷ μετ' ἄλλου σημείου τῶν δυσκόλων, κινδυνῶδες.

271. Ἐν πυρετοῖσιν ἐξαίφνης πνίγεσθαι, καὶ καταπίνειν μὴ δύνασθαι, χωρὶς οἰδήματος, κακόν.

272. Τράχηλον ἐπιστραφῆναι μὴ δύνασθαι, μηδὲ
καταπίνειν, θανάσιμον ὡς τὰ πολλά.

273. Ὑποχόνδριον δὲ χρὴ μαλθακὸν εἶναι καὶ ἄπο-
νον καὶ ὁμαλόν· φλεγμαῖνον δέ, ἢ ὀδύνην ἔχον ἢ
644   ἀνωμάλως διακείμενον, σημεῖον ἀρρωστίης ἐστὶν οὐκ
εὐήθεος.

274. Οἴδημα δὲ ἐν ὑποχονδρίῳ, σκληρόν τε ὂν καὶ
ἐπώδυνον, κάκιστον μέν, εἰ παρὰ πάντων εἴη τῶν
μερέων· τῶν δ' ἐκ τοῦ ἑνὸς μέρεος, ἀκινδυνότερον τὸ ἐκ
τῶν ἀριστερῶν· σημαίνει δὲ ἐν ἀρχῇ μὲν τὰ τοιαῦτα
θάνατον σύντομον, ὑπερβάλλοντα δὲ τὰς εἴκοσι, τοῦ
πυρετοῦ μένοντος, ἐμπύησιν προσδέχεσθαι· γίνεται
δὲ τούτοισιν ἐν τῇ πρώτῃ περιόδῳ[70] ῥῆξις αἵματος διὰ
ῥινῶν, καὶ κάρτα ὠφελέει· τὰ γὰρ πολλὰ κεφαλὴν
οὗτοι πονέουσι, καὶ ὄψις ἀμαυροῦται, καὶ μᾶλλον εἰς
ταῦτα προσγίνεσθαι προσδέχου τὴν ῥῆξιν, ἡλικίῃσι
δὲ πέντε καὶ τριήκοντα ἐτέων, τοῖσι δὲ πρεσβυτέροισιν ἧσσον.

275. Τὰ μαλθακὰ δὲ καὶ ἀνώδυνα τῶν οἰδημάτων,
χρονιώτερα δὲ τὰς κρίσιας ποιέεται, καὶ ἧσσόν ἐστιν
ἐπικίνδυνα· τὰς δὲ ἑξήκοντα καὶ ταῦτα ὑπερβάλλοντα,
τοῦ πυρετοῦ μένοντος, ἐμπυοῦται. παραπλήσια δὲ σημαίνει τοῖσιν ἐν ὑποχονδρίοισι καὶ τὰ περὶ κοιλίην,
πλὴν ἧσσον ἐκπυοῦται ταῦτα ἐκείνων, ἥκιστα δὲ ὑπ'

270. For the throat to be ulcerated in a fever in association with any other of the signs indicating trouble signifies danger.

271. Suddenly to suffocate during fevers, and to be unable to swallow fluids, when no oedema is present, is a bad sign.

272. Not to be able to turn the neck or to swallow fluids is generally a sign of death.

273. The hypochondrium should be soft, free of pain, and flat. But if it is inflamed, has a pain, or lies in an irregular way, this indicates an illness that is not benign.

274. A swelling in the hypochondrium which is hard and painful is a very bad sign, if it involves both sides; of swellings that come from only one side, those from the left side are less dangerous. At the beginning of diseases, these signs indicate a rapid death, but if the swelling lasts beyond twenty days and the fever persists, expect internal suppuration. In the first phase, these patients have a haemorrhage through the nostrils, and this helps considerably. For generally such patients have pain in the head, and their sight is dimmed: expect these things to progress to the haemorrhage more in the ages up to thirty-five years, in older persons, less often.

275. Swellings that are soft and painless have their crisis rather late and are not too dangerous; if they extend beyond sixty days, with fever still present, they form internal suppurations. Swellings in the area of the cavity have similar signs to those in the hypochondria, except that the former tend to suppurate less than the latter, and those below

---

[70] The text in I ends at this point, leaving A as the sole independent witness for the remainder of the treatise.

ὀμφαλόν· καὶ γίνεται δὲ ταῦτα μὲν ἐν χιτῶνι, τὰ δ' ἄνω
κεχυμένα· θανάσιμα δ' ἐστὶν αὐτῶν, ὅσα ἂν εἴσω
ῥαγῇ· τῶν δὲ λοιπῶν ἐμπυημάτων τὰ μὲν ἔξω ῥηγνύμενα, βέλτιστον μὲν ὡς εἰς ἐλάχιστον καὶ ὀξύτατον
συλλέγεσθαι· τὰ δὲ εἴσω, μήτε ὄγκῳ, μήτε πόνῳ, μήτε
χρώματι διάδηλον ἔξω ποιέειν[71] τὸ δὲ ἐναντίον κάκι-
στον· τινὰ δὲ τούτων διὰ πάχος πύου οὐ διασημαίνει.

Τὰ δὲ πρόσφατα τῶν ἐν τοῖσιν ὑποχονδρίοισιν
ἐπαρμάτων, ἢν μὴ σὺν φλεγμονῇ ᾖ, καὶ τοὺς ἀπ'
646 αὐτῶν πόνους λύει βορβορυγμὸς γενόμενος ἐν | ὑπο-
χονδρίῳ, καὶ μάλιστα μὲν διεκπεσὼν δι' οὔρων καὶ
διαχωρημάτων· εἰ δὲ μή, καὶ αὐτὸς διαπεραιωθείς·
ὠφελεῖ δὲ καὶ ὑποκαταβὰς ἐς τὰ κάτω χωρία.

276. Σφυγμὸς ἐν ὑποχονδρίῳ μετὰ θορύβου, παρα-
κρουστικόν, καὶ μᾶλλον ἢν αἱ ὄψιες πυκνὰ κινέωνται.

277. Καρδίης πόνος καὶ σφυγμὸς ὑποχονδρίῳ, πυρετοῦ περιψυχθέντος, κακόν, ἄλλως τε κἢν ἐφιδρῶσιν.

278. Ἐς ὑποχόνδριον ἐμπίπτοντα ἀλγήματα, ἄλλως τε πονηρόν, καὶ ἢν κοιλίας καθυγραίνῃ· κακίω δέ,
ἐν ὀλίγῳ γινόμενα· καὶ τὰ παρ' οὖς τε ἀνιστάμενα ἐκ
τούτων, κακοήθεα, καὶ τὰ ἄλλα ἐκπυήματα.

279. Καρδιαλγικὰ καὶ μετὰ στρόφου, κοιλίης θηρία καταρρήγνυται.

280. Καρδίης ἄλγημα, πρεσβυτέρῳ πυκνὰ ἐπιφοιτέον, θάνατον ἐξαπίναιον σημαίνει.

281. Οἷσιν ὑποχόνδρια μετεωρίζεται, κοιλίης ἐπι-

71 Littré: -έει A.

the navel least; these last arise in a tunic, whereas those higher up are diffuse. Deadly are swellings whose suppuration ruptures internally. Of other suppurations, some rupture externally—in the best instance forming a very small and pointed collection of pus—, and others inwards, best without becoming noticeable on the outside by any mass or pain or coloration: signs opposite to these are very bad. Some suppurations do not show any signs, due to the thickness of their pus.

New swellings in the hypochondria, if they are without an inflammation, have their pains resolved by intestinal rumbling in the hypochondrium, and especially if the rumbling passes off with the urines and stools: but if not so, then even if it comes to an end by itself. It also helps if the rumbling descends into the lower parts.

276. Pulsation in the hypochondrium accompanied by confusion points towards derangement, especially if the eyes have rapid movements.

277. Pain of the cardia and pulsation in the hypochondrium, after a fever has cooled off, are a bad sign, especially if they are followed by sweating over the body.

278. Pains invading the hypochondrium are an evil sign, especially if they cause diarrhoea; it is even worse if this happens over a short time. Swellings beside the ear that develop in such cases are malignant, as are other suppurations.

279. Pains in the cardia together with colic announce the evacuation of worms.

280. A pain in the cardia, if it recurs frequently in an older person, announces a sudden death.

281. Persons in whom the hypochondria swell up, while

στάσης,[72] κακόν· μάλιστα δὲ ἐν φθινώδεσι τῶν μακρῶν, καὶ οἷς κοιλίαι ὑγραίνονται.

282. Ἐν ὑποχονδρίῳ φλεγμονὴ ἀποπυητική, ἔστιν
οἷς πρὸ τῶν θανάτων μέλανα διαχωρέει.

283. Ὑποχονδρίων σύντασις, μετὰ καύματος ἀσώδεος, κεφαλαλγικῷ, τὰ παρ' οὖς ἐπαίρει.

284. Μετὰ ὑποχονδρίων ἔπαρσιν, τοῖσι χολώδεσι,
πνεῦμα μέγα καὶ πυρετὸς ὀξὺς τὰ παρ' οὖς ἐπαίρει.

285. Ἐν ὑποχονδρίων ἀλγήματι, ὑποβορβορύζον-
τι, ὀσφύος ἄλγημα ἐπιγενόμενον ἐν πυρετοῖς κοιλίας
ἐπὶ τὸ πολὺ καθυγραίνει, ἢν μὴ φῦσα καταρραγῇ, ἢ
οὔρου πλῆθος ἔλθῃ.

648   286. Ἐπὶ ὑποχονδρίῳ χρονίῳ καὶ | κοιλίῃ δυσώδει,
παρ' οὖς ἀπόστημα κτείνει.

287. Τοῖσιν ἀπὸ ὑποχονδρίων ἀλγήμασι κοιλίη
κατὰ μικρὸν ὑπόγλισχρα διαδιδοῦσα βραχέα κοπρώδεα, ἐκχλοιοῖ· ἆρα καὶ αἱμορραγεῖ;

288. Οἷσιν ἐξαίφνης ἀπυρέτοισιν ἐοῦσιν ὑποχονδρίου καὶ καρδίης πόνος, καὶ περὶ σκέλεα καὶ τὰ κάτω
μέρεα, καὶ κοιλίη ἐπῆρται, λύει φλεβοτομίη καὶ κοιλίης ῥύσις· πυρέξαι βλαβερὸν τούτοισιν· μακροὶ γὰρ
οἱ πυρετοὶ καὶ ἰσχυροὶ γίνονται, καὶ βῆχες καὶ πνεῦμα καὶ λυγμοὶ γίνονται· λύεσθαι δὲ μελλόντων τούτων, πόνος ἰσχυρὸς ἰσχίων ἢ σκελέων, ἢ πύου πτύσις,
ἢ ὀφθαλμῶν στέρησις ἐπιγίνεται.

289. Οἷσι πόνοι ὑποχονδρίων, καρδίης, ἥπατος,

---

[72] ἐπιστ. Lind. from Foes' note: ὑποστ. A

the cavity is stopped, are in a bad way, especially in chronic consumptions, and if they have diarrhoea.

282. During suppurative inflammation in the hypochondria, in some patients dark stools pass before their deaths.

283. Contraction of the hypochondria together with a nauseating burning, in a person with a headache, provokes swelling beside the ear.

284. After a swelling of the hypochondria in persons suffering from bile, deep breathing and an acute fever lead to swelling beside the ear.

285. In a person with pain of the hypochondria and slight rumbling sounds, the arrival of a pain in the loins during fevers generally causes diarrhoea, unless flatulence is expelled or a quantity of urine passes.

286. If it occurs in a chronic disorder of the hypochondrium accompanied by the evacuation of ill-smelling stools, swelling beside the ear is fatal.

287. For the cavity to excrete a scanty amount of somewhat sticky, fecal material, a little at a time while there are pains of the hypochondria, makes the patient turn sallow: will he also have a haemorrhage?

288. Persons in whom, together with a sudden remission of fever, there is pain of the hypochondrium and the cardia, and also in the legs and lower parts, and the cavity is raised, are relieved by phlebotomy or a flux of the cavity. To become febrile is bad for them, since such fevers are long and intense, and coughing, difficult breathing, and hiccups occur. When these patients are about to recover, an intense pain attacks the hips or legs, or they cough up pus, or their eyes become blind.

289. Persons with pains of the hypochondria, the

τῶν περὶ ὀμφαλὸν μερῶν, αἵματος διαχωρήσαντος, σῴζονται, μὴ διαχωρήσαντος δέ, θνήσκουσιν.

290. Οἷσιν ὑποχόνδρια λαπαρά, πρόσωπον ἐρρωμένον, οὐ λύεται χωρὶς αἵματος ῥύσιος ἐκ ῥινῶν πολλοῦ, ἢ σπασμοῦ, ἢ ὀδύνης ἰσχίων.

291. Αἱ πρὸς ὑποχόνδρια ἐν πυρετῷ ὀδύναι ἀναύδῳ, ἀνιδρωτὶ λυόμεναι, κακόν· τούτοισιν ἐς ἰσχία ἀλγήματα.

292. Οἱ κατὰ κοιλίην ἐν πυρετῷ παλμοὶ ἐκστάσιας ποιέουσιν· αἱμορροίη δὲ φρικώδης.

293. Αἱ ἐς ὑποχόνδρια ἐν πυρετῷ ὀδύναι ἀναΐσσουσαι, ἀνιδρωτὶ λυόμεναι, κακοήθεες, τούτοισιν ἐς ἰσχία ἀλγήματα, ἅμα πυρετῷ καυσώδεϊ, κοιλίη καταρραγεῖσα, ὀλέθριον.

294. Οἱ περὶ ὀμφαλὸν πόνοι παλμώδεες ἔχουσι μέν τι καὶ γνώμης παράφορον· περὶ κρίσιν δ᾽ οὖν τούτοισι
650 φλέγμα ἅλες | συχνὸν σὺν πόνῳ διέρχεται.

295. Μετὰ κοιλίης ἐπίστασιν[73] ὑποχόνδρια μετέωρα, κακόν· μάλιστα δὲ τοῖσι φθινώδεσι τῶν μακρῶν, καὶ οἷσι κοιλίαι ὑγραίνονται.

296. Τοῖσιν ἀλυσμώδεσιν ἐν ὑποχονδρίῳ τὰ παρ᾽ οὖς ἐπαρθέντα κτείνει.

297. Τὰ κατὰ κοιλίην σκληρύσματα μετὰ πόνου, πυρετοῖσι φρικώδεσιν, ἀποσίτοισι, σμικρὰ ἐφυγραινομένης, κάθαρσιν οὐ διδόντα, ἐς ἐμπύησιν ἥξει.

[73] ἐπίστ. Lind.: ὑπόστ. A.

cardia, the liver, or the area around the navel are saved if they pass blood in their stools, but die if they do not.

290. Persons whose hypochondria are slack,[6] while their faces are healthy, do not recover unless there is a copious haemorrhage from their nostrils, or spasm or pain in their hips.

291. During a fever, for pains moving towards the hypochondria, in association with speechlessness, to be relieved without sweating is a bad sign: these patients will have pains in their hips.

292. In a fever, trembling over the abdomen causes the patient to lose his mind; a haemorrhage is likely to produce shivering.

293. In a fever, pains rising up towards the hypochondria, if relieved without sweating, bode ill: if, in such persons, there are pains in the hips, together with an ardent fever, and the cavity evacuates, they are doomed.

294. Pains about the navel in association with trembling are likely to derange the mind. About the time of the crisis, these patients pass a great mass of phlegm with pain.

295. Swelling of the hypochondria after stoppage of the cavity is a bad sign, especially in chronic consumptives and patients with diarrhoea.

296. In persons troubled in the hypochondrium, a swelling beside the ear is fatal.

297. Indurations through the abdomen with pain, shivering fevers, anorexia, and the evacuation of small amounts of fluid stools which do not bring about an adequate cleaning, will arrive at suppuration.

[6] Littré accepts J. Opsopoeus' introduction of a negation on the basis of a parallel text in ch. 125 above.

298. Ὑπὲρ ὀμφαλὸν πόνος, καὶ ὀσφύος ἄλγημα, φαρμακείη μὴ λυόμενα, ἐς ὑδρωπιῶδες ξηρὸν ἀποτελευτᾷ.

299. Τὰ δὲ ἐξ ὀσφύος ἀλγήματα, χρονιώτερα, πυρετῷ παροξυνόμενα τριταιογενῶς, ποιέει τὰ θρομβώδεα αἵματα διαχωρέειν.

300. Τὰ ἐν ὀσφύϊ ἀλγήματα, αἱμορροϊκά.

301. Αἱ ἐξ ὀσφύος ἀλγήματος αἱμόρροιαι, λάβραι.

302. Οἷσιν ἐξ ὀσφύος ἀλγήματος ἀναδρομὴ ἐς κεφαλήν, καὶ χεῖρες ναρκώδεες, καὶ καρδιαλγικά, καὶ ἠχώδεα, αἱμορροϊκὰ λάβρως, καὶ κοιλίαι καταρρήγνυνται τούτοισι, καὶ γνῶμαι ταραχώδεες ἐπὶ τὸ πολύ.

303. Αἱ ἐκ νώτου ἀλγήματος ἀρρωστίης ἀρχαί, δύσκολοι.

304. Ἐν ὀσφύος ἀλγήματι συντόνῳ καὶ ὑποφορῇ πλέονι, ἀπ' ἐλλεβόρου ἐμέσαι ἀφρώδεα συχνά, ὠφελεῖ.

305. Ῥάχιος διαστροφὴν καὶ δύσπνοιαν αἵματος ῥύσις λύει.

306. Ἐν ὀσφύϊ ἐπωδύνῳ καρδιαλγικὰ προσελθόντα, σημεῖα αἱμορροώδη, ἢ καὶ προγεγενημένα.

307. Τὰ ἐξ ὀσφύος ἐς τράχηλον καὶ κεφαλὴν ἀναδιδόντα, παραλύοντα παραπληκτικὸν τρόπον, σπασμώδεα, παρακρουστικά· ἆρα καὶ λύεται τὰ | τοιαῦτα σπασμοῖσιν; ἢ τῶν τοιούτων κοιλίαι νοσέουσι, διὰ τῶν αὐτῶν ἰόντων;

652

176

298. Pain above the navel and an ache in the loins, if not relieved by drinking a purgative medication, terminate in a dry dropsy.

299. Pains arising from the loins, when they become more chronic and have a paroxysm of the kind arising from tertian fever, cause clotted blood to pass with the stools.

300. Pains in the loins tend to lead to haemorrhoidal fluxes.

301. Haemorrhoids arising from pains in the loins have violent fluxes.

302. If persons with a migration upward to the head from a pain in the loins, and with numbness of their arms, pains in the cardia, and ringing in the ears, have a violent haemorrhoidal flux, they will also have a discharge of their cavity, and usually a disturbance of their mind.

303. The beginning stages of a disease arising from a pain of the back are difficult.

304. In the case of an intense pain of the loins and a more copious purging below, to vomit up copious frothy material as the result of taking hellebore helps.

305. Twisting of the back and difficult breathing are relieved by a haemorrhage.

306. In the case of a pain in the loins, the arrival of pain in the cardia announces a haemorrhoidal flux, or that one has occurred before.

307. Migration from the loins to the neck and head of paralytic signs, disabling in the manner of apoplexy, indicate convulsions and delirium: are these resolved by the convulsions? Or are the cavities of such persons ill, passing through the same stages?

308. Ἐξ ὀσφύος ἀναδρομὴ πόνου, ὀφθαλμῶν ἴλλωσις, κακόν.

309. Πόνος ἐς στῆθος ἱδρυθεὶς νωθρότητι, κακόν· ἐπὶ πυρετῷ οὗτοι ὀξέως ἀπόλλυνται.

310. Ἐξ ὀσφύος ἀλγήματος ἀναδρομαὶ ἐς καρδίην, πυρετώδεες, φρικώδεες, ἀνεμέοντες λεπτά, ὑδατώδεα, παρενεχθέντες ἄφωνοι, ἐμέσαντες μέλανα, τελευτῶσιν.

311. Τὰ κατ᾽ ὀσφὺν καὶ τὸ λεπτὸν χρόνια ἀλγήματα, καὶ πρὸς ὑποχόνδρια πόνοι, ἀπόσιτοι, ἅμα πυρετῷ, τούτοισιν ἐς κεφαλὴν ἄλγημα σύντονον ἐλθὸν κτείνει ὀξέως τρόπον σπασμώδεα.

312. Οἷσιν ὀσφύος ἄλγημα, οὗτοι κακοί· ἆρα τούτοισι τρομώδεα γίνεται, καὶ φωνὴ δ᾽ ὡς ἐν ῥίγει;

313. Ἆρα τοῖς ὀσφυαλγέσιν, ἀσώδεσιν, ἀνημέτοισιν, ὀλίγα θρασέως παρακρούσασιν, ἐλπὶς μέλανα διελθεῖν;

314. Ὀσφύος πόνος, καρδιαλγικῷ, μετὰ ἀναχρέμψιος βιαίης, ἔχει τι σπασμῶδες.

315. Ὑπάφωνον ἅμα κρίσει ῥῖγος.

316. Ὀσφύος ἄλγημα, ἄνευ προφάσιος πυκνὰ ἐπιφοιτέον, κακοήθεος ἀρρωστίης σημεῖον.

317. Ὀσφύος ἄλγημα μετὰ καύματος ἀσώδεος, πονηρόν.

318. Ὀσφύος σύντασις ἐκ γυναικείων πλήθεος, ἐκπυητικόν· καὶ τὰ ποικίλως ἰόντα, γλίσχρα, δυσώ-

308. A migration of pain upward from the loins in conjunction with distortion of the eyes is a bad sign.

309. Pain settling in the chest at the same time that indolence is present is a bad sign; in a fever, such patients die rapidly.

310. Upward migrations of pain from the loins to the cardia, that are accompanied by fever, shivering, the vomiting of thin, watery material, and loss of speech, end with the vomiting of dark material.

311. Chronic pains in the loins and the small intestine, and pains extending to the hypochondria with anorexia at the same time as fever: in such patients, an intense pain coming to their head kills them rapidly with convulsions.

312. Persons with pain in the loins are in a bad way: do they experience trembling and have a voice like a person with a chill?

313. In patients with pains of the loins, nausea but no vomiting, and a slight mental derangement toward the over-bold: are they likely to pass dark stools?

314. Pain of the loins accompanied by pain in the cardia and violent expectoration has a tendency towards convulsions.

315. A chill tends to be accompanied by a partial loss of speech at its crisis.

316. A pain of the loins occurring frequently without a cause is the sign of a malignant disease.

317. A pain of the loins in association with a nauseating burning bodes ill.

318. Contraction of the loins after the passage of copious menses announces the expulsion of pus. Also the passage of many-coloured viscous evil-smelling menses, in

δεα, πνιγώδη, ἐπὶ τοῖσι προειρημένοισιν, ἐκπυητικόν·
οἶμαι δὲ καὶ παρακρούειν τι τὰς τοιαύτας.

319. Οἷσιν ὀσφύος ἄλγημα καὶ πλευροῦ ἄνευ προφάσιος, ἰκτεριώδεες γίνονται.

320. Αἱ ἐν κρισίμοισιν ἐκ τῶν αἱμορραγιῶν περι-
ψύξιες νεανικαί, κάκισται.

654 321. Τὸ ἀνάπαλιν αἱμορραγέειν, πονηρόν, | οἷον
ἐπὶ σπληνὶ μεγάλῳ ἐκ τῶν δεξιῶν· καὶ κατὰ ὑποχόνδρια ὡσαύτως.

322. Τὰ αἱμορραγεῦντα, ἐπιρριγοῦντα τρώματα,
κακοήθεα· διαλεγόμενοι λαθραίως τελευτῶσιν.

323. Τὰ πεμπταῖα αἱμορραγοῦντα λάβρως, ἕκτῃ
ἐπιρριγώσαντα, ἑβδόμῃ περιψυχθέντα, ἀναθερμαν-
θέντα ὀξέως, τούτοισι κοιλίαι πονηρεύονται.

324. Μεθ᾽ αἱμορραγίην μελάνων διαχώρησις, κα-
κόν· πονηρὸν δὲ καὶ τὰ ἐξερυθρώδεα· τεταρταίοισιν
αἱ τοιαῦται αἱμορραγίαι· κωματώδεες, ἐκ τοιούτων
σπασθέντες θνήσκουσι, μελάνων προδιελθόντων, καὶ
κοιλίης ἐπαρθείσης.

325. Μεθ᾽ αἱμορροίας καὶ μελάνων διαχωρήσιας ἐν
ὀξεῖ κώφωσις, κακόν· αἵματος διαχώρησις τούτοισιν
ὀλέθριον, κώφωσιν δὲ λύει.

326. Οἷσιν αἱμορραγίαι πλείους, προϊόντος χρόνου, κοιλίαι πονηρεύονται, ἢν μὴ οὖρον πέπον ἔλθῃ·
ἆρά γε τὸ ὑδατῶδες οὖρον τοιοῦτόν τι σημαίνει;

327. Οἷσιν ἐπὶ αἱμορραγίῃ λάβρῳ πυκνῇ μετὰ
μελάνων συχνὴ διαχώρησις, ἐπιστάσης δὲ αἱμορροεῖ,

association with suffocation, in the conditions described above, indicates the expulsion of pus. I also believe that such women will become somewhat delirious.

319. Persons with pain of the loins and side, when no cause is apparent, become jaundiced.

320. During crises vehement chills arising from haemorrhages are a very bad sign.

321. Contralateral haemorrhages are a bad sign, as for example in splenomegaly for the right side to bleed; also in the hypochondria, in the same way.

322. Injuries with haemorrhages and subsequent chills are insidious: these patients die from no apparent cause, in the act of speaking.

323. In persons who have a violent haemorrhage on the fifth day, followed by chills on the sixth day, cooling on the seventh day, and then rapid restoration of warmth, the cavities become disordered.

324. The excretion of dark material after a haemorrhage is a bad sign: reddish stools also bode ill. This kind of haemorrhage occurs on the fourth day: the patients are comatose, and then they have convulsions and die with dark excretions and swelling of the cavity.

325. After haemorrhages and dark excretions in an acute disease, deafness is a bad sign: an excretion of blood in these cases is a fatal sign, but it ends the deafness.

326. In persons with frequent haemorrhages, as time passes the cavities become disordered, unless concocted urine passes: does watery urine signify something of this kind?

327. Persons who, after a continuous violent haemorrhage, suffer an intense diarrhoea with dark stools, and, when this stops, haemorrhage again, have pains in their

οὗτοι κοιλίας ὀδυνώδεες, ἅμα δέ τῇσι φύσῃσιν εὔφο-
ροι· ἆρα οἱ τοιοῦτοι ἐφιδροῦσι πολλοῖσι ψυχροῖσιν; τὸ
ἀνατεταραγμένον οὖρον ἐν τούτοισιν οὐ πονηρόν, οὐδὲ
τὸ ἐφιστάμενον γονοειδές· ἐπὶ τὸ πολὺ δὲ οὗτοι ὑδατώδεα οὐρέουσιν.

328. Οἷσιν ἐκ ῥινῶν ἐπὶ κωφώσει καὶ νωθρότητι
μικρὰ ἀποστάζει, ἔχει τι δύσκολον· ἔμετος τούτοισι
συμφέρει καὶ κοιλίης ταραχή.

329. Αἱ ἐν ἀρχῇσι μεγάλαι αἱμορραγίαι περὶ ἀνα-
κομιδὴν κοιλίας καθυγραίνουσιν.

330. Τὰ ἐκ ῥινῶν λάβρα βίῃ ἀποληφθέντα, ἔστιν
656 ὅτε σπασμὸν ἐπικαλεῖται, φλεβοτομίῃ λύει.

331. Αἱ ἑνδεκαταῖαι στάξιες, δύσκολοι, ἄλλως τε
καὶ ἢν δὶς ἐπιστάξῃ.

332. Ἐπὶ αἵματος ῥύσει πολλῇ, ἢ λυγμὸς ἢ σπασμός, κακόν.

333. Τοῖσι νέοις ἐτῶν ἑπτὰ ἀδυναμίη μετὰ ἀχροίης,
καὶ πνεῦμα ἁλιζόμενον ἐν τῇσιν ὁδοῖσι, καὶ γῆς ἐπι-
θυμίη, αἵματος φθορὴν καὶ ἔκλυσιν σημαίνει.

334. Ἐν τοῖσι μακροῖσι τὰ μικρὰ ἐπιφαινόμενα
αἱμορροώδεα, ὀλέθρια.

335. Τὰ σκοτώδεα ἐξ ἀρχῆς αἱμορροίη ῥινὸς λύει.

336. Τὰ ἐκ ῥινῶν σμικροῖς ἱδρῶσι περιψυχόμενα,
κακοήθεα.

337. Αἵματος ἀφαίρεσις ἐν καταψύξει νενωθρευ-
μένῃ, κακόν.

338. Ὅσοι, κοιλίης ἐπιστάσης, αἱμορροέουσι, καὶ
ἐπιρριγοῦσιν ἅμα τῷ αἱμορροεῖν, τούτοισι κοιλίην

cavity; but these are relieved simultaneously with flatulence: do such cases have a subsequent sweat with many chills? Disturbed urine in these does not bode ill, nor a scum of material resembling seed: generally these patients pass watery urines.

328. Persons who, after deafness and torpor, have small haemorrages from their nostrils have a rather difficult course: vomiting or a disturbance of the cavity benefits them.

329. Copious haemorrhages at the beginning (sc. of diseases) cause diarrhoea around the time of recovery.

330. Violent discharges from the nostrils, if halted by force, sometimes provoke convulsions: phlebotomy resolves these.

331. Epistaxis on the eleventh day is difficult, especially if it happens twice.

332. After a great flow of blood, either hiccups or a convulsion is a bad sign.

333. In children of seven years, weakness together with a loss of natural colour, rapid breathing when on the road, and a desire to eat earth indicate a spoiling and dissolution of the blood.

334. In long diseases, the occurrence of short haemorrhagic episodes is a fatal sign.

335. Dizzy spells are resolved at the beginning by a flow of blood from the nose.

336. In patients chilled by minor sweats, epistaxes are a malignant sign.

337. In chills accompanied by torpor, to draw blood is bad.

338. In persons who, during a bout of constipation, haemorrhage and suffer chills at the time of the haem-

λειεντεριώδεα ποιέει καὶ ἐπίσκληρον, ἢ[74] ἀσκαρίδας, ἢ ἀμφότερα.

339. Τὰ τεταγμένοισι χρόνοισιν αἱμορροώδεα, διψώδη, μὴ αἱμορραγήσαντα, ἐπιληπτικῶς θνήσκει.

340. Ἐξ αἱμορροΐδος ὅσον ἐπιφανείσης σκοτώδεα ἐλθόντα, παραπληγικὸν μικρὸν καὶ ἐπ᾽ ὀλίγον σημαίνει· λύει φλεβοτομίη· καὶ πᾶν τὸ οὕτως ἐπιφαινόμενον κακόν τι σημαίνει.

341. Οἱ παλμώδεες δι᾽ ὅλου, ἆρα καὶ ἄφωνοι τελευτῶσιν;

342. Τὰ τρομώδεα, σπασμώδεα γενόμενα, ἐφιδροῦσι, φιλυπόστροφα· τούτοισι κρίσις ἐπιρριγώσασιν· ἐπιρριγέουσι δ᾽ οὗτοι ἐπὶ κοιλίην καύματι προκληθέντες· ὕπνος πολὺς ἐν τούτοισι, σπαῖσμῶδες, καὶ τὰ ἐς μέτωπον βάρεα, καὶ οὔρησις δυσκολαίνουσα.

343. Οἱ ἐν ὑστερικοῖσιν ἄπυροι σπασμοί, εὐχερεῖς.

344. Τὰ σπασμώδεα ἀνιδρωτί, πτύελα παρρέοντα, πυρετώδει ἐόντι εὐήθεα· τούτοισιν, ἐπεὶ κοιλίαι τι καθυγραίνονται, τάχα δέ τι καὶ ἐς ἄρθρα ἀποστήσονται.

345. Οἷσιν ἐν σπασμώδεσιν ὀφθαλμοὶ ἐκλάμπουσιν[75] ἀτενέως, οὔτε παρ᾽ ἑωυτοῖσίν εἰσι, διανοσέουσί τε μακροτέρως.

346. Τὰ σπασμώδεα τρόπον παροξυνόμενα κατόχως, τὰ παρ᾽ οὖς ἐπαίρει.

---

[74] ἢ Ermerins, cf. *Prorrhetic I* 138: καὶ A.   [75] Littré after Cornarius' ms. note: ἐκλειπάνουσιν (μ above the first π) A.

orrhage, the cavity becomes lienteric and indurated, or passes worms, or both.

339. If persons with haemorrhages at regular times suffer thirst and the haemorrhages do not take place, they die of epilepsy.

340. After a haemorrhoid that has just appeared, the arrival of vertigo announces a minor, gradual paralysis: phlebotomy relieves this. And in fact, everything that appears in that way announces something bad.

341. Persons with tremors through their whole body: do they lose their speech at the end?

342. A disorder with trembling that becomes convulsive, if sweats supervene, has a tendency to relapse. In such patients the crisis arrives after a bout of shivering; the shivering is called forth by a burning sensation over the cavity. Much sleep in such patients leads to convulsions, and they have a sensation of heaviness in the forehead, and difficult urination.

343. In women with uterine conditions, convulsions without fever are not dangerous.

344. Convulsions without sweating, if accompanied by expectoration, are benign in a person with fever. In such cases, since they have some diarrhoea, there will also probably be a deposit in the joints.

345. Persons whose eyes are bright and stare fixedly during convulsions are not in their senses, and they remain ill for a considerable time.

346. A condition of convulsions, that has its paroxysm in the form of catalepsy, causes swelling beside the ear.

347. Τρομώδεσιν, ἀσώδεσι, μικρὰ τὰ παρ' οὖς ἐπάρματα σπασμὸν σημαίνει, κοιλίης πονηρευομένης.

348. Τὰ σπασμώδεα καὶ τετανώδεα πυρετὸς ἐπιγενόμενος λύει.

349. Σπασμὸς ἐπὶ τρώματι, θανάσιμον.

350. Σπασμὸς ἐπὶ πυρετῷ γενόμενος, ὀλέθριον, ἥκιστα δὲ παιδίοισιν.

351. Οἱ πρεσβύτεροι ἑπτὰ ἐτέων ἐν πυρετῷ οὐχ ἁλίσκονται ὑπὸ σπασμοῦ· εἰ δὲ μή, ὀλέθριον.

352. Σπασμοῦ λυτικὸν πυρετὸς ἐπιγενόμενος ὀξύς, μὴ πρότερον γεγονώς· εἰ δὲ εἴη πρότερον γεγονώς, παροξυνθείς· ὠφελέει δὲ[76] καὶ οὔρου διέξοδος ὑαλοειδὴς πολλή, καὶ ῥύσις κοιλίης, καὶ ὕπνοι· τῶν δὲ ἐξαπίνης σπασμῶν λυτικόν, πυρετός, κοιλίης ῥύσις.

353. Ἐν τοῖσι σπασμοῖσιν ἀναυδίη ἐπὶ πολύ, κακόν· τὸ δὲ ἐπὶ μικρόν, ἤτοι γλώσσης ἀποπληξίην, ἢ βραχίονος καὶ τῶν ἐπὶ δεξιὰ σημαίνει· λύεται δὲ οὔροισιν ἐξαπίνης ἐλθοῦσι, πολλοῖσιν, ἀθρόοισιν.

354. Ἱδρῶτες δέ, οἱ μὲν κατὰ μικρόν, ὠφελέουσιν· οἱ δὲ ἀθρόοι, καὶ αἱ τῶν αἱμάτων ἀφαιρέσιες αἱ ἀθρόοι, βλάπτουσιν.

355. Ἐν τοῖσι τετάνοισι καὶ ὀπισθοτόνοισι γέννες
660 λυόμεναι, | θανάσιμον· θανάσιμον δὲ καὶ ἱδροῦν ἐν ὀπισθοτόνῳ καὶ τὸ σῶμα διαλύεσθαι, καὶ ἀνεμεῖν ὀπισθοτόνῳ διὰ ῥινῶν, ἢ ἐξ ἀρχῆς ἄφωνον ἐόντα βοᾶν ἢ φλυηρεῖν· ἐς γὰρ τὴν ὑστεραίην θάνατον σημαίνει.

---

[76] ὠφ. δὲ Foes: δὲ ὠφ. A.

347. In persons with tremors and nausea, small swellings beside the ear indicate a convulsion, if the cavity is disordered.

348. When fever comes on in cases of convulsions and tetanus, it resolves them.

349. A convulsion after a wound is lethal.

350. A convulsion occurring after a fever is a mortal sign, but least so in children.

351. Persons older than seven years are not usually taken by convulsions in a fever, but if they are, it is a fatal sign.

352. The arrival of an acute fever resolves a convulsion, unless the fever had existed before; if it had arisen beforehand, it brings a paroxysm. An excretion of copious clear urine is also of benefit, and also a flux of the cavity, and sleep. Sudden convulsions are resolved by fever or a flux of the cavity.

353. In convulsions, a longer loss of speech is bad; a shorter loss of speech foretells either a paralysis of the tongue, or of an arm and the right parts. It is resolved by copious urines being suddenly excreted in a mass.

354. Sweats, when they proceed a little at a time, are beneficial; when they occur all at once—just like extractions of blood that are made all at the same time—they harm.

355. In cases of tetanus and opisthotonus, for the jaws to be relaxed is a fatal sign. It is also a fatal sign in opisthotonus to sweat and for the body to be relaxed, and to vomit in opisthotonus through the nostrils, or, after losing one's speech at the beginning, to shout and talk nonsense, for this indicates death on the following day.

187

356. Πυρετώδεα ὀπισθοτονώδεα γονοειδεῖς οὐρήσιες λύουσιν.

357. Τὰ κυναγχικὰ τὰ μήτε ἐν τῷ τραχήλῳ μήτε ἐν
τῇ φάρυγγι μηδὲν εὔδηλον ποιέοντα, πνιγμὸν δὲ νεα-
νικὸν καὶ δύσπνοιαν παρέχοντα, αὐθημέρους καὶ τρι-
ταίους κτείνει.

358. Τὰ δὲ ἐπάρματα καὶ ἔρευθος ἐν τῷ τραχήλῳ
λαμβάνοντα, τὰ μὲν λοιπὰ παραπλήσια, χρονιώτερα
δέ.

359. Ὅσοισι δὲ συνεξερευθείη ἥ τε φάρυγξ καὶ ὁ
αὐχὴν καὶ τὸ στῆθος, χρονιώτερα· καὶ μάλιστα ἐξ
αὐτῶν σῴζονται, ἢν μὴ παλινδρομέῃ τὰ ἐρυθήματα·
ἢν δὲ ἀφανίζηται, μήτε φύματος συστραφέντος ἔξω,
μήτε πύου ἀναχρεμπτομένου πρηέως καὶ ἀπόνως,
μήτε ἐν ἡμέρῃσι κρισίμῃσιν, ὀλέθρια γίνεται· ἆρά γε
ἔμπυοι γίνονται; ἀσφαλέστατον δὲ τὸ ἔρευθος καὶ τὰς
ἀποστάσιας ὅτι μάλιστα ἔξω τρέπεσθαι.

360. Ἐρυσίπελας δὲ ἔξωθεν μὲν ἐπιγίνεσθαι, χρή-
σιμον· εἴσω δὲ τρέπεσθαι, θανάσιμον· τρέπεται δέ,
ὅταν, ἀφανιζομένου τοῦ ἐρυθήματος, βαρύνηται τὸ
στῆθος, καὶ δυσπνοώτερος γίνηται.

361. Οἷς δὲ κυνάγχη ἐς τὸν πλεύμονα τρέπεται, οἱ
μὲν ἐν τῇσιν ἑπτὰ ἡμέρῃσιν ἀπόλλυνται· οἱ δὲ διαφυγόντες ἔμπυοι γίνονται, μὴ γενομένης αὐτοῖς ἀνα-
γωγῆς φλεγματώδεος.

362. Οἷσι διὰ σφοδρότητος σφυγμοῦ κόπριον ἐξαπίνης διαχωρέει, θανάσιμον.

356. Febrile opisthotonus is resolved by the passage of urine that looks like seed.

357. Cases of angina that produce no visible changes in either the neck or the throat, but provoke an intense choking and difficulty in breathing, kill patients on the same day or on the third day.

358. When swelling and redness set in in the throat, the cases are otherwise similar, only longer.

359. In persons in whom the throat, the neck, and the chest all become red, the swellings are of longer duration, and in most cases these patients are saved, unless the redness recedes (sc. without a crisis). But if it disappears without a growth forming on the outside, or pus being coughed up gently and easily, and not on the critical days, then such cases turn out to be fatal. Do they suppurate internally? The safest thing is for the redness and the apostasis to be turned mostly outward.

360. For erysipelas to develop externally is advantageous, but for it to turn inward is deadly; it is turning when, after the redness disappears, the chest becomes heavy and the person has more trouble breathing.

361. Of patients in whom angina turns towards the lung, some die in seven days; others escape with their lives but suppurate internally, unless they have an expectoration of phlegmy sputum.

362. If there is a sudden excretion of fecal material as the result of a violent pulsation, this is a deadly sign.

363. Ἐν τοῖσι κυναγχικοῖσι τὰ ὑπόξηρα πτύσματα ἰσχνῶν, κακόν.

364. Τὰ κυναγχικὰ ἐν γλώσσαις οἰδήματα, ἀσήμως ἀφανιζόμενα, ὀλέθρια· καὶ τὰ ἀλγήματα | ἀφανιζόμενα χωρὶς προφάσιος, ὀλέθρια.

365. Ἐν τοῖσι κυναγχικοῖσιν οἱ μὴ παχὺ ἀναπτύοντες πέπονα, ὀλέθριοι.

366. Ἐν κυνάγχῃ ἀσήμως εἰς κεφαλὴν ἀλγήματα μετὰ πυρετοῦ, ὀλέθρια.

367. Ἐν κυνάγχῃ ἀσήμως ἐς σκέλη ἀλγήματα μετὰ πυρετοῦ, ὀλέθρια.

368. Ἐκ κυναγχικῶν ἀκρίτως ὑποχονδρίου ἄλγημα, μετὰ ἀκρασίης καὶ νωθρότητος γενόμενον, κτείνει λαθραίως, εἰ καὶ πάνυ δοκοῖεν ἐπιεικῶς ἔχειν.

369. Ἐκ κυναγχικῶν ἀσήμως ἰσχνανθέντων ἐς στῆθος ἄλγημα καὶ ἐς κοιλίην ἐλθὸν σύντονον, ποιέει πυῶδες διαχωρέειν, ἄλλως δὲ λυομένου τὸ τοιοῦτον.

370. Ἐκ κυναγχικῶν πάντα ὀλέθρια, ὅσα μὴ ἔκδηλον ἐποίησεν ἄλγημα· ἀτὰρ καὶ ἐς σκέλεα ἀλγήματα χρόνια φοιτᾷ, καὶ ἐκπυοῦται δυσκόλως.

371. Τὰ ἐκ κυνάγχης πτύαλα γλίσχρα, παχέα, ἔκλευκα, βιαίως ἀναγόμενα, κακόν, καὶ πᾶς ὁ τοιοῦτος πεπασμός, κακόν· κάθαρσις πολλὴ κάτω τοὺς τοιούτους παραπληκτικῶς ἀπόλλυσιν.

372. Ἐκ κυνάγχης ὑπόξηρα πυκνὰ πτύελα, βηχώδεα, πλευροῦ ὀδυνώδεα, ὀλέθρια· καὶ τὰ ἐν τοῖσι ποτοῖσιν ὑποβήσσοντα, καὶ κατάποσις βιαία, πονηρόν.

363. In angina, the somewhat dry sputa of reduced swellings are a bad sign.

364. When swellings of the tongue in angina disappear without leaving a mark, it is a fatal sign; also if pains disappear for no reason.

365. In anginas patients who do not cough up thick, concocted sputa are doomed.

366. In an angina pains moving to the head, in the absence of other signs but with fever, foretell death.

367. In an angina pains moving to the legs, in the absence of other signs but with fever, foretell death.

368. A pain of the hypochondrium arising from anginas, that remain without a crisis but are accompanied by weakness and torpor, kills without a visible cause, even if the case seems to be very favourable.

369. From anginas that are drying up without signs, intense pain coming to the chest and cavity causes purulent material to pass in the stools, and this tends to resolve the condition.[7]

370. Anything that supervenes after angina without producing a conspicuous pain is fatal; chronic pains lancinate to the legs and difficult suppurations occur.

371. Sputa from angina that are viscous, thick, very white, and coughed up violently are a bad sign, and every such coction is also bad. Copious downward cleaning leads to death by apoplexies in such patients.

372. Sputa from angina that are somewhat dry and frequent, and that are associated with coughing and pains of the sides, are a fatal sign; slight coughs associated with drinking and forced swallowing are also bad.

---

[7] The meaning of the final clause is very uncertain.

373. Τῶν πλευριτικῶν οἷσιν ἐν ἀρχῇ πάμπυοι αἱ
πτύσιες, τριταῖοι θνήσκουσιν, ἢ πεμπταῖοι· φυγόντες
δὲ ταύτας, μὴ πολὺ ῥᾷον ἔχοντες, τῇ ἑβδόμῃ, ἢ
ἐννάτῃ, ἢ ἑνδεκάτῃ, ἄρχονται ἐμπυοῦσθαι.

374. Οἷσι δὲ ἐν νώτῳ ἔρευθος, τῶν πλευριτικῶν, |
664 καὶ ὦμοι[77] θερμαίνονται, καὶ κοιλίη ταράσσεται χολώδεα καὶ δυσώδεα, εἰκοστῇ καὶ μιῇ κινδυνεύουσι,
φυγόντες δὲ ταύτας σῴζονται.

375. Αἱ ξηραὶ τῶν πλευριτίδων καὶ ἄπτυστοι, χαλε-
πώταται· φοβεραὶ δέ, ἐν οἷσιν ἄνω τὰ ἀλγήματα.

376. Αἱ ἄνευ σπασμάτων πλευρίτιδες χαλεπώτεραι
τῶν μετὰ σπασμάτων.

377. Τῶν πλευριτικῶν οἷσιν ἐν ἀρχῇ γλῶσσα χολώδης γίνεται, ἑβδομαῖοι κρίνονται· οἷσι δὲ τρίτῃ ἢ
τετάρτῃ, περὶ τὴν ἐννάτην.

378. Πομφόλυγος δὲ ὑποπελίου γινομένης ἐπὶ τῆς
γλώσσης ἐν ἀρχῇ, οἵη σιδηρίου βαφέντος ἐς ἔλαιον,
χαλεπωτέρη ἡ ἀπόλυσις γίνεται, καὶ ἡ μὲν κρίσις ἐς
τὴν ιδ´ ἀφικνεῖται· αἷμα δὲ ὡς ἐπὶ τὸ πολὺ πτύουσιν.

379. Πτύαλον δ᾽ ἐν τῇσι πλευρίτισι, τρίτῃ μὲν
ἀρχόμενον πεπαίνεσθαι καὶ πτύεσθαι, θάσσους ποιέει
τὰς ἀπολύσιας, ὕστερον δέ, βραδυτέρας.

380. Τὰ δὲ ἀλγήματα τοῖσι πλευριτικοῖσι χρήσιμον κοιλίην μαλάσσεσθαι, πτύαλα χρωματίζεσθαι,
ψόφους ἐν τῷ στήθει μὴ γίνεσθαι, τὸ οὖρον εὐωδεῖν·
τὰ δὲ τούτων ἐναντία δυσχερέα, καὶ πτύαλον γλυκαι-
νόμενον.

---

[77] ἐν νώτῳ . . . ὦμοι Opsopoeus: ἄνω ὠτὸς . . . ὁμοίως A.

373. Patients with pleurisy whose expectorations are totally purulent at the beginning die on the third or fifth day; if they escape that period, unless they become much better, they begin to suppurate internally on the seventh, ninth, or eleventh day.

374. Patients with pleurisy who become red on their back, whose shoulders become warm, and whose cavity is disturbed with bilious foul-smelling stools are in danger on the twenty-first day, but after that are safe.

375. Dry pleurisies with no expectorations are the most difficult to bear; those are to be feared in which pains arise in the upper regions.

376. Pleurisies without ruptures are more difficult than those with ruptures.

377. Pleurisies in which the tongue becomes bilious at the beginning have their crisis on the seventh day; if it becomes bilious on the third or fourth day, then the crisis will be on the ninth day.

378. If at the beginning a somewhat livid blister, such as forms on a cautery iron dipped in olive oil, comes up on the tongue, resolution will be more difficult, and the crisis extends to the fourteenth day; such patients usually expectorate blood.

379. For the sputum in pleurisies to begin to be concocted and expectorated on the third day makes them resolve more quickly, if later, then more slowly.

380. In the presence of pains, it is favourable in pleurisies for the cavity to be softened, the sputa to become coloured, no sounds to arise in the chest, and the urine to smell healthy. Signs opposite to these are unfavourable, as it also is for the sputum to become sweet.

381. Αἱ δὲ χολώδεες ἅμα καὶ αἱματώδεες πλευρίτιδες, ὡς ἐπὶ τὸ πολὺ κρίνονται ἐναταῖαι, ἢ ἐνδεκα-
ταῖαι, καὶ μάλιστα ὑγιάζονται· οἷσι δὲ τῶν πλευριτικῶν ἐν ἀρχῇ μὲν οἱ πόνοι μαλθακοί, πέμπτῃ δὲ
ἢ ἕκτῃ παροξύνονται, μᾶλλον πρὸς τὰς δυοκαίδεκα
ἀφικνέονται, καὶ οὐ πάνυ σῴζονται, κινδυνεύουσι δὲ
μάλιστα ἑβδομαῖοι καὶ δωδεκαταῖοι, τὰς δὲ δὶς ἑπτὰ
φυγόντες, σῴζονται.

382. Ὅσοισι τῶν πλευριτικῶν ψόφος τοῦ πτυάλου
πολὺς ἐν τῷ στήθει, καὶ πρόσωπον κατηφές, καὶ
ὀφθαλμὸς ἰκτεριώδης καὶ ἀχλυώδης, ἀπόλλυνται.

383. Οἱ ἐκ πλευριτικοῦ ἔμπυοι γενόμενοι, ἐν τῇσι
τεσσαράκοντα ἡμέρῃσιν ἀναπτύουσιν ἀπὸ τῆς ῥήξιος.

384. [Περὶ πτυέλου ἐν τοῖς πλευριτικοῖς.][78] πτύαλον
δὲ χρὴ πᾶσι τοῖσι πλευριτικοῖσι καὶ περιπλευμονικοῖσιν εὐπετέως τε καὶ ταχέως ἀναπτύεσθαι, μεμίχθαι
τε τὸ ξανθὸν τῷ πτυάλῳ· τὸ δ᾽ ὕστερον πολλῷ τῆς
ὀδύνης ἀναγόμενον ξανθόν, ἢ μὴ μεμιγμένον, καὶ
πολλὴν βῆχα παρέχον, πονηρόν· πονηρὸν δὲ πάντως
καὶ τὸ ξανθὸν ἄκρητον, καὶ τὸ γλίσχρον καὶ λευκόν,
καὶ τὸ στρογγύλον, καὶ τὸ χλωρὸν σφόδρα, καὶ τὸ
ἀφρῶδες, καὶ τὸ πελιῶδες καὶ ἰῶδες· χεῖρον δέ τε τὸ
οὕτως ἄκρητον, ὥστε μέλαν φαίνεσθαι· αἵματι δὲ μὴ
πολλῷ συμμεμιγμένον τὸ ξανθόν, ἐν ἀρχῇ μὲν σωτήριον, ἑβδομαίῳ δ᾽ ἢ παλαιοτέρῳ ἧσσον ἀσφαλές·
αἱματῶδες δὲ λίην, ἢ πέλιον εὐθέως ἐν ἀρχῇ, κιν-
δυνῶδες· πονηρὰ δὲ καὶ τὰ ἀφρώδεα, καὶ τὰ ξανθά,

381. Pleurisies that are simultaneously bilious and sanguine generally have their crises on the ninth or eleventh day, and in most cases the patients recover; but pleurisies in which the pains are mild at the beginning, but then on the fifth or sixth day become sharper, tend to arrive at the twelfth day, and such patients are not likely to be saved; they are most subject to danger on the seventh and the twelfth days, but, if they make it beyond two weeks, they have reached safety.

382. Cases of pleurisy in which the sputum causes much sound in the chest, the countenance is downcast, and the eyes are jaundiced and cloudy, are fatal.

383. Persons with internal suppurations arising from pleurisy cough them up in forty days from the rupture.

384. The sputum in all pleurisies and pneumonias should be coughed up easily and quickly, and yellow material should be mixed through it. For it to be brought up long after the pain has begun, pure yellow or not mixed, and with the accompaniment of much coughing, is bad. Absolutely bad are sputa that are unmixed yellow, or viscous and white, or globular, or very green, or frothy, or livid and rusty; and even worse, so unmixed that they appear dark. At the beginning, yellow material mixed with a little blood is a sign of safety, but on the seventh day or later this is less sure. Excessively bloody or livid sputum right at the beginning indicates danger. Difficulty is also indicated by

---

[78] Del. Aldina.

καὶ μέλανα, καὶ ἰώδεα, καὶ ἰξώδεα, καὶ ὅσα ταχέως
χρωματίζεται· τὰ δὲ μυξώδεα καὶ λιγνυώδεα καὶ χρωματίζεται ταχέως, καί ἐστιν ἀσφαλέστερα· τὰ δ' ἐντὸς
πέμπτης ἐς πέψιν χρωματιζόμενα, βελτίω.

385. Πᾶν δὲ πτύαλον μὴ λύον τὴν ὀδύνην, πονηρόν·
λύον δέ, χρήσιμον.

668   386. Ὅσοι δὲ μετὰ τοῦ χολώδεος πυῶδες ἀνάγουσιν, ἢ χωρίς, ἢ μεμιγμένον, ὡς ἐπὶ τὸ πολὺ τεσσαρεσκαιδεκαταῖοι θνήσκουσιν ἢν μή τι κακὸν ἢ ἀγαθὸν
ἐπιγένηται τῶν προγεγραμμένων· εἰ δὲ μή, κατὰ λόγον, μάλιστα δὲ οἷσιν ἑβδομαίοισιν ἄρχεται τὸ τοιοῦτον πτύαλον.

387. Ἔστι δὲ ἀγαθὸν μὲν καὶ τούτοισι καὶ πᾶσι
τοῖσι περὶ πλεύμονα, φέρειν ῥηϊδίως τὸ νούσημα, τῆς
ὀδύνης ἀπηλλάχθαι, τὸ πτύαλον εὐπετέως ἀνάγειν,
εὔπνοον εἶναι καὶ ἄδιψον, τὸ σῶμα ἅπαν ὁμαλῶς
θερμαίνεσθαι καὶ μαλθακὸν εἶναι, καὶ πρὸς τούτοισιν
ὕπνους, ἱδρῶτας, οὖρον, διαχώρησιν χρηστὴν γίνε-
σθαι· κακὰ δὲ τἀναντία τούτων. εἰ μὲν οὖν πάντα
προσγένοιτο τῷ πτύσματι τούτῳ τὰ χρήσιμα, σῴζοιτ'
ἄν· εἰ δὲ τὰ μέν, τὰ δὲ μή, <οὐ>[79] πλείους τῶν τεσσαρεσκαίδεκα βιώσας· τῶν δ' ἐναντίων σημείων ἐπιγενομένων, συντομώτερον.

388. Ὅσα δὲ τῶν ἀλγημάτων ἐν τοῖσι τόποισι
τούτοισι μὴ παύσηται μήτε πρὸς τὰς ἀναπτύσιας,
μήτε πρὸς τὰς φλεβοτομίας τε καὶ διαίτας, ἐμπυοῦται.

389. Ὅσοισι δὲ ἐκ περιπλευμονίης ἀποστάσιες
παρ' οὖς ἢ ἐς τὰ κάτω γίνονται, καὶ ἐκπυοῦσί τε, καὶ

frothy sputum, and by yellow, black, rusty, and sticky sputa, and those that take on colour quickly; but mucous and sooty sputa also take on colour quickly, and they point more to safety. Sputa which take on the colour of coction within five days are quite good.

385. Any sputum that does not resolve the pain is a bad sign, whereas one that does is a favourable sign.

386. Any patients that bring up pus, either together with bilious material or without it, or both mixed together, generally die on the fourteenth day—unless one of the bad or good signs described above appears; if it does, the significance is commensurate, especially in patients who began on the seventh day to have sputum of this kind.

387. It is good in these cases and in all conditions of the lung, for patients to bear the disease with ease, to be free of pain, to bring their sputum up easily, to breathe well, to be without thirst, for their entire body to be evenly warmed and supple, and besides this for their sleep, sweating, urine, and stools to be favourable: signs opposite to these are bad. Now if all the favourable signs are present in addition to the expectoration described above, the person will be saved; if some but not others are present, he will die not having survived more than fourteen days; if opposite signs appear, the patient will die sooner.

388. Patients with pains in these locations that are relieved neither by expectoration, nor by phlebotomy, nor by regimen, suppurate internally.

389. Patients who, out of pneumonia, develop apostases beside the ear or to the lower parts which expel

---

[79] Opsopoeus, after Cornarius' *non ultra*.

ἐκσυριγγοῦνται, περιγίνονται· γίνονται δέ, οἷσιν ἂν ὅ
τε πυρετὸς καὶ ὁ πόνος παρακολουθῇ, καὶ τὸ πτύαλον
μὴ χωρέῃ κατὰ λόγον, μηδὲ χολώδεες αἱ διαχω-
670 ρήσιες, εὔλυτοί | τε καὶ ἄκρητοι γίνωνται, μηδὲ οὖρον
παχύ τε σφόδρα καὶ πολλὴν ὑπόστασιν ἔχον, τά τε
ἄλλα σωτηρίως ἔχοιεν· γίνονται δέ, αἱ μὲν ἐς τὰ κάτω,
οἷσιν ἂν περὶ ὑποχόνδρια φλεγμονὴ γίνηται, αἱ δὲ ἐς
τὰ ἄνω, οἷσιν ἂν τὸ μὲν ὑποχόνδριον λαπαρόν τε καὶ
ἀνώδυνον ᾖ, δύσπνοοι δέ τινα χρόνον γενόμενοι παύσωνται χωρὶς προφάσιος.

390. Αἱ δὲ ἐς τὰ σκέλεα τῶν ἀποστασίων[80] ἐν τῇσιν
ἐπικινδύνοισι περιπλευμονίῃσι, λυσιτελέες μὲν πᾶ-
σαι, βέλτισται δὲ αἱ τοῦ πτυέλου πυώδεος ἀντὶ ξανθοῦ
γενομένου· μὴ χωρέοντος δὲ τοῦ πτυέλου κατὰ λόγον,
μηδὲ τοῦ οὔρου χρηστὴν ὑπόστασιν ἔχοντος, κίνδυνος χωλωθῆναι τὸν ἄνθρωπον, ἢ καὶ πολλὰ πρήγματα
παρασχεῖν· ἢν δὲ παλινδρομέωσιν αἱ ἀποστάσιες,
πυρετοῦ παρακολουθοῦντος, καὶ τοῦ πτυάλου μὴ χωρέοντος, κίνδυνος θανεῖν καὶ παραφρονῆσαι. ὅσοι δὲ
τῶν περιπλευμονικῶν μὴ ἀνεκαθάρθησαν ἐν τῇσι κυ-
ρίῃσιν ἡμέρῃσιν, ἀλλὰ παρακόψαντες διέφυγον τὰς
τεσσαρεσκαίδεκα, κίνδυνος ἐμπύους γενέσθαι.

391. Τῶν περιπλευμονιῶν αἱ ἐκ πλευριτικοῦ μετα-
στᾶσαι, τῶν ἐξ ἀρχῆς γενομένων ἀσφαλέστεραι.

392. Τῶν δὲ σωμάτων τὰ γεγυμνασμένα καὶ πυκνὰ
θᾶσσον ὑπὸ τῶν πλευριτικῶν καὶ περιπλευμονικῶν
ἀπόλλυνται τῶν ἀγυμνάστων.

393. Κορύζας καὶ πταρμοὺς τοῖσι περὶ πλεύμονα

pus and form fistulae, survive. These arise in patients in whom fever and pain follow, the sputum does not pass as it should, the stools are not bilious, easily passed, and unmixed, and the urine is not very thick with a copious sediment, but the other signs suggest safety. Some apostases develop in the lower parts—in patients with inflammation around their hypochondrium—and some in the upper parts —in patients whose hypochondrium is slack and painless, and in whom difficult breathing is present for a certain time and then ceases for no reason.

390. Apostases to the legs in dangerous pneumonias are all advantageous, but best when the sputum becomes putrid rather than yellow. If the sputum does not move the way it should, and the urine does not have favourable sediments, there is a danger the person will become lame, or that many things will happen. If the apostases recede, fever follows, and the sputum does not move, there is a danger that the person will die or lose his mind. In patients with pneumonia who are not cleaned upwards on the decisive days, but who escape the fourteenth day in a state of madness, there is a danger that they will suppurate internally.

391. Pneumonias which develop out of pleurisy are less dangerous than those which begin by themselves.

392. Trained and hard bodies die sooner from pleurisies and pneumonias than untrained ones.

393. For colds and sneezing to occur before or after

---

80 Opsopoeus, after Cornarius' *abscessus*: προφασίων A.

καὶ προγενέσθαι καὶ ἐπιγενέσθαι, πονηρόν· τοῖσι δὲ
λοιποῖσι πταρμὸς οὐκ ἀλυσιτελής.

394. Τοῖσι περιπλευμονικοῖσιν, οἷσι γλῶσσα πᾶσα λευκὴ καὶ τρηχεία γίνεται, ἀμφότερα φλεγμαίνει
τὰ μέρεα τοῦ πλεύμονος· οἷσι δὲ τὸ ἥμισυ, ἐν καθ᾽ ὃ
φαίνεται· καὶ οἷσι μὲν πρὸς τὴν μίαν κληῖδα ὁ πόνος
γίνεται, ἡ ἄνω πτέρυξ τοῦ πλεύμονος ἡ μία νοσέει·
672 οἷσι δὲ | πρὸς ἄμφω τὰς κληῖδας ὁ πόνος γίνεται, αἱ
ἄνω πτέρυγες τοῦ πλεύμονος ἄμφω νοσέουσιν· οἷσι δὲ
κατὰ μέσην τὴν πλευρήν, ἡ μέση· οἷσι δὲ πρὸς τὴν
διάτασιν, ἡ κάτω· οἷσι δὲ πᾶν τὸ ἓν μέρος πονέει,
πάντα τὰ κατὰ τοῦτο μέρος νοσέει. ἢν μὲν οὖν σφόδρα
φλεγμαίνωσιν αἱ ἀορταί, ὥστε προσκαθῆσθαι πρὸς
τὸ πλευρόν, παραλύονται τὸ κατὰ τοῦτο τὸ μέρος τοῦ
σώματος, καὶ πελιώματα περὶ τὴν πλευρὴν ἔξω γίνεται, τούτους δὲ ἐκάλεον οἱ ἀρχαῖοι βλητούς· ἢν δὲ μὴ
σφόδρα φλεγμαίνωσιν, ὥστε μὴ προσκαθῆσθαι, ἀλ-
γηδὼν μὲν γίνεται παρ᾽ ὅλον, οὐ μὴν παραλύονταί γε,
οὐδὲ πελιώματα ἴσχουσιν.

395. Οἷσι δ᾽ ἅπας ὁ πλεύμων φλεγμήνῃ μετὰ τῆς
καρδίης, ὥστε καὶ προσπεσεῖν πρὸς τὴν πλευρήν,
παραλύεται πᾶς ὁ νοσέων, καὶ κεῖται ψυχρὸς ὁ νοσέων
ἀναίσθητος· θνῄσκει δὲ δευτεραῖος ἢ τριταῖος· ἢν δὲ
καὶ χωρὶς τῆς καρδίης συμβῇ καὶ ἧσσον, πλείονα
χρόνον ζώσιν, ἔνιοι δὲ καὶ διασῴζονται.

396. Τοῖς ἐμπύοις γινομένοισι, μάλιστα δὲ ἐκ πλευριτικοῦ καὶ περιπλευμονικοῦ, θέρμαι παρακολουθοῦσι, τὴν μὲν ἡμέρην λεπταί, τὴν δὲ νύκτα συντονώτε-

conditions of the lung gives a bad indication; in other cases, sneezing is not without advantage.

394. In pneumonias, if the whole tongue becomes white and rough, then both sides of the lungs become inflamed; where this happens in half of the tongue, then the swelling appears on that one side. In cases where the pain moves towards one clavicle, the superior lobe of the lung on that side is ill; where the pain is felt in the direction of both clavicles, the superior lobes of the lungs on both sides are ill. Where the pain is located towards the middle of the side, the middle lobe is ill, where it is located towards the diaphragm, the lower lobe. Where one whole side is painful, the whole region on that side is ill. Now if the aortas are very inflamed, so that they hang down against the side, that side of the body becomes paralysed, and livid marks appear on the exterior of the side, which the early writers called "strokes"; if the aortas are not so inflamed, and thus do not hang down against the side, a pain is present through the whole chest, but such patients are not paralysed, nor do they have livid marks.

395. In cases where the whole lung is inflamed, together with the heart, so that they fall against the side, the patient is completely paralysed, and lies in his bed cold and insensible; he dies on the second or third day. If the heart is not involved and less happens, such patients live longer and some even survive.

396. Patients who suppurate internally, in most cases from pleurisy and pneumonia, are beset by fevers—mild during the day, but more intense at night—fail to cough

ραι, καὶ πτύουσιν οὐδὲν ἄξιον λόγου, ἱδροῦσί τε περὶ
τράχηλον καὶ κληῖδα, καὶ τοὺς μὲν ὀφθαλμοὺς κοιλαί-
νονται, τὰς δὲ γνάθους ἐρεύθονται, χειρῶν δὲ θερμαίνονται μὲν δακτύλους ἄκρους καὶ τραχύνονται, γρυ-
ποῦνται δὲ ὄνυχας, καὶ καταψύχονται, περί τε τοὺς
πόδας ἐπάρματα ἴσχουσι, καὶ κατὰ τὸ σῶμα φλυκται-
νίδια, σίτων τε ἀφίστανται. τὰ μὲν οὖν χρονίζοντα
τῶν ἐμπυημάτων ἴσχει τὰ σημεῖα ταῦτα. τὰ δὲ[81]
674 συντόμως ῥηγνύμενα σημειοῦσθαι τούτων | τε τοῖσιν
ἐπιγενομένοισι, καὶ τοῖσιν ἐν ἀρχῇ πόνοισι, ἅμα δὲ
καὶ ἤν τι δυσπνοώτερος γίνηται· ῥήγνυνται δὲ τὰ
πλεῖστα τῶν ἐμπυημάτων, τὰ μὲν εἰκοσταῖα, τὰ δὲ
τεσσαρακοσταῖα, τὰ δὲ πρὸς τὰς ἑξήκοντα. οἷσι μὲν
οὖν ὁ πόνος ἐν ἀρχῇ ἔγκειται σύντονος καὶ δύσπνοια
καὶ βὴξ μετὰ πτυαλισμοῦ, πρὸς τὰς εἴκοσιν ἢ συντομώτερον προσδέχου τὴν ῥῆξιν· οἷσι δὲ ἐλαφρότερα
ταῦτά ἐστι, κατὰ λόγον. λογίζεσθαι δὲ τὸν χρόνον,
ἀφ' οὗ πρῶτον ἤλγησεν, εἰ ἐβαρύνθη, εἰ ἐπύρεξεν, ἢ εἴ
ποτε ῥῖγος ἔλαβεν· προγίνεσθαι δὲ ἀνάγκη καὶ πόνον
καὶ δύσπνοιαν καὶ πτυαλισμὸν πρὸ τῆς ῥήξιος. οἷσι
μὲν οὖν ὅ τε πυρετὸς εὐθέως ἀπογίνεται μετὰ τὴν
ῥῆξιν, καὶ σιτίων ἐπιθυμέουσι, καὶ τὸ πῦον ἀνάγεται
ῥηϊδίως λευκὸν ἐὸν καὶ ἄνοσμον καὶ λεῖον καὶ ὁμό-
χροον καὶ ἀφλέγμαντον, κοιλίη τε μικρὰ συνεστηκότα
ὑποχωρέει, ὡς ἐπὶ τὸ πολὺ σῴζονται συντόμως. οἷσι
δὲ πυρετοί τε παρακολουθοῦσι καὶ δίψα καὶ ἀποσιτίη,
καὶ τὸ πῦον πελιὸν ἢ χλωρὸν ἢ φλεγματῶδες ἢ ἀφρῶδες, κοιλίη τε ἐφυγραίνεται, τελευτῶσιν. οἷσι δὲ τὰ

up anything worth mentioning, sweat about the neck and clavicle, and become hollow around their eyes and red in their cheeks; their hands become warm, the ends of their fingers become thick, and their nails are hooked. These patients have chills, swellings in their feet, and blisters over their body; they reject food. Internal suppurations that are chronic have these signs. Those, however, which rupture quickly are recognized both by the appearance of these signs, and at the beginning by pains, and by more difficulty in breathing at that time. Most internal suppurations rupture on the twentieth, fortieth, or sixtieth day; but in cases where the pain attacks intensely at the beginning, and difficult breathing and cough with expectoration follow, expect the rupture toward the twentieth day or sooner; when these symptoms are lighter, proportionately later. Reckon the time from when the patient first had pain, and whether he felt weighed down, or had fever, or a chill was ever present: pain, difficult breathing, and expectoration must precede the rupture. Now patients in whom the fever remits immediately after the rupture, there is a desire for food, the pus is brought up easily and is white, odourless, smooth, even in colour, and bereft of phlegm, and the cavity passes small compacted stools, generally recover quickly. But those in whom fevers persist along with thirst and anorexia, the pus is livid, green, phlegmy, or frothy, and diarrhoea occurs, die. Of patients in whom some of the

---

81 σημεῖα—δὲ Opsopoeus, after Cornarius' *signa haec habent. Quae vero . . .*: σημεῖα· ταῦτα δὲ A.

μὲν ἐπιγίνεται ἐκ τῶν προειρημένων, τὰ δὲ μή, τούτων
οἱ μὲν ἀποθνήσκουσιν, οἱ δὲ πολλῷ χρόνῳ σῴζονται.

397. Οἱ δὲ μέλλοντες ἔμπυοι γίνεσθαι, πτύουσι, τὸ
μὲν πρῶτον ἁλμυρόν, εἶτα γλυκύτερον.

398. Οἷσι δ᾿ ἐν πλεύμονι φύματα γίνεται, τὸ πῦον
ἀνάγουσιν ἐς τεσσαράκοντα ἡμέρας μετὰ τὴν ῥῆξιν·
ταύτας δὲ ὑπερβάλλοντες, ὡς τὰ πολλὰ φθισικοὶ γίνονται.

399. Ἐπὶ πλευροῦ ἀλγήματι στάξις ἀπὸ ῥινῶν
αἵματος, κακόν.

400. Οἷσιν ἐμπύοισιν ἐπιεικέστερον ἔχουσι δυσω-
δίαι τῶν πτυσμάτων παρακολουθοῦσι, τούτους ὑπο-
τροπὴ κτείνει.

401. Οἱ ἐν πλευριτικοῖσιν ἀναπτύοντες πυώδεα,
ὑπόχολα, στρογγύλα, ἢ πυώδεα ὕφαιμα, προεληλυ-
θότος χρόνου, ὀλέθριοι· ὀλέθριοι δὲ καὶ οἱ τὰ μέλανα
λιγνυώδεα πτύοντες, ἢ οἷσιν | ⟨οἷον⟩[82] ἀπὸ οἴνου
μέλανος γίνεται πτύσματα.

402. Ὅσοι αἷμα ἀφρῶδες πτύουσι, πονέοντες ὑποχόνδριον δεξιόν, ἀπὸ τοῦ ἥπατος πτύουσι, καὶ οἱ
πολλοὶ ἀπόλλυνται.

403. Οἷσι καιομένοισι πῦον βορβορῶδες ἔρχεται
καὶ δυσῶδες, ἀπόλλυνται ὡς τὰ πολλά.

404. Οἷσιν ἀπὸ τοῦ πύου ἡ μήλη χρωματίζεται
καθάπερ ἀπὸ πυρός, ἀπόλλυνται ὡς τὰ πολλά.

405. Μετὰ πλευροῦ ἀλγήματος, μὴ πλευριτικοῦ δέ,
καὶ ταραχωδέων λεπτῶν ἐπιεικῶν, οὗτοι φρενιτικοὶ
ἀποβαίνουσιν.

signs named occur, but others do not, some die, and others recover after a long time.

397. Patients who are about to suppurate internally expectorate first salty and then sweetish sputum.

398. Patients who develop growths in their lung bring up pus until forty days after the rupture; if they get beyond that term, they generally become consumptive.

399. Bleeding from the nostrils, in conjunction with pain of the side, is bad.

400. If ill-smelling sputa persist in patients with internal suppuration, even though they are in a quite favourable state, a relapse will kill them.

401. Patients with pleurisy, who cough up sputa that are purulent, somewhat bilious, globular, or purulent and somewhat bloody, as time passes, are doomed. Doomed also are those who expectorate dark sooty sputa, or whose sputa look like they came from dark wine.

402. Patients who expectorate frothy blood, while having pains in their right hypochondrium, are expectorating from the liver, and many die.

403. Patients who have cautery, and who pass foul-smelling pus and have rumbling in their abdomen, generally die.

404. In cases where a probe becomes coloured from the pus, as if from fire, patients generally die.

405. If persons with pain in the side, who do not have pleurisy, evacuate favourable thin stools, they turn out to have phrenitis.

---

[82] Opsopoeus, after Cornarius' *velut*.

406. Ἐν τοῖσι κατὰ πλεύμονα αἱ λίην ἐξέρυθροι ἀποστάξιες, πονηρόν.

407. Μετὰ βράγχου[83] πτύελα γλίσχρα, ἁλμυρώδεα, κακόν· ἢν δέ τι καὶ ἐπαίρηται κατὰ στῆθος, ἐπὶ τούτοισι κακόν· τὰ ἐς τράχηλον ἀλγήματα, τούτων ἰσχνανθέντων, ὀλέθριον.

408. Βράγχος μετὰ βηχὸς καὶ κοιλίης ὑγρῆς, πῦον ἀνάγει.

409. Οἷσιν ἐν περιπλευμονίῃ οὖρα παχέα ἐν ἀρχῇ, εἶτα πρὸ τῆς τετράδος λεπτύνεται, θανάσιμον.

410. Οἱ ἐν ξηροῖσι περιπλευμονικοῖσιν ὀλίγα πέπονα ἀνάγοντες, φοβεροί· τὰ ἐν τοῖσι στήθεσιν ἐρυθήματα ὑποπλάτεα, γίνεται τοῖς τοιούτοισιν ὀλέθρια.

411. Πλευροῦ ‹ἄλγημα›[84] ἐν πτύσει χολώδει ἀλόγως ἀφανισθέν, ἐξίστανται.

412. Οἱ δι' ἐμπύησιν πυρετοὶ διαλείποντες, ἐφιδροῦντες οἱ πολλοί εἰσιν.

413. Τοῖσιν ἐμπύοισι κώφωσις γενομένη αἱματώδεα διαχώρησιν σημαίνει· τούτοισι πρὸς τὴν τελευτὴν μέλανα διαχωρέει.

414. Πλευροῦ ἄλγημα μετὰ πυρετοῦ χρονίου σημαίνει πῦον ἀνάξειν.

415. Οἱ φρικώδεες πυκνὰ ἐς ἐμπύησιν ἔρχονται· ἀτὰρ καὶ πυρετὸς τὸν τοιοῦτον ἄγει ἐς ἐμπύησιν. |

678   416. Οἷσιν ἐκ πλευροῦ ἀλγήματος ἀσιτίαι παρακολουθοῦσιν, ὑπό τι καρδιαλγικοί, ἱδρώδεες, ἔχοντος δὲ

---

[83] Froben: βρόγχου A.

206

406. In conditions of the lung, excessively red nose bleeds are bad.

407. Viscous salty sputa, in association with hoarseness, are bad; also bad is if there is some swelling of the chest besides; pains invading the neck when these swellings go down are a fatal sign.

408. Hoarseness in association with coughing and diarrhoea presages the expectoration of pus.

409. Patients with pneumonia whose urines are thick at the beginning, but then become thin before the fourth day, are doomed.

410. Patients with dry pneumonia, who bring up slightly concocted sputa, are to be feared for; flattish red excrescences on the chest are a fatal sign in such cases.

411. If, in association with bilious expectoration, a pain of the side disappears for no reason, such patients will become deranged.

412. Many fevers that remit because of an internal suppuration are followed by sweating over the body.

413. When deafness arises in cases of internal suppuration, it foretells the excretion of bloody stools; towards the end these patients pass dark stools.

414. Pain of the side, in association with a chronic fever, indicates that pus will be brought up.

415. Frequent shivering indicates there will be an internal suppuration. Fever, too, may bring such a case to internal suppuration.

416. Persons in whom anorexia follows from a pain in the side, who have a degree of heartburn and sweating,

---

84 Opsopoeus, after Cornarius' *Lateris dolor*.

προσώπου ἄνθη, καὶ κοιλίης ὑγροτέρης, ἐκπυήματα κατὰ πλεύμονα ἴσχουσιν.

417. Τὰ ὀρθοπνοϊκὰ ποιέει ὑδρωπιώδεα σκληρά.

418. Τὰ σπάσματα μὲν πάντα ὀχληρὰ γίνεται, καὶ πόνους τε ἐν ἀρχῇ συντόνους παρέχει, καὶ ἐξ ὑστέρου ἐνίους ὑπομιμνήσκει· δυσκολώτατα δὲ τὰ περὶ θώρηκα, μάλιστα δὲ κινδυνεύουσιν, οἷσιν ἔμετος αἵματος, πυρετὸς πολύς, καὶ πόνος περὶ μαζὸν καὶ θώρηκα καὶ μετάφρενον· οἷσι γὰρ γίνεται πάντα ταῦτα, συντόμως θνῄσκουσιν· οἷσι δὲ μὴ πάντα, μηδὲ σφόδρα, βραδύτερον· φλεγμαίνει δὲ τὸ μακρότατον ἡμέρας τεσσαρεσκαίδεκα.

419. Τοῖσιν αἷμα πτύουσιν ἀπυρέτοις εἶναι συμφέρει, καὶ βήσσειν καὶ πονέειν ἐλαφρῶς, καὶ τὸ πτύαλον λεπτύνεσθαι πρὸς τὰς δὶς ἑπτά· πυρέσσειν δὲ καὶ βήσσειν καὶ πονέειν συντόνως, καὶ αἷμα πρόσφατον αἰεὶ πτύειν, ἀσύμφορον.

420. Ὅσοισι τὸ πλευρὸν μετέωρον καὶ θερμότερον, ὅταν ἐγκεκλιμένοισιν ἐπὶ θάτερον βάρος ἐξηρτῆσθαι δοκέῃ, τούτοισι τὸ πῦον ἐκ τοῦ ἑνὸς μέρεός ἐστιν.

421. Τοῖσιν ἐμπύοισι τὸν πλεύμονα, κατὰ κοιλίην πῦον ὑποχωρέειν, θανάσιμον.

422. Ὅσοι, τρωθέντες ἐς τὸν θώρηκα, τὸ μὲν ἐκτὸς τοῦ τρώματος ὑγιάσθησαν, τὸ δ' ἐντὸς μή, κινδυνεύουσιν ἔμπυοι γενέσθαι· ὅσοις δ' ἂν ἀσθενὴς ἔνδο-
680 θεν ἡ οὐλὴ γένηται, ῥηϊδίως | ἀναρρήγνυται.

whose face has an eruption, and whose excretions are fluid, will have productive suppurations from the lung.

417. Orthopnoea produces dropsies with indurations.

418. All ruptures in the lung are troublesome, and provoke intense pains at the beginning, of which some persist later. Most difficult are the pains about the thorax, and such patients are in most danger who have vomiting of blood, great fever, and pain in the breast, thorax, and back, for if they have all of these, they die rapidly. Patients who do not have all of these signs, and not intensely, die later. The inflammation lasts, at the longest, for fourteen days.

419. For patients expectorating blood, it is beneficial to be without a fever, to have gentle coughing and mild pains, and for their sputum to become thin towards two weeks. If the fevers, coughing, and pains are intense, and the patients continue to expectorate fresh blood, it is unhelpful.

420. Patients in whom one side is raised and very warm, and who, when they lean on the opposite side, seem to feel a weight hanging down, have pus on the one (i.e. raised) side.

421. In patients with a suppuration in the lung, for pus to be passed down through the cavity is a fatal sign.

422. Persons wounded in the thorax who heal at the external part of the wound, but not at the internal part, are in danger of suppurating internally. Those in whom the scar is weak on the inside are liable to have it torn open.

423. Ἀπόλλυνται δὲ ἐκ μὲν τῶν περιπλευμονικῶν ἐμπυημάτων οἱ γεραίτεροι μᾶλλον· ἐκ δὲ τῶν λοιπῶν οἱ νεώτεροι.

424. Τῶν ἐμπύων οἷσι σειομένοισιν ἀπὸ τῶν ὤμων πολὺς γίνεται ψόφος, ἔλασσον ἔχουσι πῦον, ἢ οἷσιν ὀλίγος δυσπνοωτέροισιν ἐοῦσι καὶ εὐχροωτέροισιν· οἷσι δὲ ψόφος μὲν μηδὲ εἷς ἐγγίνεται, δύσπνοια δὲ ἰσχυρή, καὶ ὄνυχες πέλιοι, πλήρεις οὗτοί εἰσι πύου καὶ ὀλέθριοι.

425. Ὅσοι ἀφρῶδες αἷμα ἐμέουσι, πόνου μὴ ἐόντος κάτω τοῦ διαφράγματος, ἀπὸ τοῦ πλεύμονος ἐμέουσιν· καὶ οἷσι μὲν ἡ μεγάλη φλὲψ ἐν αὐτῷ ῥήγνυται, πολύ τε ἐμέουσι καί εἰσιν ἐπικίνδυνοι· οἷσι δὲ ἡ ἐλάσσων, ἔλασσόν τε ἀνάγουσι, καί εἰσιν ἀσφαλέστεροι.

426. Τῶν φθισικῶν οἷσιν ἐπὶ τοῦ πυρὸς ὄζει τὸ πτύαλον κνίσης βαρύ, καὶ αἱ τρίχες ἐκ τῆς κεφαλῆς ῥέουσιν, ἀπόλλυνται.

427. Τῶν φθισικῶν οἷσιν ἐπὶ θάλασσαν πτύουσιν ἐς τὸν πυθμένα βαδίζει τὸ πῦον, ὀλέθριον συντόμως· ἔστω δὲ ἐν χαλκῷ ἡ θάλασσα.

428. Ὅσοισι τῶν φθισικῶν αἱ τρίχες ἐκ τῆς κεφαλῆς ῥέουσιν, ὑπὸ διαρροίης ἀπόλλυνται· καὶ ὅσοισι φθισικοῖσιν ἐπιγίνονται διάρροιαι, θνήσκουσιν.

429. Αἱ ἐν φθινώδεσιν ἐπισχέσιες πτυάλων ἐξιστᾶσι ληρωδῶς· αἱμορροΐδα τούτοισιν ἐλπὶς ἐπιφανῆναι.

423. From the internal suppurations of pneumonia, older persons are more liable to die, from other internal suppurations, younger persons.

424. Patients with internal suppurations who, on being shaken by the shoulders, make a loud sound, have less pus than those who, with more difficulty in breathing and a better colour, make less sound. Those who do not make any sound, who have great difficulty in breathing, and whose nails are livid, are filled with pus and given over to death.

425. Patients who vomit up frothy blood, but have no pain below the diaphragm, are vomiting from the lung; those whose great vessel in the lung is torn, vomit much blood and are in danger, whereas if it is the lesser vessel, they bring up less blood and are safer.

426. Patients with consumption, whose sputum, when cast on a fire smells with a heavy meat odour, and whose hair falls out of their heads, die.

427. Patients with consumption, whose pus, when they expectorate into brine (let the brine be in a bronze vessel), goes to the bottom, are soon doomed.

428. Persons with consumption whose hair falls out die from diarrhoea; in fact any patients with consumption, whom diarrhoea befalls, die.

429. The holding up of expectoration in persons with consumption makes them become deranged and talk nonsense; expect haemorrhoids to appear in these.

430. Φθίσιες ἐπικινδυνόταται, αἵ τε ἀπὸ ῥήξιος φλεβῶν τῶν παχειῶν, καὶ ἀπὸ κατάρρου τοῦ ἀπὸ κεφαλῆς.

431. Τῶν δὲ ἡλικιῶν ἐπικινδυνόταται πρὸς φθίσιν ἀπὸ ιη΄ ἐτέων μέχρι ε΄ καὶ λ΄.

432. Τὰ κνησμώδεα σώματα μετὰ κοιλίης στάσιν ἐν φθισικοῖσι, κακόν.

433. Ἐπὶ τῇσι φθισινώδεσιν ἕξεσι μετὰ πυρετοῦ ἐς οὖλα καὶ ὀδόντας ῥεύματα ἐπιφαινόμενα, κακόν.

434. Ἐπὶ πᾶσιν ὑποχόνδρια μετέωρα, κακόν· κάκιστον δὲ ἐπὶ τοῖσι φθισικοῖσι· τῶν μακρῶν, ἐπὶ τοῖσι τετηκόσιν ὀλεθρίοισιν, ἔνιοι πρὸ τῶν τελευτῶν ἐπιρριγοῦσιν.

435. Τὰ ἀμυχώδεα ἐξανθήματα φθίσιν ἕξιος σημαίνει.

436. Οἱ δύσπνοοι ξηρῶς, ἢ πολλὰ ἄπεπτα ἀνάγοντες ἐν φθίσει, ὀλέθριοι.

437. Οἷσιν ἡπατικοῖσι πολὺ πτύαλον αἱματῶδες, εἴ τε ἐνυπόσαπρον, εἴ τε χολῶδες ἄκρητον, ὀλέθριον εὐθέως.

438. Ἐφ' ἡπατικῷ τῆξις ἅμα βράγχῳ, κακόν, ἄλλως τε κἢν ὑποβήσσῃ.

439. Οἱ καθ' ἧπαρ ὀδυνώδεες, καρδιαλγικοί, καρώδεες, ῥιγώδεες, κοιλίαι ταραχώδεες, λεπτοί, ἀπόσιτοι, ἐφιδροῦντες πολλῷ, πυώδεα κατὰ κοιλίην προΐενται.

440. Τοῖσιν ἧπαρ ἐξαπίνης περιωδυνοῦσι πυρετὸς ἐπιγενόμενος λύει.

430. The most dangerous of consumptions are those arising from a rupture of the wide vessels, or from a downward flux from the head.

431. The most dangerous ages with regard to consumption are from eighteen to thirty-five years.

432. In consumptions, for the body to suffer from itching after constipation has been present is bad.

433. In continuing consumptions with fever, for fluxes to the gums and teeth to occur, is bad.

434. In all diseases, raised hypochondria are a bad sign, and in consumptions they are worst; in lengthy cases with terminal melting away (sc. of flesh), some patients have chills before their deaths.

435. Superficial skin eruptions indicate that the condition of consumption is present.

436. In consumption, patients with dry, difficult breathing, or who bring up much unconcocted sputum, are doomed.

437. Patients with liver disease who cough up much bloody sputum, which is either partly putrid or completely bilious, will soon die.

438. In a liver condition, the melting away of the flesh in association with hoarseness is a bad sign, especially if the patient has a mild cough.

439. Patients with a liver condition, who have pains, heartburn, stupor, chills, disordered cavities, thinness, anorexia, and frequent sweats over their body, evacuate purulent stools through their cavity.

440. Patients who suddenly suffer severe liver pains are relieved by a fever coming on.

441. Ὅσοι δὲ ἀφρῶδες αἷμα πτύουσι, πονέοντες ὑποχόνδριον δεξιόν, ἀπὸ τοῦ ἥπατος πτύουσι, καὶ θνήσκουσιν.

442. Οἷσιν ἧπαρ καυθεῖσιν οἷον ἀμόργη ἔρχεται, θανάσιμον.|

443. [Περὶ ὑδρώπων.][85] οἱ δὲ ὕδρωπες οἱ ἐκ τῶν ὀξέων νοσημάτων, ἐπίπονοι γίνονται καὶ ὀλέθριοι· ἄρχονται δὲ οἱ πλεῖστοι μὲν ἀπὸ τῶν κενεώνων, οἱ δὲ καὶ ἀπὸ τοῦ ἥπατος. τοῖσι μὲν οὖν ἀπὸ τῶν κενεώνων ἀρχομένοισιν οἱ πόδες οἰδέουσι, καὶ διάρροιαι πολυχρόνιοι παρακολουθοῦσιν, οὐ λαπάσσουσαι κοιλίην, οὐδὲ τὰς ὀδύνας λύουσαι τὰς ἐξ ὀσφύος καὶ κενεώνων. ὅσοι δὲ ἀπὸ τοῦ ἥπατος, βῆξαί τε θυμὸς ἐγγίνεται, καὶ οἱ πόδες οἰδέουσι, καὶ ἡ κοιλίη σκληρὰ διαδίδωσι καὶ πρὸς ἀνάγκην, οἰδήματά τε περὶ αὐτὴν γίνεται, τὰ μὲν ἐπὶ δεξιά, τὰ δ' ἐπ' ἀριστερά, καὶ πάλιν καταπαύεται.

444. Ἐπὶ τοῖσι ξηροῖσιν ὑδρωπιώδεσι τὰ στραγγουρικά, μοχθηρόν· φλαῦρα δὲ καὶ τὰ μικρὰς ὑποστάσιας ἔχοντα.

445. Τοῖσιν ὑδρωπιώδεσιν ἐπιληπτικὰ ἐπιγενόμενα, ὀλέθριον, ἀλλήλων τε σημεῖον μοχθηρόν, καὶ κοιλίας ἐξυγραίνουσιν.

446. Ἐν τοῖσι χολώδεσι κοιλίη ταραχώδης,[86] διαδιδοῦσα σμικρὰ γονώδεα, μυξώδεα, καὶ πόνον περὶ ἦτρον ἐμποιέοντα, καὶ οὖρα οὐκ εὐλύτως ἰόντα, ἐς ὕδρωπα ἀποτελευτᾷ ἐκ τῶν τοιούτων.

441. Patients who expectorate frothy blood, and suffer pain in their right hypochondrium, are expectorating from the liver, and they die.

442. Patients, who on being cauterized in the liver have a fluid resembling olive water come out, are doomed.

443. Dropsies arising from acute diseases are painful and deadly. Most begin from the flanks, but others from the liver. In the ones beginning from the flanks, the feet swell up and chronic diarrhoeas follow, which do not empty the cavity or resolve the pains arising from the loins and flanks. In dropsies arising from the liver, a desire to cough comes on, the feet swell up, the cavity secretes hard material and only when it is stimulated, and swellings come up in the abdomen—sometimes on the right side, at other times on the left side—which then go down again.

444. Strangury occurring in dry hydropsies is troublesome; to have meagre sediments (sc. in the urines) is also an indifferent sign.

445. When epileptic symptoms come on in dropsies, they are a fatal sign, and each of the conditions represents a troublesome sign for the other one: diarrhoeas follow.

446. In bilious states, if the cavity is disturbed and produces a few mucous stools that resemble seed, if pain invades the region of the lower abdomen, and if urines pass but not freely, after these events the condition ends by turning into dropsy.

---

85 Del. Aldina.
86 Aldina: -ώδεσι A.

447. Ὑδεριῶντι πυρετώδει οὖρον μικρὸν καὶ τεταραγμένον, ὀλέθριον.

448. Ἐπὶ δὲ ὑδέρῳ ἀρχομένῳ διάρροια γενομένη ὑδατώδης, χωρὶς ἀπεψίης, λύει τὸ νόσημα.

449. Τοῖσι ξηροῖσιν ὑδρωπιώδεσι, προσημαίνουσι στρόφοι περὶ τὸ | λεπτὸν ἐμπίπτοντες, κακόν.

450. Τὰ ἐξ ὑδρωπικῶν ἐπιληπτικά, ὀλέθρια.

451. Ὕδερος πρὸς θεραπείην ἐνδιδούς, παλινδρομέων, ἀνέλπιστον.

452. Τοῖσιν ὑδρωπιώδεσι, κατὰ φλέβας ἐς κοιλίην ῥαγέντος τοῦ ὕδατος, λύσις.

453. Δυσεντερίη ἀκαίρως ἐπιστᾶσα ἀπόστασιν ἐν πλευροῖσιν, ἢ σπλάγχνοισιν, ἢ ἐν ἄρθροισι ποιέει· ἆρα ἡ μὲν χολώδης ἐν ἄρθροισιν, ἡ δὲ αἱματώδης ἐν πλευροῖσιν, ἢ σπλάγχνοισιν;

454. Δυσεντερικοῖσιν ἔμετος χολώδης ἐν ἀρχῇ, κακόν.

455. Οἷσιν ἐκ δυσεντερίης ὀξείης ἐς πυώδεα ἥκει τὸ ὑγρόν, τὸ ἐφιστάμενον ἔλευκον ἔσται καὶ πολύ.

456. Τὰ δυσεντεριώδεα, ὑπέρυθρα, ἰλυώδεα, λάβρα διαχωρήματα, ἐπὶ φλογώδεσιν ἐξερύθροισι χρώμασι λυόμενα, ἐλπὶς ἐκμανῆναι.

457. Δυσεντερίη σπληνώδεσι μὴ μακρή,[87] χρήσιμον, μακρὴ δέ, πονηρόν· ληγούσης γάρ, ‹εἰ›[88] ὕδρωπες ἢ λειεντερίαι γίνονται, θανάσιμον.

458. Ἐν λειεντερικοῖσι μετὰ θηρίων, ὀδύναι στρό-

447. In a person with dropsy and fever, a little disturbed urine is a fatal sign.

448. In a dropsy just starting, the arrival of watery diarrhoea without unconcocted material resolves the disease.

449. In dry dropsies, colic invading the region of the small intestine presages evil.

450. Epilepsy following dropsy is a fatal sign.

451. Dropsy that abates with treatment, but relapses, is hopeless.

452. For dropsies, water breaking through the vessels into the cavity means resolution.

453. A dysentery that remits in an untimely way provokes an apostasis in the sides, internal parts, or joints: is it bilious in the joints, but bloody in the sides or internal parts?

454. In dysenteries, vomiting bile at the beginning is a bad sign.

455. If the fluid (sc. excreted) in patients with acute dysentery becomes purulent, it will have a very white, copious scum on it.

456. If dysenteric, reddish, slimy, violent excretions are relieved by becoming a flaming bright red colour, you may expect the patient to become delirious.

457. In persons with a condition of the spleen, a dysentery that is not long is favourable, whereas one that is long bodes ill; for if, when the dysentery ends, dropsies or lienteries come on, the condition is fatal.

458. In lienteries with intestinal worms, pains relieved

---

87 μ. μ. Foes, after Cornarius' *Dysenteria non longa*: μα-κρῆσι A.   88 Aldina.

φῳ λυόμεναι τὰ περὶ ἄρθρα μετεωρίζουσιν· ἐκ τοιούτων λέπια ἐξέρυθρα, φλυκταινούμενα· ἐφιδρώσαντες οὗτοι διαφοινίσσονται οἷα μάστιξιν.

459. Οἱ ἐν λειεντεριώδεσι μακροῖσιν ἅμα θηρίοισι στροφώδεες, ὀδυνώδεες, λυομένων, ἐποιδέουσι· τὸ ἐπιρριγοῦν τούτοισι κακόν.

460. Λειεντερικὰ μετὰ δυσπνοίης καὶ πλευροῦ τῆς κνήσιος,[89] ἐς φθίσιν ἀποτελευτᾷ.

461. Εἰλεώδεσιν ἔμετος καὶ κώφωσις, κακόν.

462. Κύστιες δὲ σκληραί τε καὶ ἐπώδυνοι, πάντως |
688 μὲν κακόν, κάκιστον δὲ πυρετῷ συνεχεῖ· καὶ γὰρ οἱ ἀπ' αὐτῶν πόνοι, ἱκανοὶ ἀνελεῖν· καὶ κοιλίαι τούτοισιν οὐ πάνυ διαχωρέουσιν· λύει δὲ τούτους οὖρον πυῶδες ἐλθόν, λευκὴν καὶ λείην ἔχον ὑπόστασιν· μὴ λυομένων δὲ τούτων, μηδὲ τῆς κύστιος λαπασσομένης, ἐν τῇσι πρώτῃσι περιόδοισιν ἐλπὶς ἀπολέσθαι τὸν νοσέοντα· μάλιστα δὲ γίνεται τοῦτο τοῖσιν ἀπὸ ἑπτὰ ἐτέων μέχρι πεντεκαίδεκα.

463. Οἱ λιθιῶντες, σχηματισθέντες ὥστε τὸν λίθον μὴ προσπίπτειν πρὸς τὸν οὐρητῆρα, ῥηϊδίως οὐρέουσιν· οἷσι δὲ φῦμα περὶ τὴν κύστιν ἐστὶ τὸ παρέχον τὴν δυσουρίην, παντοίως σχηματισθέντες ὀχλέονται· λύσις δὲ τούτου γίνεται, πύου ῥαγέντος.

464. Οἷσι λανθάνει τὸ οὖρον προσπῖπτον, καὶ τὸ αἰδοῖον ἕλκονται, ἀνέλπιστοι.

465. Ἐπὶ στραγγουρίῃ εἰλεὸς ἐπιγενόμενος ἑβδο-

---

[89] Littré in *app. crit.*: τι κεινήσει Α.

by colic cause the joints to swell; from these conditions a red efflorescence is formed with blisters; if these patients have an attack of sweating over their body, they become quite red, as if they had been whipped.

459. In chronic lienteries with intestinal worms, patients who have colic and pains, if relieved, swell up; for chills to follow in these cases is a bad sign.

460. Cases of lientery with difficult breathing and a degree of irritation in the side terminate in consumption.

461. In cases of ileus, vomiting and deafness are bad signs.

462. For bladders to be hard and painful is always a bad sign; it is worst in conjunction with a continuous fever, for the pains arising in these cases are sufficient to carry the patient off. The cavities in such cases tend not to evacuate. These are relieved by the arrival of purulent urine with a fine, white sediment, but if they are not relieved and the bladder is not emptied, you may expect the patient to die in the first periods of the disease. This condition occurs most frequently in persons between seven and fifteen years.

463. Patients with stones situated such that they do not fall against the urethra, pass water easily, but those with a growth around the bladder which provokes dysuria will have difficulties no matter how the stones are situated; these are relieved when pus breaks out.

464. Patients who pass urine unawares, and whose genital parts are retracted, are in a hopeless state.

465. If ileus comes on in strangury, it kills such patients

μαίους ἀπόλλυσιν, ἢν μή, πυρετοῦ ἐπιγενομένου, ἀθρόον οὖρον ἔλθῃ.

466. Νάρκαι καὶ ἀναισθησίαι γινόμεναι παρὰ τὸ ἔθος, ἀποπληκτικῶν συμβησομένων σημεῖον.

467. Ὅσοι ἐκ τρώματος ἀκρατέες γίνονται τοῦ σώματος, πυρετοῦ μὲν ἐπιγενομένου χωρὶς ῥίγεος, ὑγιάζονται· μὴ γενομένου δέ, ἀποπληκτικοὶ γίνονται τὰ δεξιὰ ἢ τὰ ἀριστερά.

468. Ἀποπληκτικοῖσιν αἱμορροΐδες ἐπιγενόμεναι, χρήσιμον· ψύξιες δὲ καὶ ναρκώσιες, πονηρόν.

469. Ἐν τοῖσιν ἀποπληκτικοῖσιν ἐπὶ τῇ δυσφορίῃ τοῦ πνεύματος ἱδρὼς ἐπιγενόμενος, θανάσιμον· ἐν αὐτοῖσι δὲ πάλιν τούτοισιν ἢν πυρετὸς ἐπιγένηται, λύσις.

470. Τὰ ἐξαίφνης ἀποπληκτικὰ λελυμένως ἐπιπυρετήναντα, χρόνῳ ὀλέθρια.

471. Οἷσιν ἔκ τινος ἀρρωστίης ἐς ὕδερον περιίσταται, τούτοισι κοιλίαι ξηραὶ σπυραθώ|δεες ἔρχονται μετὰ περιτήξιος μυξώδεος καὶ οὔρου οὐ καλοῦ· διατάσιές τε περὶ ὑποχόνδρια, καὶ πόνοι καὶ ἐπάρματα περὶ κοιλίην, καὶ πόνοι περὶ κενεῶνας, καὶ περὶ τοὺς ῥαχιαίους μύας προσπίπτουσι, πυρετοί τε καὶ δίψαι καὶ βῆχες ξηραὶ παρακολουθοῦσι, καὶ δύσπνοια περὶ τὰς κινήσιας, καὶ σκελέων βαρύτης, σιτίων τε ἀφιστᾶσι, καὶ προσενεγκάμενοι μικρὰ πληροῦνται.

472. Τοὺς λευκοφλεγματοῦντας διάρροια παύει· αἱ μετὰ σιγῆς ἀθυμίαι καὶ ἀπανθρωπίαι, ἐπιεικῶς αὐτῶν κατεργαστικαί.

on the seventh day, unless fever supervenes and copious urine is passed.

466. If numbness and anaesthesias occur against habit, they indicate that strokes are imminent.

467. Persons who lose command over their body, as the result of an injury, recover if a fever without chills comes on; if no fever comes on, they are paralysed in the right or left parts of the body.

468. In cases of stroke, for haemorrhoids to supervene is favourable; chills and numbnesses are evil signs.

469. In cases of stroke, for sweating to supervene during difficult breathing is a deadly sign, but if in these same cases a fever supervenes, it means resolution.

470. Sudden apoplexies resolved when fever comes on, in time, become mortal.

471. In patients who change from some disease or other to dropsy, dry evacuations like sheep droppings follow, together with a mucous discharge and unfavourable urine. There are contractions in the hypochondria, pains and swellings in the cavity, and pains set in in the flanks and the muscles of the spine. Fevers, thirst, and dry coughing follow, and difficult breathing during movements, heaviness of the legs, rejection of food, and fulness after eating little.

472. Cases of white phlegm are halted by diarrhoea. Silent depressions of the spirit and unsociability are likely to wear such patients down.

473. Ὅσοι ἐκ φόβου μετὰ καταψύξιος ἐξίστανται, πυρετοὶ μεθ᾽ ἱδρώτων, καὶ ὕπνοι οἱ πάννυχοι ταῦτα λύουσιν.

474. Ἐκ μανίης ἐς βράγχον μετὰ βηχὸς ἀπόστασις.

475. Ἐν τοῖσι μανιώδεσι σπασμὸς προσγινόμενος ἀμαύρωσιν ἴσχει.

476. Αἱ σιγῶσαι ἐκστάσιες, οὐχ ἡσυχάζουσαι, ὅμμασι περιβλέπουσαι, πνεῦμα ἔξω ἀναφέρουσαι, ὀλέθριαι· ποιοῦσαι δὲ παραπληκτικὰ χρόνια· ἀτὰρ καὶ ἐκμαίνονται οὗτοι· ὅσοι δὲ ἐπὶ ταραχῇ κοιλίης οὕτω παροξύνονται, περὶ κρίσιν μέλανα διέρχεται.

477. Οἷσιν ὑγιαίνουσι, χειμῶνος ἐόντος, περὶ τὴν ὀσφὺν ψυχρότης καὶ βάρος ἀπὸ βραχείης προφάσιος, 692 καὶ κοιλίης | ἐπίστασις, τῆς ἄνω καλῶς ὑπηρετούσης, ἰσχιάς, ἢ νεφρῶν πόνος ἢ στραγγουρίη τάχα ἂν συμβαίη.

478. Οἷσι τὰ κάτω κακοῦται, κνησμῶν ἐγγενομένων ἔμπροσθεν ἰσχυρῶν, τούτοισιν ἀμμῶδες οὖρον γίνεται, καὶ ἐφίσταται· τοῖσι δὲ ὀλεθρίοισιν αὐτῶν ἡ διάνοια ἀπαναρκοῦται.

479. Οἱ τὰ ἄρθρα φλυκταινούμενοι ἐξερύθροισιν ἐπιπολαίοις, ἐπιρριγώσαντες, οὗτοι κοιλίας καὶ βουβῶνας διαφοινίσσονται, οἷα πληγῇσιν ἐπωδύνοισι, καὶ ἀποθνήσκουσιν.

480. Τὰ ἰκτερώδεα, οὐ πάνυ τι ἐπαισθανόμενα, οἷσι λύγγες, κοιλίαι καταρρήγνυνται· ἴσως δὲ καὶ ἐπίστασις· οὗτοι ἐκχλοιοῦνται.

473. Patients who become deranged from fear, when they have a great chill, are relieved by fevers with sweating, and sleeping through the whole night.

474. To pass from mania to hoarseness with coughing represents an apostasis.

475. In cases of mania the addition of a convulsion dims the vision.

476. Patients with silent derangements of the mind, who are not at peace, who gaze about with their eyes, and who expel their breath forcefully, are doomed; these conditions bring about chronic paralyses, and also mania. Patients that have paroxysms of this kind after a disturbance of the cavity, pass dark stools around the time of their crisis.

477. In healthy persons, who during winter have coldness and heaviness about the loins from a trivial cause, and are constipated while the upper cavity functions well, sciatica, pain of the kidneys, or strangury is likely to develop.

478. In patients whose the lower parts are afflicted after intense irritations have already set in, the urine becomes sandy and stops: in the cases destined to die, the mind becomes torpid.

479. In patients who form very red, superficial blisters at their joints, an attack of chills also makes their cavities and inguinal regions become red, as if they had received painful blows; such patients die.

480. In cases of jaundice that are not particularly in their senses, if hiccups occur, stools are passed violently downward; probably there will also be a retention. These cases become sallow.

481. Τὰ κατὰ πλευρὸν ἀλγήματα ἐν πυρετοῖσιν ἰσχνῶς ἑστηκότα, ἄσημα, φλεβοτομίη βλάπτει, κἢν ἀπόσιτος ᾖ, κἢν ὑποχόνδριον μετέωρον· καὶ ἐν καταψύξει οὐκ ἀπύρους νενωθρευμένους αἵματος ἀφαίρεσις βλάπτει· καὶ δοκέοντες δὲ ἐπιεικέστερον ἔχειν, οὗτοι θνήσκουσιν.

482. Κεφαλὴν καὶ πόδας καὶ χεῖρας κατεψῦχθαι, κοιλίης καὶ πλευρῶν θερμῶν ἐόντων, κακόν· βέλτιστον δὲ πᾶν ὁμοίως τὸ σῶμα θερμόν τε εἶναι καὶ μαλθακόν.

483. Στρέφεσθαι δὲ ῥηϊδίως χρὴ τὸν νοσέοντα, καὶ ἐν τοῖσι μετεωρισμοῖσιν ἐλαφρὸν εἶναι· βαρύτης δὲ ὅλου τοῦ σώματος καὶ χειρῶν καὶ ποδῶν, πονηρόν· εἰ δὲ καὶ πρὸς τῷ βάρει πέλιοι γίνονται οἱ δάκτυλοι καὶ οἱ ὄνυχες, πλησίον ὁ θάνατος· μελαινόμενα δὲ παντελῶς, ἧσσον ὀλέθρια τῶν πελίων· ἀλλὰ τὰ λοιπὰ θεωρεῖν· ἢν γὰρ εὐπετέως φέρῃ τὸ νόσημα, καὶ ἄλλο | τι τῶν χρησίμων ὑποδεικνύῃ, τὸ νόσημα ἐς ἀπόστασιν τρέπεται, καὶ τὰ μελανθέντα τοῦ σώματος ἀποπίπτει.

484. Ὄρχιες καὶ αἰδοῖον ἀνεσπασμένα πονηρὸν σημαίνει.

485. Φῦσαν δὲ ἄνευ ψόφου καὶ πραδήσιος διεξιέναι, βέλτιστον· κρέσσον δὲ καὶ σὺν ψόφῳ διελθεῖν, ἢ αὐτοῦ ἀνειλέεσθαι· καίτοι τὸν τοιοῦτον τρόπον διελθοῦσα σημαίνει πονηρὸν καὶ παραφροσύνην, ἢν μὴ ἑκὼν οὕτω ποιέηται τὴν ἄφεσιν τῆς φύσης.

694

481. Pains in the side occurring without swelling during fevers, in the absence of other signs, are made worse by phlebotomy, even if the person is off food, and if the hypochondrium swells up. The removal of blood also harms torpid patients who have fever during a chill: although these appear to be in quite a good state, they die.

482. For the head, feet, and hands to have a chill, while the cavity and the sides are warm, is a bad sign; best is for the whole body to be equally warm and supple.

483. A patient must be able to turn himself easily and seem light when he is lifted: heaviness of the whole body and the arms and legs is a bad sign. If in addition to the heaviness, the digits and nails become livid, death is near —for them to turn altogether dark is less deadly than being livid—: but you must observe the other signs, for if the patient bears his disease easily, and another favourable sign appears, the disease is turning toward an apostasis, and the parts that turned dark will fall off.

484. Retraction of the testicles and the penis is a bad sign.

485. For flatulence to pass out without any sound or breaking of wind is best, but it is still better for it to come out with a sound than to be pent up inside: indeed, coming out with a sound it indicates that the patient is in a bad way and deranged in his mind, unless such an expulsion of wind is caused intentionally.

486. Ἕλκος πέλιον καὶ ξηρὸν ἢ χλωρὸν γινόμενον, θανάσιμον.

487. Ἀνάκλισις βελτίστη μέν, ὡς εἴθισταί τις ὑγιαίνων· ὕπτιον δὲ κεῖσθαι, τὰ σκέλεα ἐκτεταμένον, οὐκ ἀστεῖον· εἰ δὲ καὶ καταρρέοι προπετὴς ἐπὶ πόδας, χεῖρον· θανάσιμον δὲ καὶ κεχηνέναι καὶ καθεύδειν ἀεί· καὶ τὰ σκέλεα ὑπτίου κειμένου συγκεκαμμένα τε εἶναι ἰσχυρῶς καὶ διαπεπλεγμένα· τὸ δ' ἐπὶ γαστέρα κεῖσθαι οἶσι μὴ σύνηθες, παραφροσύνην σημαίνει καὶ πόνους περὶ κοιλίην· πόδας δὲ γυμνοὺς ἔχειν καὶ χεῖρας, μὴ θερμὸν ἐόντα ἰσχυρῶς, καὶ τὰ σκέλεα διερρῖφθαι, κακόν, ἀλυσμὸν γὰρ σημαίνει· ἀνακαθίζειν δὲ βούλεσθαι, κακὸν ἐν τοῖσιν ὀξέσι, κάκιστον δὲ ἐν περιπλευμονικοῖσι καὶ πλευριτικοῖσιν. καθεύδειν δὲ χρὴ τὴν νύκτα, τὴν δὲ ἡμέρην ἐγρηγορέναι· τὸ δ' ἐναντίον, πονηρόν· ἥκιστα δ' ἂν βλάπτοι τὸ πρωῒ κοιμώμενος ἕως τοῦ τρίτου τῆς ἡμέρης· οἱ δὲ μετὰ ταῦτα ὕπνοι, πονηροί· κάκιστον δὲ μὴ καθεύδειν μήτε ἡμέρης, μήτε νυκτός, ἢ γὰρ ὑπὸ ὀδύνης τε καὶ πόνου ἀγρυπνοίη ἄν, ἢ παραφρονήσει ἀπὸ τούτου τοῦ σημείου.|

488. Ὁκόσοισι κρόταφος τάμνεται, σπασμὸς ἐκ τῶν ἐναντίων τῆς τομῆς ἐπιγίνεται.

489. Ὅσοισιν ἂν ὁ ἐγκέφαλος σεισθῇ, καὶ πονέσῃ πληγεῖσιν ἢ ἄλλως, πίπτουσι παραχρῆμα, ἄφωνοι γίνονται, καὶ οὔτε ὁρῶσιν, οὔτε ἀκούουσι, καὶ τὰ πολλὰ θνήσκουσιν.

490. Οἷς ὁ ἐγκέφαλος τιτρώσκεται, πυρετὸς ὡς ἐπὶ

696

486. For an ulcer to become livid and dry, or green, is a mortal sign.

487. The best position in bed is the one a person habitually takes when he is healthy, but to lie on the back with the legs outstretched is not good; if the patient slips down in bed towards the foot, that is a very bad sign. It is a mortal sign to yawn and sleep all the time; also for the legs of a person lying on his back to be forcefully bent together and twisted. For patients to lie on their stomachs, if this is not their habit, indicates derangement of the mind and pains about the cavity. For a patient to keep his feet and hands naked, when he is not strongly heated, and to toss his legs about, is a bad sign, for it indicates restlessness. To want to sit up is a bad sign in acute states, and very bad in pneumonias and pleurisies. A person should sleep at night and be awake during the day, the opposite being a bad indication. Least harmful would be to sleep from dawn through the first third of the day: patients who sleep after that are in an evil way. Worst of all is to sleep neither at day nor at night, for either the person is kept awake by his pain and distress, or he will become deranged in his mind after this sign.

488. A person who is cut in the temple will have a spasm on the opposite side of his body.

489. Persons whose brain is shaken and who suffer pain, either as the result of blows or otherwise, immediately fall down, lose their speech, can neither see nor hear, and in most cases succumb.

490. Persons whose brain is injured are generally be-

τὸ πολὺ καὶ χολῆς ἔμετος ἐπιγίνεται, καὶ ἀποπληξίη
σώματος, καὶ ὀλέθριοι οἱ τοιοῦτοι.

491. Τῶν ῥηγνυμένων ἐν κεφαλῇ ὀστέων, χαλεπώ-
τατον γνῶναι τὰ κατὰ τὰς ῥαφὰς ῥηγνύμενα· ῥήγνυται δὲ ὑπὸ τῶν βαρέων καὶ στρογγύλων βελέων μάλιστα, καὶ ἐκ τῶν ἐξ ὑπεναντίου φερομένων, καὶ μὴ ἐξ
ἰσοπέδου. τὰ δ᾽ ἀπορεύμενα, πότερον ἔρρωγεν ἢ οὔ,
κρίνειν δεῖ, διαμασᾶσθαι διδόντα ἐφ᾽ ἑκατέρην τὴν
σιηγόνα ἀνθέρικον ἢ νάρθηκα, καὶ προσέχειν κελεύειν, εἴ τι ψοφεῖν αὐτῷ δοκέει τὸ ὀστέον· τὰ γὰρ
κατεηγότα δοκέει ψοφεῖν. προϊόντος δὲ τοῦ χρόνου, τὰ
ἐρρωγότα μὲν ἑβδομαῖα, τὰ δὲ τεσσαρεσκαιδεκαταῖα,
τὰ δὲ καὶ ἄλλως διασημαίνει· τῆς τε γὰρ σαρκὸς
ἀπόστασις ἀπὸ τοῦ ὀστέου γίνεται, καὶ τὸ ὀστέον
πελιόν, καὶ πόνοι, ἰχώρων ὑπορρεόντων· γίνεται δὲ
ταῦτ᾽ ἤδη δυσβοήθητα.

492. Ὅσοισιν ἐπίπλοον ἐκπίπτει, ἀνάγκη ἀποσα-
πῆναι.

493. Ἢν ἔντερον διακοπῇ τῶν λεπτῶν, οὐ συμφύεται.

494. Νεῦρον διακοπέν, ἢ γνάθου τὸ λεπτόν, ἢ
ἀκροποσθίη, οὐ συμφύεται.

495. Ὅ τι ἂν ἐν τῷ σώματι ὀστέον ἀποκοπῇ, ἢ
χόνδρος, οὐκ αὔξεται. |

496. Ἐπὶ τρώματι σπασμὸς ἐπιγενόμενος, κακόν.

497. Ἐπὶ τρώματι χολῆς ἔμετος ἐπιγενόμενος, κακόν, καὶ μάλιστα ἐπὶ τοῖσι κεφαλικοῖσιν.

498. Νεῦρα ὅσα παχέα τιτρώσκεται, ὡς ἐπὶ τὸ πολὺ

fallen by fever and the vomiting of bile, and by paralysis of the body; these are doomed.

491. Of fractures in the skull, the most difficult to discern are those along the sutures. The bone is usually fractured by heavy and rounded missiles, and from those being thrown from the opposite side, and also not from an equal height. Cases that are unclear as to whether or not they are fractured must be judged by giving a stalk of asphodel or fennel to the patient to chew by each mandible in turn, and ordering him to pay attention to whether his bone seems to make a noise, for mandibles that are broken are noticed to make a noise. As time passes, the broken bones will reveal themselves in seven or fourteen days, or at another time: for a separation of the flesh from the bone occurs, the bone becomes livid, pains supervene, and sera collect in the tissues. When this happens, the case is already difficult to help.

492. Cases where the omentum falls out must form apostases.

493. If the small intestine is cut through, it does not unite.

494. A severed cord, the narrow part of the mandible, or the tip of the foreskin does not unite.

495. If any part of the body which is bone or cartilage is severed, it does not grow back.

496. A spasm following upon a wound is a bad sign.

497. The vomiting of bile following upon a wound is bad, especially if the wound is of the head.

498. Persons wounded in the thick cords generally be-

χωλοῦνται, καὶ λοξὰ τιτρωσκόμενα μάλιστα, καὶ τῶν
μυῶν αἱ κεφαλαί, μάλιστα τῶν ἐν μηροῖσιν.

499. Ἀποθνήσκουσι δὲ μάλιστα ἐκ τῶν τρωμάτων,
ἤν τις ἐγκέφαλον τρωθῇ ἢ ῥαχίτην μύελον ἢ ἧπαρ ἢ
φρένας ἢ καρδίην ἢ κύστιν ἢ φλέβα τῶν παχειῶν·
θνήσκει δέ, κἢν ἐς ἀρτηρίην καὶ πλεύμονα μεγάλαι
σφόδρα αἱ πληγαὶ γένωνται, ὥστε, τοῦ πλεύμονος
πληγέντος, ἔλασσον προερχόμενον πνεῦμα κατὰ στόμα γίνεσθαι, ἢ τὸ ἐκπῖπτον ἐκ τοῦ τρώματος· θνήσκουσι δὲ καὶ οἱ ἐς τὰ ἔντερα,⁹⁰ ἤν τέ τι τῶν λεπτῶν
τρωθῶσιν, ἤν τε τῶν παχέων, ἢν ἐπικάρσιος ἡ πληγὴ
γένηται καὶ μεγάλη· εἰ δὲ μικρὴ καὶ εὐθεῖα, περι-
γίνονται ἔνιοι· ἥκιστα δὲ θνήσκουσιν οἱ τιτρωσκό-
μενοι, ἐν οἷσι ταῦτα μὴ ἔνι τῶν τοῦ σώματος μερῶν, ἢ
τούτων προσωτάτω.

500. Τὴν δὲ ὄψιν ἀμαυροῦνται ἐν τοῖσι τρώμασι
τοῖσιν ἐς τὴν ὀφρὺν καὶ μικρὸν ἐπάνω· ὅσῳ δ' ἂν τὸ
τρῶμα νεώτερον ᾖ, μάλιστα βλέπουσι, χρονιζομένης
δὲ τῆς οὐλῆς, ἀμαυροῦσθαι μᾶλλον συμπίπτει.

501. Αἱ σύριγγες χαλεπώταταί εἰσιν, ὅσαι ἐν τοῖσι
χονδρώδεσί τε καὶ ἀσάρκοισι τόποισι πεφύκασιν, εἰσί
τε κοῖλαι, μέλαιναι⁹¹ καὶ ἰχωρορροοῦσιν αἰεί, σαρκίον
τε ἐπὶ τῷ στόματι ἔπεστιν αὐταῖς· εὐθεραπευτότεραι
δέ, ὅσαι ἐν τοῖσι μαλθακοῖσι τόποισι καὶ σαρκώδεσί
τε καὶ ἀνεύροισι πεφύκασιν. |

700   502. Τάδε πρὸ ἥβης οὐ γίνεται νοσήματα, περι-
πλευμονικά, πλευριτικά, ποδαγρικά, νεφρῖτις, κιρσὸς
περὶ κνήμην, ῥοῦς αἱματηρός, καρκίνος μὴ σύμφυτος,

230

come lame, and especially when the oblique ones are wounded; also if they are wounded in the heads of the muscles, especially at the thighs.

499. Patients usually die from wounds to the brain, the spinal marrow, the liver, the diaphragm, the heart, the bladder, or one of the wide vessels. A patient also dies if especially great blows strike the trachea and the lung, so that, with the lung injured, there is less breath coming through the mouth than escaping from the wound. Persons wounded in the intestines—either part of the small ones or of the large ones—also die if the blow was at an angle and great; if the blow was minor and oriented lengthwise, some survive. Patients die least often if they are wounded in parts not located among the ones mentioned, or which are farthest from them.

500. Patients lose their sight in wounds to the eyebrow and the region above it; as long as the wound is fresh, they generally still see, but as the scar becomes chronic, their loss of sight increases.

501. The most difficult fistulae are those which form in the cartilaginous and fleshless parts: they are hollow, dark, and continually exude serum, and a small piece of tissue grows at their mouth. Most easily treatable are fistulae in the soft and fleshy parts without cords.

502. The following diseases do not arise before puberty: pneumonia, pleurisy, gout, nephritis, varicosities in the lower leg, a bloody flux, cancer (unless it is congenital),

---

⁹⁰ ἔντ. Opsopoeus, after Cornarius' *intestina*: ἐντὸς νεῦρα A.
⁹¹ Potter: μ*λοῦντι A.

λεύκη μὴ συγγενής, κατάρρους νωτιαῖος, αἱμορροΐς, μὴ σύμφυτος χορδαψός· τούτων τῶν νοσημάτων πρὸ ἥβης οὐ χρὴ προσδέχεσθαι γενησόμενον οὐδέν. ἀπὸ τεσσαρεσκαίδεκα μέχρι δύο καὶ τεσσαράκοντα ἐτέων πάμφορος ἡ φύσις νοσημάτων ἤδη τοῦ σώματος γίνεται. πάλιν δὲ ἀπὸ ταύτης τῆς ἡλικίης μέχρι ξγ´ ἐτέων οὐ γίνονται χοιράδες, οὐδὲ λίθος ἐν κύστει, ἢν μὴ τύχῃ πρότερον ὑπάρχων, οὐδὲ κατάρρους νωτιαῖος, οὐδὲ νεφρῖτις, ἢν μὴ παρακολουθῶσιν ἐξ ἄλλης ἡλικίης, οὐδὲ αἱμορροΐδες, οὐδὲ ῥοῦς αἱματηρός, ἢν μὴ πρότερον τύχῃ γεγενημένος· ταῦτα μέχρι γήρως ἀπέχεται νοσήματα.

503. Ἐν γυναικείοισι τὰ πρὸ τῶν τόκων ἰόντα ὑδατώδεα, κακόν.

504. Στόματα ἀφθώδεα, τῇσιν ἐπιφόροισιν οὐ χρηστόν· ἆρα καὶ κοιλίαι καθυγραίνονται;

505. Ἐκ κενεώνων μεθιστάμενα ἀλγήματα ἐς τὸ λεπτὸν ἐν μακροῖσιν, ἐκ διαφθορῆς καὶ μὴ λίην καθαρθείσης, ὀλέθριον.

506. Τὰ ἐκ τόκου καὶ διαφθορῆς πολλὰ ὀξέως ὁρμήσαντα, ἐπιστάντα, δύσκολα· ῥῖγος ταύτῃσι πολέμιον, καὶ κοιλίης ταραχή, ἄλλως τε καὶ ὑποχονδρίου ὀδυνώδεες.

507. Τῇσιν ἐπιφόροισι κεφαλαλγικὰ καρώδεα, μετὰ βάρεος γινόμενα | καὶ σπασμοῦ, φλαῦρα ὡς ἐπὶ τὸ πολύ.

508. Ἧισιν ἐκ γυναικείων περὶ τὸ ἄνω καὶ τὸ λεπτὸν πόνοι σύντονοι, κοιλίας καθυγραίνουσιν, ὑπ-

leuce (unless it is congenital), a downward flux in the back, haemorrhoids (unless they are congenital), and chordapsus;[8] one should not expect any of these diseases to occur before puberty. From the fourteenth to the forty-second year, the nature of the body is apt to bear all diseases. But then again, from that age until sixty-three years, scrofula does not occur, nor stone in the bladder (unless it happens to have existed before), nor a downward flux in the back, nor nephritis (unless the cases are carried over from an earlier time), nor haemorrhoids, nor a bloody flux (unless it happens to have arisen before): these diseases stay away until old age.

503. In women passage of the waters before their deliveries is a bad sign.

504. Thrush in the mouth is not a favorable sign in pregnant women: will they also have diarrhoea?

505. The migration of pains from the flanks to the small intestine, in long diseases after an abortion and when the woman is not adequately cleaned, is a deadly sign.

506. Copious fluxes after a delivery or an abortion, that start up rapidly and then stop, are a bad sign. A chill in these women is harmful, and also disturbance of their cavity, especially if they have pains in the hypochondrium.

507. In pregnant women, for a headache with torpor to arise, in conjunction with a feeling of heaviness and a convulsion, is generally an indifferent sign.

508. Women who after their menses have intense pains in the upper regions and the small intestine, suffer diar-

8 "Old term for a painful colic, in which the intestines seem tied in knots." *New Sydenham Society Lexicon* (London, 1882).

ασώδεες, ταύτῃσι περὶ κρίσιν καταφοραί, καὶ ἀδύνατοι κενεαγγικῶς ἐφιδροῦσι καὶ περιψύχουσιν· αἱ τοιαῦται ὑποστροφαὶ τῇσι πλείστῃσι γενόμεναι μετὰ τὴν ἄφεσιν, ταχέως κτείνουσιν.

509. Τὰ μετὰ μυχθισμοῦ ἔξω ἀναφερόμενα πνεύματα, καὶ τῆξις παράλογος, τῇσιν ἐπιφόροισιν ἐκτιτρώσκει· ὀδύνη κοιλίης μετὰ τόκον, ἐπὶ ταύτῃσι πυώδεα καθαίρει.

510. Αἱ ναρκώδεες καὶ μάλιστα ἐν τῇσι κινήσεσι μετὰ ἀδυναμίης κατακεκλασμέναι, περὶ κρίσιν ἐνοχληθεῖσαι, ἀσώδεες, ἐφιδροῦσι πολλῷ· κοιλίαι καθυγρανθεῖσαι ταύτῃσι, κακόν.

511. Τὰ δὲ γυναικεῖα μὴ ἐπιστῆναι, χρήσιμον· ἐπιληπτικὰ ἐκ τῶν τοιούτων, οἶμαι, ἐνίῃσι δὲ ὑποφοραὶ μακραί, ἐνίῃσι δὲ αἱμορροΐδες.

512. Τῇσιν ἐπιφόροισιν ὑποχονδρίου ἄλγημα, κακόν· καὶ κοιλίαι ταύτῃσι φερόμεναι, κακόν· καὶ τὸ ἐπιρριγοῦν ταύτῃσι κακόν· ὀδύνη κοιλίης ἐν τῇσι τοιαύτῃσιν, ἧσσον κακόν, ἢν ἰλυώδεα καθαίρῃ· ᾗσι ῥηϊδίως τῶν τοιούτων τίκτεται, μετὰ τόκον δύσφορα σφόδρα.

513. Τῇσι κυούσῃσι φθινώδεσιν, ᾗσιν ἔρευθος ἐπὶ προσώπου γίνεται, αἱ ἀπὸ ρινῶν ἀποστάξιες τοῦτο ἀποτρέπουσι γινόμεναι.

514. Ἧισιν ἐκ τόκου λευκά, ἐπιστάντων δὲ ἅμα πυρετῷ κώφωσις καὶ ἐς πλευρὸν ὀδύνη ὀξεῖα, ἐξίστανται ὀλέθριοι.

rhoea and a degree of nausea; around the time of their crisis, they are subject to lethargy and weakness like that experienced by a person whose vessels have been emptied, and they sweat over their whole body and have a generalized cooling. Relapses of this kind befall most of them after their remission, and lead quickly to death.

509. Breaths drawn up and expelled with snorting, and emaciation for no reason cause abortion in pregnant women; in these cases, pain of the cavity after delivery cleans out purulent material.

510. Women who have numbness and feel broken down and weak, especially in their movements, and around their crisis are troubled and nauseated, sweat copiously over their whole body: diarrhoea in such women is a bad sign.

511. For the menses not to stop is a favourable sign. If they do stop, epilepsies result, I think, and in some cases there are prolonged diarrhoeas, while in others there are haemorrhoids.

512. In pregnant women pain in the hypochondrium is a bad sign; for their cavities to be moved is also bad, and for chills to attack. Pain of the cavity in such women is less bad, if they are cleared of slimy material. Any of these women who give birth with ease will still have great difficulties after the birth.

513. Redness appearing in the face of pregnant women suffering from consumption is averted if bleeding from the nostrils occurs.

514. Women who, after giving birth, have a white flux which ceases with a fever and is succeeded by deafness and sharp pain in the side, become delirious in a fatal manner.

515. Τὰ ἐν τῇσιν ἐπιφόροισιν ἁλμυρώδεα σημαίνει μετὰ τόκον δύσκολα λευκοῖσι δακνώδεσιν· αἱ τοιαῦται καθάρσιες ἀποσκληρύνουσιν· λὺγξ ἐπὶ τούτοισι

704 φλαῦρον, καὶ πτύξις ὑστερῶν, καὶ συντείνει.

516. Ἐς πόδας καὶ ἐς ὀσφὺν συντάσιες ἐκ γυναικείων, ἐκπυητικόν, καὶ τὰ ἀπὸ κοιλίης γλίσχρα, δυσώδεα ἐπιπόνως ἰόντα· πνιγμοὶ ἐπὶ τοῖσι προγεγραμμένοισιν, ἐκπυητικόν.

517. Τὰ ὑστερικὰ ἐν κοιλίῃσι σκληρύσματα ἐπώδυνα, ὀξέως ὀλέθριον.

518. Τῇσιν ἐπιφόροισιν ἤδη ἀφθώδεα ρεύματα ἐπώδυνα, πονηρόν· αἱμορροῖς ταύτῃσι, κάκιστον.

519. Ἧισι, κοιλίης ἐπαρθείσης, ἐς αἰδοῖον ἔρευθος ἦλθε, γυναικείων λευκῶν ὑγρῶν κατελθόντων ἐξαπίνης, ἐν μακροῖσι πυρετοῖσι τελευτῶσιν.

520. Σπασμῷ, γυναικείων ἐν ἀρχῇσι φανέντων, πυρετοῦ μὴ ἐπιγενομένου, λύσις.

521. Οὖρα λεπτὰ ὑπονέφελα ἐν μέσῳ αἰωρεύμενα, ρῖγος σημαίνει.

522. Ἢν ἀπὸ τῆς τετράδος αἵματος ρύσις γένηται, χρόνια σημαίνει, καὶ κοιλίη καταρρήγνυται, καὶ σκελέων οἰδήματα.

523. Τῇσιν ἐπιφόροισι κεφαλαλγικὰ καρώδεα μετὰ βάρους γενόμενα, φλαῦρα· ἴσως δὲ ταύτῃσι καὶ ἅμα σπασμῶδές τι παθεῖν ὀφείλει.

524. Αἱ προαλγήσασαι τρόπον χολερώδεα πρὸ τῶν

515. Salty fluxes in pregnant women indicate that after the birth they will be troubled with white, irritating lochia. Such cleanings cause indurations; hiccups in such cases are an indifferent sign, and also folding of the uterus, and contraction.[9]

516. Contractions in the feet and loins after the menses is a sign of suppuration; so too are viscous, foul-smelling excretions of the cavity that are passed with pain; suffocation together with the signs listed also indicate suppuration.

517. Painful uterine indurations in the cavity are quickly fatal.

518. Painful aphthous fluxes in pregnant women are a bad sign; a haemorrhoidal flux in them is a very bad sign.

519. If redness occupies the genital parts, in women with a raised cavity, and watery, white menstrual fluxes suddenly pass down, they die of chronic fevers.

520. The appearance of the menses unaccompanied by fever, at the beginning of a convulsion, indicates resolution.

521. Thin urines with somewhat cloudy material suspended in the centre indicate an incipient chill.

522. If a haemorrhage occurs after the fourth day, it indicates chronicity: the cavity has a violent discharge and there are swellings of the legs.

523. In pregnant women, for torpid headaches to arise in conjunction with a feeling of heaviness is an indifferent sign, and they will probably suffer some sort of convulsion at the same time.

524. Women who have pains of a choleric kind before

---

[9] The text here is difficult to understand.

τόκων, τίκτουσι μὲν ῥηϊδίως, πυρέξασαι δέ, κακοήθεες, ἄλλως τε κἤν τι κατὰ φάρυγγα ὀχλῇ, ἤ τι τῶν ἐν
πυρετῷ κακοήθων ἐπιφανῇ σημείων.

525. Τὰ πρὸ τῶν τόκων ῥηγνύμενα ὑδατώδεα,
φλαῦρα.

526. Τῇσιν ἐπιφόροισι κατὰ φάρυγγα ἁλμυρώδεες
ῥύσιες, πονηρόν.

527. Τὸ πρὸ τῶν τόκων ἐπιρριγοῦν, καὶ τὰ ἀνωδύ-
νως τικτόμενα, κινδυνώδεα.

528. Τῇσιν ἐπιφόροισι τὰ ἀφθώδεα ῥεύματα, πονη-
ρόν· σπασθεῖσαι, ἐκλυθεῖσαι, μετακαταψυχθεῖσαι, ἐκ-
θερμαίνονται ὀξέως· καὶ μέντοι καὶ δύσκολα ἀποβαί-
706 νει τῇσιν | ἐπιφόροισι τὰ περὶ τὸ λεπτὸν οἰδήματα,
οἷα τὰ περὶ τὰ ἰσχία[92] γίνεται, ἀπολαμβανόμενα ὀρ-
θοπνοίησιν· ἆρα τὰ τοιαῦτα οἰδήματα διδυμοτοκεῖ·
ἆρα καὶ σπασμῶδες τὰ τοιαῦτα οἰδήματα ποιέει;

529. Τὰ μυχθῶδες ἐξαναφέροντα πνεῦμα ἐν πυρετοῖσιν, ἐκτιτρώσκονται.

530. Φρικώδεσι, κοπιώδεσι, καρηβαρικῇσι, γυναικεῖα καταρρήγνυνται.

531. Αἱ πρὸς χεῖρα νωθραί, κατάξηροι, ἄδιψοι,
γυναικεῖα πολλὰ χαλῶσαι, ἐκπυητικαί.

532. Τὰ ἐξαίφνης λευκὰ κατατρέχοντα ἐπὶ τρω-
σμῷ, ἤν τι ῥιγῇ, καὶ ἐς μηρὸν ὁρμᾷ τρόμος, δύσκολον.

533. Τὰ ἀφθώδεα στόματα τῇσιν ἐπιφόροισι κοιλίας καθυγραίνει.

[92] τὰ ἰσχία Potter: τὰς ὀσχίας A.

their delivery, give birth easily, but then become febrile and enter a malignant state, especially if there is some trouble in their throat, or if any of the malignant signs of fever appear.

525. For the waters to break out before the birth process is an indifferent sign.

526. In pregnant women, salty fluxes in the throat are an evil sign.

527. To have chills before giving birth, and to give birth without pains, are dangerous signs.

528. In pregnant women aphthous fluxes are an evil sign; they have convulsions which go away, after that severe chills, but then quickly warm up. In pregnant women, swellings about the small intestine indicate trouble, as do swellings about the hips accompanied by orthopnoea.[10] Do such swellings announce the birth of twins? Do they provoke convulsions?

529. In fevers, expirations accompanied by snorting breaths indicate abortions.

530. Shivering, weariness, and heaviness of the head are signs that the menses will break out.

531. Women who are insensible to touch, who are very dry but have no thirst, and who pass copious menses, are likely to suppurate.

532. For a white flux suddenly to come down after an abortion, if it is accompanied by any chill or trembling that starts up into the thigh, is a difficult sign.

533. Aphthae of the mouth in pregnant women provoke diarrhoea.

---

[10] Littré comments: "la phrase entière est fort obscure."

534. Αἱ[93] δὲ τῶν κυουσέων προνοσέουσαι πρὸ τῶν τόκων ἐπιρριγοῦσιν.

535. Αἱ ναρκώδεες ἐκλύσιες, δύσκολοι μὲν ἐκ τῶν τόκων ἀποβαίνουσι καὶ παρακρουστικαί, οὐ μέντοι ὀλέθριοι· ἀτὰρ καὶ πλῆθος γυναικείων προσημαίνουσιν.

536. Αἱ ἐν τόκῳ καρδίην προαλγήσασαι, ὀλίγῳ ὕστερον ἀποβάλλουσιν.

537. Τὰ φρικώδεα, κοπιώδεα, καρηβαρικά, τραχήλου ὀδυνώδεα, γυναικεῖα καταρρήγνυσιν· περὶ κρίσιν τὸ τοιοῦτον γινόμενον μετὰ βηχίου ἐπιρριγεῖ.

538. Ἧισι κόρησιν ὀρθοπνοϊκὰ συμβαίνει, ἐν τῆισιν ἐπιφορῆισι τιτθοὺς ἐκπυοῦνται· γυναικεῖα ἐπιφαίνεσθαι ἐν ἀρχῇ, κακόν.

539. Τὰ μανικὰ πυρετοὺς ὀξεῖς ταραχώδεας ἀχόλῳ καρδιαλγικῷ λύουσιν.

540. Τῆισιν ἀτόκοισιν αἵματος ἔμετος πρὸς τὸ συλλαβεῖν ὠφελεῖ.

541. Τὰ ἀχλυώδεα, γυναικείων συχνῶν ἐπιφανέντων, λύεται.

542. Ὅσηισι γυναιξὶν ἐκ πυρετῶν ἄλγημα τιτθῶν γίνεται, πτύσις αἱμάλωπος οὐ τρυγώδης ἐγγενομένη λύει τοὺς πόνους.

543. Οἱ ἐν ὑστερικῆισιν ἀπύροισι σπασμοί, εὐχερέες, οἷον καὶ Δορκάδι.

<hr>

93 Αἱ Foes, after Cornarius' *Praegnantes praeaegrotantes*: εἰσὶ Α.

534. Pregnant women suffering from some previous disease will have a chill over their whole body before giving birth.

535. After giving birth numbness and faintness indicate trouble and foretell delirium, but such signs do not threaten life. They do, however, indicate that there will be a good amount of lochial discharge.

536. Women who while giving birth have a continual pain in their cardia will expel the child before long.

537. Chills, weariness, heaviness of the head, and pains in the neck are signs that the menses will break out. If this occurs around the time of the crisis and is accompanied by a little cough, there will be chills.

538. Those women who chance to have had orthopnoea as girls, when they are pregnant will have abscesses in their breasts; if their menses appear at the beginning, it is a bad sign.

539. Mania resolves disordered, acute fevers in persons with non-bilious heartburn.

540. In women who have no children, the vomiting of blood helps towards conception.

541. Cloudiness before the eyes is resolved by the appearance of copious menses.

542. In women who, after fevers, have pains in their breasts, the expectoration of a blood clot that does not become thick like wine lees resolves the pains.

543. Convulsions in afebrile, hysterical disorders are easy to manage, as for example in the case of Dorcas.

544. Ἦισιν ἐκ ῥίγεος πυρετὸς κοπιώδης, γυναικεῖα κατατρέχει· τράχηλος ἐν τούτοις ὀδυνώδης, αἱμορραγικόν.

545. Ἔμετος δὲ ἀλυπότατος, φλέγματος καὶ χολῆς συμμεμιγμένος, μὴ πολὺς δὲ κάρτα ἐμείσθω· τὰ δὲ ἀκρητέστερα τῶν ἐμουμένων, κακίω· πρασοειδὴς δὲ ἔμετος, καὶ μέλας, καὶ πελιός, πονηρόν· εἰ δὲ καὶ πάντα τὰ χρώματα ὁ αὐτὸς ἐμέοι, ὀλέθριον· τάχιστον δὲ θάνατον σημαίνει ὁ πελιὸς καὶ κακώδης· ἐστὶ δὲ θανάσιμος ὁ ἐρυθρὸς ἔμετος, καὶ μάλιστα εἰ μετὰ ἀνάγκης ἐμέοιτο ἐπωδύνου.

546. Οἱ ἀσώδεες ἀνημέτως παροξυνόμενοι, κακόν, καὶ οἱ σπαρασσόμενοι ἀνημέτως.

547. Τὰ μικρὰ ἐμέσματα, χολώδεα, κακόν, ἄλλως τε κἢν ἀγρυπνέωσιν.

548. Ἐπὶ μελάνων ἐμέτων κώφωσις οὐ βλάπτει.

549. Οἱ κατὰ μικρὰ ταχεῖς, χολώδεες, ἄκρητοι ἔμετοι, κακὸν ἐν ὑποφορῇ[94] πλείονι καὶ ὀσφύος ἀλγήματι συντόνῳ.

550. Τὰ ἐξ ἐμέτων ἀσώδεα, κλαγγώδεα, | ὄμματα ἐπίχνουν ἴσχοντα, μανικά· ὀξέως μανέντες θνήσκουσιν ἄφωνοι.

551. Ἐν ἐμέτῳ διψώδεα ἐόντα, ἄδιψον γενέσθαι, κακόν.

552. Ἐν ἀσώδεσιν ἀγρύπνοις, τὰ παρ' οὖς μάλιστα.

---

[94] Duretus, after Cornarius' *subductione*: –φθορῇ A.

544. In women who have a wearisome fever after a chill, the menses run down; if their neck is sore, it suggests that a haemorrhage will follow.

545. The least harmful vomitus is a mixture of phlegm and bile, and should not be vomited in too great a quantity. More unmixed vomitings are a worse sign; leek-coloured, dark, and livid vomitus bode ill; if the same person vomits all these colours, it indicates death; the most rapid death is indicated by livid and fetid vomitus. Red vomitus is also deadly, and most especially if it is vomited with a painful retching.

546. Exacerbations of nausea unaccompanied by vomiting are a bad sign, and also to retch without vomiting.

547. Scanty, bilious vomitus is a bad sign, especially if patients cannot sleep.

548. Deafness occurring after the vomiting of dark material is not harmful.

549. Frequent vomiting a little at a time of bilious, unmixed material is bad when it occurs in association with copious evacuations and an intense pain in the loins.

550. If, subsequent to vomiting, there is nausea, the voice is shrill, and the eyes develop a wool-like covering, this announces mania: such patients rage violently and die after losing their voice.

551. To be thirsty while vomiting, but then to have no thirst, is a bad sign.

552. In persons with nausea and sleeplessness, swelling beside the ear is especially common.

553. Τοῖς ἀσώδεσι, κοιλίης ταραχώδους⁹⁵ ἐπίστασις διὰ ταχέων ἐξανθεῖ οἷα κωνώπων κεντήματα, καὶ ἐς ὄμματα δακρυώδης ἀπόστασις ἔρχεται.

554. Ἐπὶ ἀκρήτοις ἐμέτοις λυγμός, κακόν· κακὸν δὲ καὶ σπασμός· ὁμοίως δὲ καὶ ἐν τῇσιν ὑπερκαθάρσεσι τῇσιν ἐκ τῶν φαρμακειῶν.

555. Οἱ μέλλοντες ἐμεῖν πτυαλίζουσιν ἔμπροσθεν.

556. Ἐπὶ ἐλλεβόρῳ σπασμός, ὀλέθριον.

557. Ἐπὶ πάσῃ καθάρσει πλεοναζούσῃ ψύξις μεθ' ἱδρῶτος, ὀλέθριον· καὶ οἱ ἐπανεμέοντες διψώδεες ἐν τούτοισι, κακόν· οἱ δὲ ἀσώδεες ὀσφυαλγέες κοιλίην καθυγραίνονται.

558. Αἱ ἐξερύθρων, μελάνων ὑπὸ ἐλλεβόρου, καθάρσιες, πονηραί· καὶ ἔκλυσις δὲ μετὰ τοιούτων, κακόν.

559. Ἀπὸ ἐλλεβόρου ἐμέσαι ἐρυθρά, ἀφρώδεα, ὀλίγα, ὠφελέει· ποιέει μέντοι σκληρύσματα, καὶ ἐμπυήσιας μεγάλας ἀφίστησιν· εἰσὶ δὲ οἱ τοιαῦτα ἐμέοντες ἄλλως τε καὶ στῆθος ἐπώδυνοι, καὶ ἐν τοῖς ῥίγεσιν ἐφιδροῦντες, καὶ ὄρχιας ἐπαίρονται· τούτου προσγενομένου, ἐπιρριγοῦσι καὶ ἰσχναίνονται.

560. Αἱ πυκναὶ διὰ τῶν αὐτῶν ὑποστροφαὶ ἐμετώδεες περὶ κρίσιν μέλανα ἔμετον ποιέουσιν· γίνονται δὲ καὶ τρομώδεες. |

712 561. Ἱδρὼς ἄριστος μὲν ὁ λύων τὸν πυρετὸν ἐν ἡμέρῃ κρισίμῳ, χρήσιμος δὲ καὶ ὁ κουφίζων· ὁ δὲ

⁹⁵ Opsopoeus: -ώδεας A.

244

553. In persons with nausea, the stoppage of a disordered cavity quickly causes a skin eruption like mosquito bites, and a lachrymal apostasis enters the eyes.

554. Hiccups occurring together with the vomiting of unmixed material is a bad sign; a convulsion is also bad. It is the same with excessive cleanings by means of medications.

555. Persons who are going to vomit expectorate first.

556. A convulsion after taking hellebore is a deadly sign.

557. In every excessive cleaning, a chill with sweating is a deadly sign; in such cases, it is bad to vomit at intervals and to be thirsty. Patients who are nauseated and have pains in their loins pass fluid stools.

558. Cleanings of very red or dark material brought on by taking hellebore are an evil sign; also a faintness together with these things is bad.

559. It is helpful to vomit a little frothy, red material by taking hellebore: however, this causes indurations, although it does prevent serious suppurations. Patients who vomit like this are especially likely to have pains in their chest, and to sweat over their whole body during the chills, and their testicles swell up: when this once happens, chills come on and the swelling goes down.

560. Frequent regular relapses accompanied by vomiting produce a dark vomitus around the time of crisis; such patients also have tremors.

561. The best sweat is one that resolves fever on the critical day, but also favourable is one that brings relief. A

ψυχρὸς καὶ μοῦνον περὶ κεφαλὴν καὶ τράχηλον γινόμενος, φλαῦρος, καὶ γὰρ χρόνον καὶ κίνδυνον σημαίνει.

562. Ἱδρὼς δὲ ψυχρός, ἐν ὀξεῖ μὲν πυρετῷ θανάσιμος, ἐν πρηΰτέρῳ δὲ χρόνον σημαίνει.

563. Ἱδρὼς ἅμα πυρετῷ γενόμενος ἐν ὀξεῖ, φλαῦρον.

564. Οὖρον ἐν πυρετῷ λευκὴν ἔχον καὶ λείην ὑπόστασιν ἱδρυμένην, ταχεῖαν ἄφεσιν σημαίνει· ταχεῖαν
δὲ καὶ τὸ ἐξ ἀκρίτου λίπος ἴσχον τι ἐξυδατούμενον· τὸ
δὲ ὑπέρυθρον καὶ τὴν ὑπόστασιν ἔχον ὑπέρυθρόν τε
καὶ λείην, πρὸ μὲν τῆς ἑβδόμης γενόμενον, ἑβδομαῖον
ἀπολύει, μετὰ δὲ τὴν ἑβδόμην, χρονιώτερον ἢ πάντως
χρόνιον· τό τε ἐν τετάρτῃ λαβὸν ἐπινέφελον ὑπέρυθρον, ἑβδομαῖον ἀπολύει, τῶν λοιπῶν κατὰ λόγον
ἐχόντων. τὸ δὲ λεπτὸν καὶ χολῶδες καὶ τὸ μόλις
γλίσχρων ἔχον ὑπόστασιν, καὶ τὸ μεταβάλλον ἐπὶ τὸ
βέλτιον καὶ χεῖρον, χρόνιον· ἐπὶ πλεῖον δὲ τοῦτο
ἐπακολουθοῦν, ἢ περὶ κρίσιν χειρόνων γενομένων, οὐκ
ἀκίνδυνον.

565. Ὑδατῶδες δὲ καὶ λευκὸν διατελέως ἐν χρονί-
οισι, δύσκριτον γίνεται καὶ οὐκ ἀσφαλές.

566. Νεφέλαι δὲ ἐν οὔροισι λευκαὶ μὲν καὶ κάτω,
λυσιτελέες· ἐρυθραὶ δέ, καὶ μέλαιναι, καὶ πελιαί, δύσκολοι.

567. Κινδυνῶδες τῶν οὔρων ἐστὶ τὸ χολῶδες μὴ
714 ὑπέρυθρον ἐν τοῖσιν ὀξέσι, καὶ τὸ κριμνῶδες λευκὰς
ἔχον ὑποστάσιας, καὶ τὸ ποικίλον χροιῇ καὶ ὑπο-

cold sweat that arises only about the head and neck is an indifferent sign, and indicates chronicity and danger.

562. In an acute fever, a cold sweat is a deadly sign, whereas in a milder fever, it indicates chronicity.

563. In an acute disease, sweating during a fever is an indifferent sign.

564. In a fever, urine that sets down a fine, white precipitate indicates a swift recovery; a swift recovery is also indicated when the urine goes from being mixed to having fatty material separate somewhat from the aqueous part. Urine that is reddish and has a fine reddish precipitate, when occurring before the seventh day, resolves the disease on the seventh day, but when occurring after the seventh day, indicates that the disease will be quite chronic or even very chronic. Urine that on the fourth day acquires a reddish cloud floating on it resolves the disease on the seventh day, as long as the other signs are as they should be. Thin, bilious urine that has a very light, viscous precipitate, and urine that changes sometimes for the better and sometimes for the worse, indicate chronicity. If this kind of urine persists for a longer time, or around the crisis becomes worse, it is not without danger.

565. In chronic diseases, consistently watery and white urines indicate that there will be difficulty in reaching a crisis, and that there is danger.

566. Cloudy material in urines, that is white and collects lower down, is advantageous: if red, dark, or livid, it indicates trouble.

567. Dangerous among urines in acute diseases is a bilious urine without any redness, a farinaceous urine with white precipitates, and a urine that is variable in its col-

στάσει, καὶ μάλιστα τοῖσιν ἀπὸ τῆς κεφαλῆς ῥευμα-
τισμοῖσιν. κινδυνῶδες δὲ καὶ τὸ ἐκ μέλανος μεθι-
στάμενον ἐς λεπτὸν χολῶδες, καὶ τὸ ἐξ ὑποστάσιος
διασπώμενον, καὶ τὸ ἐκ τροφιώδεος ὑπόστασιν ἴσχον
ὑποπέλιον ἰλυῶδεα· ἆρα ἐκ τοιούτων ὑποχόνδριον ὀδυ-
νῶνται, δοκέω δεξιόν, ἢ καὶ χλοώδεες γίνονται, καὶ
τὰ παρ᾽ οὓς ὀδυνώδεες; τούτοισιν ἐπὶ βραχὺ κοιλίη
καταρραγεῖσα, ὀλέθριον.

568. Οὖρα ἐξαίφνης παραλόγως ἐπ᾽ ὀλίγον πεπαινόμενα, φλαῦρα, καὶ ὅλως τὸ παραλόγως πέπον ἐν
ὀξεῖ, φλαῦρον· φλαῦρον δὲ καὶ τὸ ἐξέρυθρον ἐκ τούτων
ἐπάνθισμα ἰώδει κατεχόμενον. Λευκὸν δὲ καὶ κατα-
χεόμενον διαφανὲς οὖρον, πονηρόν· μάλιστα ἐν φρενιτικοῖσιν ἐπιφαίνεται. πονηρὸν δὲ καὶ τὸ μετὰ ποτὸν
ταχέως διουρούμενον, καὶ μάλιστα πλευριτικοῖσι καὶ
περιπλευμονικοῖσιν. πονηρὸν δὲ καὶ τὸ πρὸ ῥίγους
ἐλαιῶδες οὐρούμενον. πονηρὸν δ᾽ ἐν τοῖσιν ὀξέσι καὶ
τὰ χλοιώδεα μὴ ἐπὶ χροιῇ ἐόντα.

569. Ὀλέθριον δέ ἐστι τῶν οὔρων τό τε μέλαιναν
τὴν ὑπόστασιν ἔχον, καὶ τὸ μέλαν· μᾶλλον δ᾽ ἐν τοῖσι
παισὶ τὸ λεπτὸν τοῦ παχέος, τοῖσι δὲ λεπτοῖσι τὸ
ἀνάπαλιν· οἷσι συνεστραμμένοις καὶ τὸ χαλαζῶδες
διαχεόμενον, τὸ δ᾽ αὐτὸ καὶ ἐπίπονον· ὀλέθριον δ᾽ ἐστὶ
καὶ πᾶν τὸ λαθραίως οὐρούμενον· περιπλευμονικοῖσι
δ᾽ ἐστὶν ὀλέθριον καὶ τὸ ἐν ἀρχῇ μὲν πέπον, μετὰ δὲ
τὴν τετράδα λεπτυνόμενον.

716   570. Πλευριτικοῖσιν οὖρον αἱματῶδες, ζοφῶδες,
μεθ᾽ ὑποστάσιος ποικίλης ἀδιακρίτου, θανάσιμον ἐν

our and precipitate, especially in patients with fluxes from their head. Also dangerous is a urine that changes from dark to thin and bilious, a urine with a dispersed precipitate, and a urine that from its coagulated material sets down a muddy somewhat livid precipitate: do patients have pain in the hypochondrium from conditions like this—I think on the right side—or do they become pale and have pains beside the ear? In these cases, for the cavity to have a violent discharge after a short time is a fatal sign.

568. For urines suddenly to become concocted over a short time for no reason, is an indifferent sign, and in general any coction for no reason in an acute disease is an indifferent sign; also indifferent is a very red scum that separates out and covers urines with rust. Urine that is passed white and transparent is a bad sign: this usually happens in phrenitis. Also evil is when urine is passed rapidly after a drink, especially in pleurisies and pneumonias. Also evil is the passage of oily urine before a chill. In acute diseases, urine with green components not on its surface also bodes ill.

569. Deadly among urines is one with a dark precipitate, and also dark urine itself. In children thin urine is more fatal than thick, and thin ones that become that way for a second time. In urines that are compacted, material like hail is excreted, and this is troublesome. Deadly is any urine which is passed unawares. In pneumonias it also bodes ill if there is coction at the beginning, but after the fourth day the urine becomes thin.

570. In pleurisies, urine that is bloody, dark, and has a mixed variegated sediment in it is generally a sign of death

τέσσαρσι καὶ δέκα ἡμέρῃσιν ὡς ἐπὶ τὸ πολύ· θανάσιμον δὲ καὶ ἐν τοῖσι πλευριτικοῖσι συντόμως καὶ τὸ
πρασοειδὲς μέλαιναν ἔχον ὑπόστασιν ἢ πιτυρώ-
δεα. καυσώδεσι δὲ κατόχως κάκιστον οὖρόν ἐστι τὸ
ἔκλευκον.

571. Οὖρον δὲ ὠμὸν πλείονα χρόνον γινόμενον, τῶν
ἄλλων σωτηρίων ἐόντων, ἀπόστασιν καὶ πόνον σημαίνει, καὶ μᾶλλον ἐν τοῖσιν ὑπὸ φρένα· ἀλγημάτων
δὲ ἐν ὀσφύϊ πλανωμένων, ἐς ἰσχίον, καὶ ἐν πυρετῷ καὶ
ἄνευ πυρετοῦ. τὸ δὲ ἐκπεμπόμενον λίπος ἴσχον οὖρον
ὑπόστασιν, σημαίνει πυρετόν· τὸ δὲ αἱματῶδες ἐν
ἀρχῇ οὐρηθέν, χρόνιον· τὸ δ' ἀνατεταραγμένον μεθ'
ἱδρῶτος, ὑποτροπήν· τὸ δὲ λευκὸν οἷον τῶν ὑποζυγίων
κεφαλαλγίην· τὸ δ' ὑμενῶδες, σπασμόν· τὸ δὲ πτυα-
λώδεας ἔχον ὑποστάσιας οὖρον ἢ ἰλυώδεας, ῥίγεος
δηλωτικόν· τὸ δὲ ἀραχνιῶδες, συντήξιος· τὰ δ' ἐν
πλανώδεσι πυρετοῖσι μέλανα νεφέλια, τεταρταίου· τὰ
δ' ἄχροα μέλασιν ἐναιωρεύμενα μετὰ ἀγρυπνίης καὶ
ταραχῆς, φρενιτικά· τὰ δὲ κονιώδεα μετὰ δυσπνοίης,
ὑδατώδεα.

572. Οὖρον ὑδατῶδες ἢ τεταραγμένον ψαφερῇ τρη-
χύτητι, κοιλίην ὑγρὴν ἐσομένην σημαίνει· τὸ δὲ
ἔκλεπτον οὖρον δασυνόμενον, ἆρα ἱδρῶτα μέλλοντα
δηλοῖ, γεγενημένον δέ, τὸ ἀφρῶδες ἐφ' αὐτὸ ἐφιστάμενον;

573. Τὰ δ' ἐν τριταίοισι μετὰ φρίκης, οἷα νεφέλια
μέλανα, φρίκης ἀκαταστάτου δηλωτικά· καὶ ὑμενώ-

in fourteen days. Also rapidly fatal in pleurisies is leek-coloured urine with a dark or farinaceous sediment. In ardent fevers accompanied by catalepsy, the worst urine is a very white one.

571. A urine that is unconcocted for a longer time, while the other signs point towards delivery, foretells an apostasis and pain, especially in the region below the diaphragm, and in cases where pains are moving about in the loins, or to a hip—this whether fever is present or not. Urine passed with a fatty sediment indicates that a fever will follow. Bloody urine passed at the beginning is the sign of a longer disease. A urine containing a stirred up precipitate, in conjunction with sweating, indicates a relapse. White urine like that of cattle presages a headache, urine containing membranous material, a convulsion; urine with sputum-like sediments and muddy urine are indicative of a chill, a urine with spider-webs, of the melting of flesh, and urines with dark cloudy material in irregular fevers, of a quartan fever. Colourless urines with dark suspended material in them, in association with sleeplessness and restlessness, indicate phrenitis, ash-like urines in association with difficult breathing, dropsical conditions.

572. Watery urine, or a urine with coarse friable material in it, indicates there will be diarrhoea. Very thin urine that becomes cloudy: does it indicate there will be sweating, and when frothy material forms on it, that there has been sweating?

573. In tertian fever with shivering, dark cloudy material in the urine indicates that the shivering will be irregular, and urines with membranous material and those which

ΚΩΑΚΑΙ ΠΡΟΓΝΩΣΕΙΣ

718 δεες οὐρήσιες, καὶ | αἱ μετὰ φρίκης ὑφιστάμεναι, σπασμώδεες.

574. Οὖρον χρηστὴν ἔχον ὑπόστασιν, ἐξαπίνης μὴ ἴσχον, πόνον καὶ μεταβολὴν σημαίνει· τὸ δὲ ὑπόστασιν ἔχον, ἐπιταραχθὲν καθιστάμενον, ῥῖγος περὶ κρίσιν, τάχα δὲ καὶ ἐς τριταῖον ἢ τεταρταῖον μετάστασιν.

575. Ἐν πλευριτικοῖς οὖρον ὑπέρυθρον, ἔχον λείην ὑπόστασιν, ἀσφαλέα κρίσιν σημαίνει· τὸ δ' ὑπόχλωρον εὐανθές, λευκὴν ἔχον ὑπόστασιν, καὶ ταχεῖαν· τὸ δὲ ἐρυθρὸν σφόδρα καὶ εὐανθές, ὑπόστασιν χλωρὴν ἔχον λείην εἰλικρινέα, πολυχρόνιον σφόδρα ταραχώδεα νοῦσον μεταβάλλουσαν ἐς ἄλλην, οὐ μὴν ὀλέθριον· τὸ δὲ λευκόν, ὑδατῶδες, κριμνώδεα πυρρὴν ἔχον ὑπόστασιν, πόνον καὶ κίνδυνον σημαίνει· καὶ τὸ χλωρὸν πυρρὴν ἔχον ὑπόστασιν κριμνώδεα, χρόνον καὶ κίνδυνον σημαίνει.

576. Οὖρα τοῖσι παρ' ὦτα ταχὺ καὶ ἐπ' ὀλίγον πεπαινόμενα, φλαῦρον· καὶ τὸ κατεψῦχθαι ὧδε, πονηρόν.

577. Κύστις ἀποληφθεῖσα, ἄλλως τε καὶ μετὰ κεφαλαλγίης, ἔχει τι σπασμῶδες· τὰ ναρκώδεα ἐν τοιούτοισιν ἐκλυόμενα, δύσκολα, οὐ μὴν ὀλέθρια· ἀρά τι καὶ παρακρούουσιν;

578. Νεφρῶν ἐξαπίναιον ἄλγημα, μετὰ οὔρου ἐπισχέσιος, λιθιδίων οὔρησιν ἢ παχέων οὔρων σημαίνει· τρομώδεα πρεσβυτέροισιν ἐν πυρετῷ καὶ οὕτως ἐπιφαινόμενα, λιθίδιά που διουρέει.

252

set down a sediment while shivering is present, that convulsive disorders will occur.

574. Urine with a favourable sediment, that then suddenly loses it, points to suffering and change. Urine with a precipitate which settles after it is stirred up, indicates a chill around the time of crisis, and perhaps a change to a tertian or quartan fever.

575. In pleurisies, a reddish urine with a fine sediment indicates a safe crisis, a florid greenish one with a white sediment, also that the crisis will be rapid. A very red and florid urine, with a fine unmixed green sediment, indicates a very lengthy disease with disturbances, changing into another disease, but not too deadly a one. A white, watery urine with a farinaceous, flame-coloured sediment points to pain and danger. A green urine with a farinaceous, flame-coloured sediment indicates chronicity and danger.

576. In patients with swellings beside the ears, urines that quickly become concocted and remain that way for a short time are an indifferent sign. To become chilly in this condition bodes ill.

577. Stoppage of the bladder, especially together with a headache, gives some indication of convulsions. In such patients numbness with resolution indicates trouble, but not of a fatal kind. Is there also a degree of delirium?

578. A sudden pain in the kidneys, occurring together with a suppression of urine, indicates the passage of pebbles or thick urines: if trembling with fever comes on in such a case in older persons, they may pass pebbles in their urine.

579. Οὔρου ἀπόληψις καὶ βάρος ἐν νειαίρῃ σημαίνει ὡς τὰ πολλὰ στραγγουρίην ἐσομένην· εἰ δὲ μή, ἄλλην ἀρρωστίην, ἣν εἴωθεν ἀρρωστεῖν.

720   580. Ἐν χολώδε|σιν οὔρου ἀπόληψις κτείνει συντόμως.

581. Οὖρον ἐν πυρετῷ δάσος ἔχον διασπώμενον, ὑποτροπικόν, ἢ ἱδρῶδες.

582. Ἐν μακροῖσι πυρετοῖσι λεπτοῖσι πλανώδεσι, λεπτῶν οὔρων οὐρήσιες, σπληνώδεες.

583. Ἐν πυρετῷ ἄλλοτε ἀλλοίων οὔρων οὐρήσιες μηκύνουσιν.

584. Τὰ οὐρούμενα, ὑπομνησάντων, ἄλλως δὲ ὀλέθρια· ἆρα τούτοισιν οὐρεῖται, οἷον εἰ τὴν ὑπόστασιν ταράξειας.

585. Οἷσιν οὖρα ὀλίγα, θρομβώδεα, οὐκ ἀπυρέτοις, πλῆθος ἐκ τούτων ἐλθὸν λεπτόν, ὠφελεῖ· ἔρχεται δὲ τοιαῦτα, οἷσιν ἐξ ἀρχῆς ἢ διὰ ταχέων ὑπόστασιν ἴσχει.

586. Οἷσιν οὖρα ταχέως ὑπόστασιν ἴσχει, ταχέως οὗτοι κρίνονται.

587. Ἐπιληπτικοῖς οὖρα λεπτὰ καὶ ἄπεπτα παρὰ τὸ ἔθος ἄνευ πλησμονῆς, ἐπίληψιν σημαίνει, ἄλλως τε κἢν τις ἐς ἀκρώμιον ἢ τράχηλον ἢ μετάφρενον πόνος ἢ σπασμὸς ἐμπεπτώκῃ, ἢ νάρκη περιγίνηται τοῦ σώματος, ἢ ταραχῶδες ἐνύπνιον ἑωράκῃ.

588. Τὸ μικρὰ ἐπιφαίνεσθαι, οἷον στάξιας, καὶ οὖρον, καὶ ἔμετον, καὶ διαχωρήματα, κακὸν μὲν πάντως, κάκιστον δέ, ἐγγὺς ἀλλήλων ἰόντα.

579. The suppression of urine, in association with a sensation of heaviness in the abdomen, generally indicates that strangury will occur: if not strangury, then some other disease from which the patient is wont to suffer.

580. In the case of a bilious condition, suppression of the urine is rapidly fatal.

581. Urine which during a fever has a cloudiness spread through it foretells a relapse or a sweat.

582. In chronic light irregular fevers, the passage of thin urines indicates a condition of the spleen.

583. In a fever, the passage of different kinds of urine at different times prolongs the state.

584. For patients to pass urine only on being reminded to do so is an especially fatal sign. Does the urine passed in such cases look as if you had stirred up a sediment in it?

585. In patients who are not afebrile, for the passage of a small amount of urine with clots to be followed by the arrival of a plentiful thin urine is helpful. These things happen in cases where the urine has a sediment from the beginning, or soon.

586. Patients whose urines quickly acquire a sediment have their crisis quickly.

587. In epileptics, thin unconcocted urines passed against custom without a feeling of fulness indicate an attack of epilepsy, especially if some pain or spasm attacks the shoulder, neck or back, or numbness comes over the body, or the person has a troubling dream.

588. For anything to appear in small amounts, such as drops of blood, or urine, or vomitus, or stools, is always a bad sign, but it is worst if these pass in rapid succession.

589. Διαχώρημα κοιλίης βέλτιστον, μαλθακόν, συνεστηκός, ὑπόπυρρον, μὴ σφόδρα δυσῶδες, διαχωρέον τὴν εἰθισμένην ὥρην, πλῆθος δὲ πρὸς λόγον τῶν εἰσιόντων· παχυνέσθω δὲ πρὸς τὴν κρίσιν· χρήσιμον δὲ καὶ ἕλμινθας στρογγύλας διεξιέναι, πρὸς κρίσιν προσάγον.

590. Ἐν ὀξέσι τὸ ἀφρῶδες περίχολον διαχώρημα, κακόν· κακὸν δὲ καὶ τὸ ἔκλευκον· ἔτι δὲ κάκιον τὸ ἀλητοειδὲς κοπριῶδες· κάρος ἐπὶ τούτοισι, κακόν, καὶ αἱματώδης διαχώρησις, καὶ | κενεαγγίη παράλογος.

591. Κοιλίης ἀπόληψις μικρὰ μέλανα σπυραθώδεα πρὸς ἀνάγκην χαλῶσα, μυκτὴρ τούτοισι ῥηγνύμενος, κακόν.

592. Γλίσχρον ἄκρητον, ἢ λευκὸν διαχώρημα, φλαῦρον· <φλαῦρον>[96] δὲ καὶ τὸ ἄλες ἐζυμωμένον ὑποφλεγματῶδες· πονηρὸν δὲ καὶ ἐκ τροφιωδέων ὑπόστασις ὑποπέλιος, πυώδης μετὰ χολώδεος.

593. Αἷμα λαμπρὸν διαχωρέειν, κακόν, ἄλλως τε κἢν τις ὀδύνη παρῇ.

594. Τὸ ἀφρῶδες περίχολον διαχώρημα, φλαῦρον· καὶ ἰκτεροῦνται δὲ ἐκ τοιούτων.

595. Ἐπὶ τοῖσι χολώδεσι τὸ ἀφρῶδες ἐπάνθισμα, κακόν, μάλιστα δὲ ὀσφὺν πεπονηκότι, καὶ παρενεχθέντι· ἀραιὰ[97] δὲ τούτοισι τὰ ἀλγήματα.

596. Λεπτὸν ἔπαφρον διαχώρημα, ὑδατόχλοον ἴσχον ὑπόστασιν, πονηρόν· πονηρὸν δὲ καὶ τὸ πυῶδες· καὶ τὸ μέλαν αἱματῶδες, πονηρὸν σὺν πυρετῷ καὶ ἄλλως· καὶ τὸ ποικίλον κατακορὲς διαχώρημα, φλαῦ-

589. The best stool passed by the cavity is soft, consistent, slightly flame-coloured, and not especially foul-smelling, for it to pass at the accustomed hour, and for the amount to be in accordance with what was consumed: as the crisis approaches, it should become thick. It is also useful if round-worms are passed at the approach of the crisis.

590. In acute diseases, a frothy stool full of bile is a bad sign; also bad is a very white stool, and even worse is a mealy, fecal one. A stupor occurring in such cases is bad, as is a bloody excretion or an unexplained inanition.

591. If the cavity is blocked, but on being forced passes small dark stools that look like basketry, this, in association with a haemorrhage from one nostril, is a bad sign.

592. Viscous, unmixed or white stools are an indifferent sign, as is a somewhat phlegmy fermenting mass. Bad is also a somewhat livid deposit from coagulated stools, which is purulent and has bilious material in it.

593. To pass bright blood in the stools is a bad sign, especially if any pain is present.

594. A frothy stool full of bile is an indifferent sign, and patients develp jaundice from this.

595. A frothy scum on bilious stools is a bad sign, especially in a patient with pain in his loins who is delirious: the pains will be intermittent.

596. A thin frothy stool, with a watery green deposit, is an evil sign. Evil is also a purulent stool; a dark bloody stool bodes ill, too, with fever and otherwise. A completely variegated stool is an indifferent sign, and the more fearful its

---

96 φλ. Froben.
97 ἀραιὰ Littré: ἆρα A.

ρον, καὶ χεῖρον ὅσῳ φοβερώτερον τῇ χροιῇ, πλὴν ἐν
φαρμακείῃσιν, ἐν δὲ ταύτῃσιν ἀκίνδυνον μὴ πλήθει
ὑπερβάλλον· καὶ τὸ ψαφαρὸν μαλθακὸν ἐν πυρετῷ
διαχώρημα, φλαῦρον· φλαῦρον δὲ καὶ τὸ ξηρόν, ψα-
φαρόν, ἄχλοον, καὶ ἄλλως καὶ ⟨ἢν⟩ κοιλίην⁹⁸ καθυ-
γραίνῃ· μελάνων δὲ προδιελθόντων, κτείνει.

597. Ὑγρὸν διαχώρημα καὶ ἀθρόον κατὰ μικρόν,
κακόν· τὸ μὲν γὰρ ἀγρυπνίην, τὸ δὲ ἔκλυσιν τάχ' ἂν
ποιήσῃ.

598. Ἔνυγρον ὑποψάφαρον διαχώρημα περιψυχό-
μενον μὴ ἀπύρῳ, φλαῦρον· τὰ ἐπὶ τούτοισι ῥίγεα
κύστιν, κοιλίην ἐπιλαμβάνει. |

724   599. Ὑδατῶδες δὲ σφόδρα διαχώρημα μὴ παυό-
μενον ἐν ὀξέσι, κακόν, καὶ μᾶλλον εἰ καὶ ἀδιψήσει.

600. Ἐξέρυθρον ἐν περιπλύσει διαχώρημα, φλαῦ-
ρον· φλαῦρον δὲ καὶ τὸ σφόδρα χλωρόν, ἢ λευκόν, ἢ
ἀφρῶδες ὑδαρές· καὶ τὸ μικρόν τε καὶ γλίσχρον, καὶ
λεῖον καὶ ὑπόχλωρον, κακόν· καὶ τὸ κωματώδεσι νενω-
θρευμένοισιν ὑγρὸν διαχώρημα, κάκιστον· θανατῶδες
δὲ καὶ αἱμορροεῖν αἱματῶδες πολὺ θρομβῶδες· λευκόν
τε καὶ ὑγρὸν μετὰ κοιλίης μετεώρου.

601. Διαχώρημα μέλαν οἷον αἷμα, καὶ σὺν πυρετῷ
καὶ ἄνευ πυρετοῦ, πονηρόν· πονηρὸν δὲ καὶ πάντα τὰ
ποικίλα· καὶ τὰ κατακορέα πονηρά.

602. Τὰ ἐς ἀφρώδεα ἄκρητα τελευτῶντα διαχωρήματα, παροξυντικὰ μὲν πᾶσι, τοῖσι δὲ σπασμώδεσι
καὶ πάνυ· ἐκ τοιούτων τὰ παρ' οὓς ἀνίσταται· τὰ δὲ
ἐξυγραινόμενα καὶ πάλιν συνιστάμενα, ἄκρητα,⁹⁹ κο-

colour, the worse it is, except in cases where medications have been taken—in these it is without danger, unless excessive in amount. In a fever, a soft, friable stool is an indifferent sign; indifferent is also a dry, friable, discoloured stool, especially if the patient has diarrhoea: if dark stools came before, this leads to death.

597. Moist stools and copious stools passed a little at a time, are a bad sign: for the former is likely to cause sleeplessness, and the latter exhaustion.

598. To pass loose moist stools that produce a generalized cooling, in a person who is febrile, is an indifferent sign; subsequent chills inhibit the bladder and the cavity.

599. In acute diseases, very watery stools that do not cease are bad, and more so if the person is without thirst.

600. Very red stools in a liquid evacuation are an indifferent sign. Indifferent is also a very green, white, or frothy liquid evacuation. A small, viscous, smooth, greenish stool is also bad. In patients with coma and torpor, a moist evacuation gives a very bad indication. It is a fatal sign for a person to pass copious bloody clotted material in the stools, and white and moist stools with a raised cavity are also deadly.

601. Stools dark like blood are a bad sign, whether fever is present or not; bad also are all variegated stools, as are deeply coloured ones.

602. Evacuations that in the end become frothy and unmixed indicate an exacerbation in all cases, but especially so in convulsive conditions; subsequent to such things, the region beside the ear swells up. Cases where the stools are

---

98 ἢν κοιλίην Aldina: κοιλίης A.
99 Foes, after Cornarius' *meracae*: κρητά A.

πρώδεα, μῆκος νόσου σημαίνει· τὸ δὲ ἐξέρυθρον ἐν
πυρετῷ, παρακοπήν· τὸ δὲ λευκὸν κοπρῶδες ἰκτέρῳ,
δύσκολον· τὸ δὲ ὑγρὸν ἐν τῷ τεθῆναι λαβὸν[100] ἔρευθος, αἱμορραγέσι.

603. Γλίσχρον διαχώρημα μέλασι διαποίκιλον,
κακόηθες, μάλιστα δὲ ἐκλεύκοις.

604. Ἔκλευκον διαχώρημα ἐν πυρετῷ, οὐκ εὐκρι-
νές.

605. Κοιλίη ταραχώδης σμικρῇσι πυκνῇσιν ἀναστάσεσι, σιηγόνας ἐντείνει· λύει δὲ καὶ ἐπὶ προσώπου
γενόμενα ἐρυθήματα.

606. Κοπρώδης μετὰ τόνου διαχώρησις, κοιλίης
πονηρίην σημαίνει· φλεγματώδης δὲ ὀξέως μετὰ καρ-
διωγμοῦ, δυσεντερίην, τάχα δὲ καὶ ὀσφυαλγίην· τοῖσι
τοιούτοισι κοιλίης περίτασις,[101] πρὸς ἀνάγκην ὑγρὰ |
χαλῶσα, ταχὺ ὀγκυλλομένη,[102] ἔχει τι σπασμῶδες· τὸ
ἐπιρριγοῦν τούτοισιν ὀλέθριον.

607. Οἷσι μέλανα διαχωρέει, ἐφιδροῦσι ψυχροῖς.

608. Οἷσι κοιλίη κατ' ἀρχὰς ταράσσεται, τὰ δὲ
οὖρα μικρά, προαγόντων κοιλίη μὲν ξηραίνεται, τὸ δ'
οὖρον πληθύει λεπτόν, τούτοισιν ἀποστάσιες ἐς ἄρθρα.

609. Αἱ κατὰ μικρὰ ἀναστάσιες, φρικώδεες ῥιγω-
τικαί, οἷς φλαῦρον διαχώρημα, δυσκολώτατον τεταρταίοισιν ἀρχόμενον.

[100] Foes, after Cornarius' *capit*: λάβρον A.
[101] Opsopoeus: -στασις A.
[102] Opsopoeus: τάχιον κυλλομένη A.

very moist, and then again solid, unmixed, and fecal, indicate the length of the disease. A very red stool in a fever indicates delirium. A white, fecal excretion in jaundice is a sign of trouble: so too, in patients with haemorrhages, is a moist evacuation which takes on redness when it is left to sit.

603. A viscous stool dappled with black is malignant, and especially malignant if it is dappled with very white material.

604. In a fever, a very white stool is a sign that the disease will not reach a crisis easily.

605. If the cavity is disturbed by frequent unproductive movements, this provokes spasms of the jaws; these are relieved by red areas appearing on the face.

606. A fecal excretion with tension of the cavity indicates trouble in it, while an acute phlegmy one with heartburn indicates dysentery and perhaps also pain in the loins. In such cases, a distension of the cavity which, on being forced, excretes moist stools and rapidly swells up, indicates a tendency to convulsions. For a chill to follow in these patients is a fatal sign.

607. Patients who pass dark stools are likely to have cold sweats over their whole body.

608. Persons in whom the cavity is stirred up at the beginning, whose urines are small in amount, and who, as time passes, become constipated and pass more thin urine, will have apostases into their joints.

609. Patients who go to stool at short intervals will suffer shivering and chills, and those whose evacuations are indifferent will have great difficulties if this begins on the fourth day.

610. Αἱ πυκναὶ κατὰ μικρὰ ἀναστάσιες ὑπόγλισχροι, ἔχουσαι μικρὰ κοπρώδεα, μεθ᾽ ὑποχονδρίου καὶ πλευροῦ ἀλγήματος, ἰκτερώδεες· ἆρα, ἐπιστάντων, οὗτοι ἐκχλοιοῦνται; οἶμαι δὲ καὶ αἱμορροεῖν τούτους· τὰ δ᾽ ἐς ὀσφὺν ἀλγήματα ἐν τούτοισιν αἱμορροεῖ.

611. Οἷσιν αἷμα διαχωρέει λαμπρὸν μετὰ κάρου καὶ κεφαλαλγίης, τὸ ἐπιχλιαίνεσθαι, ὀλέθριον.

612. Τὰ γλίσχρα χολώδεα μᾶλλόν τι τὰς ἀποστάσιας παρ᾽ οὓς ποιέει.

613. Ὅσα, κοιλίης καθυγραινομένης, οἰδήματα μετεωρίζεται μετὰ ἀλγημάτων, κακόν· κοιλίης δ᾽ ἐπιστάσης, ἄλλου δέ τινος μὴ νεωτερισθέντος, ταχέως καταρρήγνυται, καὶ κακοηθέστερον· τὰ ἐμούμενα ἐπὶ τούτοισι, πονηρὰ καὶ θηριώδεα.

614. Οἷσιν ἐπὶ φλογώδεσι καὶ ἐξερύθροις λυομένοις δυσῶδες, λάβρον, ὑπέρυθρον, ἐλπὶς ἐκμανῆναι.

615. Ὁ αὐχμώδης χρὼς σημαίνει κοιλίην πονηρευομένην· ἐπὶ τούτοισιν ἐξέρυθρα | σαρκόπυα μάλιστα δίεισιν.

616. Ἐπὶ κοιλίην χολώδη, μαλθακήν, κοπρώδη, καύματα ἐπιφανέντα παρ᾽ οὓς ἔπαρμα ποιέει.

617. Χολώδεα διαχωρήματα κώφωσις παύει· κώφωσιν δὲ παύει χολῶδες διαχώρημα.

618. Τὰ ἑρπυστικὰ ὑπεράνω βουβῶνος, πρὸς κενεῶνα καὶ ἥβην γινόμενα, σημαίνει κοιλίην πονηρευομένην.

619. Ἔκλυσις ὀδύνην λύουσα, κοιλίην μάλα καθυγραίνει.

610. Slightly viscous stools with small fecal pieces, when passed frequently a little at a time in association with pain in the hypochondrium and side, are a sign of jaundice. When these evacuations cease, do the patients become sallow? I think they may bleed, too; pains that move to the loins in such patients foretell a haemorrhage.

611. For patients who pass bright blood while they have stupor and headache to be warmed up is a deadly sign.

612. Viscous, bilious stools cause apostases to arise beside the ear in some cases.

613. Swellings with pain that come up during diarrhoea are a bad sign. If the cavity closes—especially when nothing else new happens—the swellings quickly rupture, and this is even more malignant. The vomiting that takes place in these cases is difficult and verminous.

614. In patients whose evacuations turn from being flame-coloured and very red to being foul-smelling, violent, and only slightly red, you may expect mania.

615. For the skin to become dry indicates that the cavity is coming into difficulties; in such cases very red, flesh-like, purulent stool usually passes.

616. With the cavity passing bilious, soft, fecal stools, a fever heat coming on provokes swelling beside the ear.

617. Deafness terminates bilious evacuations: bilious evacuations terminate deafness.

618. Migrating ulcers above the groin that creep towards the flanks and the pubes indicate that the cavity is coming into difficulties.

619. A faintness that resolves pain provokes severe diarrhoea.

620. Τὰ καθ᾽ ἕδρην ὀδυνώδεα ἐκπυήματα κοιλίην ἐπιταράσσει.

621. Θανατώδεά ἐστι τῶν διαχωρημάτων τὸ λιπαρόν, καὶ τὸ μέλαν, καὶ τὸ πελιὸν μετὰ δυσωδίης, καὶ τὸ χολῶδες ἔχον ἐν ἑωυτῷ φακῶν ἢ ἐρεβίνθων ἐρίγμασι παραπλήσια, ἢ οἷον θρόμβους αἵματος εὐανθεῖς, κατὰ τὴν ὀδμὴν ὅμοιον τῷ τῶν νηπίων, καὶ τὸ ποικίλον, τὸ δ᾽ αὐτὸ καὶ χρόνιον· γίνοιτο δ᾽ ἂν τοιοῦτον αἱματῶδες, ξυσματῶδες, χολῶδες, μέλαν, πρασοειδές, καὶ ὁμοῦ καὶ ἐναλλάξ. θανατῶδες δὲ καὶ πᾶν ἐστι τὸ ἀναισθήτως διεξιόν.

622. Ποτὸν χαλεπῶς καταβρογχίζοντι, πνεύματι βηχῶδει, ἐρευγμὸς ὑποσπώμενος, εἴσω κατειλούμενος, σημαίνει πόνον κοιλίης.

623. Πονηρὸν δὲ καὶ ἐξερυθρώδεα τεταρταίοισι, καὶ αἱ τοιαῦται αἱμόρροιαι, κωματώδεες· ἐκ τούτων σπασμῷ τελευτῶσι, μελάνων προδιελθόντων.

624. Οἷσι μέλανα διαχωρέει, ἐφιδροῦσι ψυχροῖς.

625. Αἱ ἐξαίφνης παράλογοι ἐκλύσιες κοιλίης ἐν τοῖσι τετηκόσι χρονίοισιν, ἅμα ἀφωνίῃ τρομώδει, ὀλέθριοι· αἱ λεπταὶ μελάνων διαχωρήσιες αἱ φρικώδεες, βελτίους τοῖσι τοιούτοισιν· αἱ τοιαῦται ὠφελοῦσι μάλιστα κατὰ τὴν ἡλικίην, ἢ προακμάζουσιν.

626. Πᾶσι τὰ κνησμώδεα μελάνων διαχώρησιν σημαίνει καὶ ἔμετον θρομβώδεα· καὶ τρομώδεα σὺν | δηγμῷ μετὰ κεφαλαλγίης, τὰ μέλανα διαχωρήματα· πρὸ τῶν τοιούτων ἔμετος διέρχεται, καὶ ἐμέσασι συχνὰ τοιαῦτα προσκατασπᾶται.

730

620. Painful abscesses in the seat cause evacuations to pass.

621. Deadly among stools are the fat, the dark, the livid with a foul smell, the bilious which contain material like pounded lentils or chick-peas, or like fresh clots of blood, and in their smell resemble the stools of infants, and the variegated (such would be bloody stools that contain shreads of flesh, and are bilious, dark, and leek-coloured—both at the same time and alternately), which incidentally also indicate chronicity. Every stool that passes unperceived foretells death.

622. When drink is swallowed with difficulty, breathing is accompanied by a cough, and an eructation is drawn back and compressed inside, this announces a pain of the cavity.

623. Very red stools passed on the fourth day are also an evil sign, and such haemorrhages are indicative of coma; patients die from these with a convulsion, if dark stools have passed beforehand.

624. Patients with dark evacuations will have cold sweats.

625. Sudden bouts of diarrhoea for no reason in chronic wasting diseases, in association with a tremulous loss of speech, are a fatal sign. Thin evacuations of dark material accompanied by shivering are a better indication in such cases: these are most helpful in the age just before maturity.

626. In all cases, itchiness indicates that there will be dark evacuations and the vomiting of clots; trembling with gnawing pains and headache point to dark evacuations. Before the evacuations patients vomit, and after that many sorts of things are expelled.

627. Οἷσι δὲ ἐπὶ ταραχῆς κοιλίης παροξύνεται περὶ κρίσιν, κάτω μέλανα διέρχεται.

628. Ἐπὶ κοιλίῃ μακρῇ, ἐμετώδεσι, χολώδεσιν, ἀποσίτοις, ἱδρὼς πολὺς μετὰ ἀδυναμίης ἐξαπίνης κτείνει.

629. Ἐν φαρμακείῃσιν ἢν περιρρέῃ λεπτὸν συχνὸν[103] αἷμα ἐξετηξομένοισι, ἔπειτα δὲ φλαῦρον.

630. Τὰ κατὰ κοιλίην σκληρύσματα μετὰ πόνου, πυρετοῖσιν ἅμα φρικώδεσιν, ἀποσίτοις, σμικρὰ ἐφυγραινομένης κοιλίης κάθαρσιν οὐ ‹διδόν›τα, ἐς ἐμπύησιν ‹ἥξει›.[104]

631. Ἅμα πυρετῷ κοιλίη ταραχώδης τρόπον ἁλμυρώδεα, κωματώδεσι νωθροῖς οὐ πάνυ παρέ‹πεται›.[105]

632. Ἐπὶ κοιλίῃ ὑγρῇ, κοπιώδει, κεφαλαλγικῷ, διψώδει, ἀγρύπνῳ, ἐξερύθρῳ χρώματι λυομένους ἐλπὶς ἐκμανῆναι.

633. Ἢν δύσπνοοι ἔωσι, πρὸς τὸ ἐκχλοιοῦσθαι εὔπνοον ἄσιτόν τε, κοιλίης ἐπεισελθούσης.

732   634. Τὰ καυματώ|δεα διαχωρήματα τόνον ἴσχοντα, κοιλίην πονηρευομένην σημαίνει.

635. Τοῖσι χολώδεσι κοιλίη ταραχώδης, μικρὰ πυκνὰ διαδιδοῦσα τονώδεα μικροῖσι μυξώδεσι, πόνον περὶ τὸ λεπτὸν ποιέουσι, καὶ οὖρον οὐκ εὐλύτως ἰόν, ἐς ὕδρωπα ἐκ τοιούτων ἀποτελευτᾷ.

[103] ἢν—συχνὸν Potter: ἐν περιρρῷ λεπτῷ (corr. to -ὸν) συχνῷ A.
[104] Littré, cf. ch. 297 above: A has blank spaces in the text.

627. In patients who, after a disturbance of the cavity, have an exacerbation around the time of their crisis, dark stools will pass.

628. In a long disorder of the cavity accompanied by vomiting, biliousness, and loss of appetite, a copious sweat accompanied by weakness brings sudden death.

629. In cases where a purgative medication has been given, if much thin blood flows around the material that has been melted out of the body, it is an indifferent sign.

630. Painful indurations in the cavity occurring at the same time as fevers, shivering, and loss of appetite, when they effect no cleaning although the cavity is slightly moist, will turn to internal suppuration.

631. The cavity being disturbed with salty substances in the course of a fever rarely happens in coma or torpor.

632. In cases of diarrhoea, weariness, headache, thirst, and sleeplessness that are resolved by stools of a very red colour, expect mania.

633. If patients have difficulty breathing, as they become sallow they both recover their breath and lose their appetite when the cavity has an evacuation.

634. Burnt stools passed with force indicate that the cavity is coming into difficulty.

635. In bilious patients, a disturbance of the cavity with small, frequent, forceful evacuations containing a little mucus is followed by a pain in the area of the small intestine and urine that does not pass freely, and the condition ends in dropsy.

---

105 Froben: A has a blank space in the text.

636. Αἱ τρομώδεες γλῶσσαι, σημεῖον ἐνίοισι κοιλίης καταρραγησομένης.

637. Οἷσι καῦμα γίνεται, ἐπάφρων[106] διελθόντων, πυρετὸς παροξύνεται.

638. Ἐπὶ κοιλίῃσιν ὑγρῇσι κατάψυξις μεθ᾽ ἱδρῶτος, φλαῦρον.

639. Ἐπὶ κοιλίῃσι ὑγρῇσι τὰ ἀπ᾽ οὔλων αἵματα ἐπιρρυέντα, θανατῶδες.

640. Διαχώρημα καθαρὸν ἐπιγενόμενον, λύει πυρετὸν ὀξὺν μεθ᾽ ἱδρῶτος.

---

[106] Foes in note: ἐφιδροῦσι A.

636. Trembling of the tongue is an indication in some patients that the cavity will have a violent discharge.

637. In patients with a burning heat, if frothy evacuations occur the fever grows virulent.

638. In cases of diarrhoea, a chill accompanied by sweating is an indifferent sign.

639. In cases of diarrhoea, blood flowing from the gums is a deadly sign.

640. The advent of clean evacuations resolves an acute fever accompanied by sweating.

CRISES AND CRITICAL DAYS

INTRODUCTION

Crises and *Critical Days* are both derivative compositions based on other Hippocratic texts, and probably dating from the period of the third to fifth centuries A.D.; no ancient writer including Galen, who wrote books of his own with these same titles, betrays any acquaintance with the Hippocratic *Crises* and *Critical Days*. Despite their shared subject matter, the treatises do not exhibit any specific resemblances of style or content that would suggest a common origin, and each is transmitted in a different branch of the manuscript tradition, *Crises* in V and *Critical Days* in M.

Crises is a somewhat loosely ordered collection of Hippocratic views on prognostic signs in crises arranged in chapters as follows:[1]

1–16:	Crises indicating early improvement.
16–19:	Crises indicating late improvement.
19–34:	Crises indicating late deterioration.
34–38:	Crises indicating early deterioration.
39–40:	Relapses.

[1] Each of the first four sections is introduced by a general statement printed in italics; since this structure was not recognized before Preiser, Littré's chapter divisions, which I preserve, occasionally appear erratic.

41–44: Mental disorders.
45–50: Fevers.
51–56: Intestinal disorders.
57–60: Disorders in the head.
61–64: Spasms.

This material is drawn from a wide range of sources, in particular *Prognostic*, *Aphorisms*, *Epidemics II* and *VI*, *Sevens*,[2] and *Regimen in Acute Diseases (Appendix)*, and is reworked to varying degrees: some passages are taken over verbatim, but most are shortened, expanded, rearranged, or adjusted grammatically to fit into their context.

The eleven extended excerpts that constitute the chapters of *Critical Days* all derive directly without reworking from Hippocratic texts:

   1: *Epidemics III* 16
   2: *Sevens* 46
   3: *Internal Affections* 48
4–6: *Internal Affections* 52–54
   7: *Diseases III* 6
   8: *Internal Affections* 51
   9: *Diseases III* 11
 10: *Diseases III* 15
 11: *Sevens* 26

Only the first sentence of ch. 7 and the body of ch. 11 (a summary listing of critical days in acute and chronic condi-

---

[2] *Sevens* has survived in its entirety only in Latin translations transmitted by the manuscripts Ambrosianus Lat. G 108 (IX c.) and Parisinus Lat. 7027 (X c.). See Kerstin Agge, *Die pseudo-hippokratische Schrift von der Siebenzahl. Edition, Übersetzung und Kommentar*, Marburg, 2004.

tions based on well documented Hippocratic principles)
were contributed by the excerptor himself.

Crises and *Critical Days* appear in the collected editions and translations of the Hippocratic Collection, as
well as in the work:

> Joh. Rod. Zwinger, *Magni Hippocratis Coi opuscula
> aphoristica, semeiotico-therapeutica . . . graece et
> latine*, Basel, 1748.

Littré gives much specific information on the sources of individual chapters, but only with the following study are the
treatises comprehensively investigated:

> Gert Preiser, *Die hippokratischen Schriften "De iudi-
> cationibus" und "De diebus iudicatoriis,"* Diss. Kiel,
> 1957. (= Preiser)

The present edition, which is indebted to Preiser's work in
many ways, is based on a collation of the independent
manuscripts from microfilm.

ΠΕΡΙ ΚΡΙΣΕΩΝ

1. [Περὶ][1] κρίσεων ξυντόμων ἐπὶ τὸ ἄμεινον τὰ μὲν πλεῖστα ταῦτ᾽ ἐστιν ἅπερ ⟨ἐς⟩[2] ὑγίην σημεῖα.

Ἱδρῶτες γὰρ ἄριστοί εἰσι καὶ τάχιστα πυρετὸν παύοντες οἱ ἐν τῆσι κρισίμησιν ἡμέρῃσι γινόμενοι καὶ τελέως τὸν πυρετὸν ἀπαλλάσσοντες· ἀγαθοὶ δὲ καὶ ὅσοι διὰ παντὸς τοῦ σώματος γενόμενοι εὐπετεστέρως τὸ νόσημα φέρειν ποιήσουσιν· οἱ δ᾽ ἂν τούτων τι μὴ ἐργάσωνται, οὐ λυσιτελέσουσι γινόμενοι.

2. Παχύνεσθαι δὲ χρὴ τὸ διαχώρημα πρὸς τὴν κρίσιν ἰούσης τῆς νόσου· ἔστω δὲ ὑπόπυρρον καὶ μὴ σφόδρα δυσῶδες· ἐπιτήδειον δὲ καὶ ἕλμινθας ἐξιέναι πρὸς τὴν κρίσιν.

3. Οὖρον δὲ ἄριστόν ἐστιν, ὃ ἂν ἔχῃ λευκότατον ὑπόστημα καὶ λεῖον καὶ ὁμαλὸν παρὰ πάντα τὸν χρόνον ἕως ἂν κριθῇ τὸ νόσημα· σημαίνει γὰρ ἀσφαλέα καὶ ὀλιγοχρονίην τὴν νοῦσον ἔσεσθαι. ἢν ἱδρῶτος ἐγγενομένου ἡ νοῦσος[3] ἐκλίπῃ, καὶ τὸ οὖρον πυρρὸν θεωρηθῇ λευκὴν ὑπόστασιν ἔχον, τούτοισιν αὐθημερὸν ὑποστροφὴ τοῦ πυρετοῦ γίνεται, οὗτος καὶ ἐν πέντε ἀκινδύνως κρίνεται.

CRISES

1. *Of early crises tending towards improvement, most signs are the same as those which indicate a return to health.*

Sweats are best and stop fever most quickly that occur on critical days and completely dispel the fever. Sweats that occur over the whole body and that make the disease easier to bear are also good, whereas sweats that do not have these characteristics bring no benefit.

2. Stools should become thicker as a disease approaches its crisis, and they should be slightly reddish-yellow in colour and not overly ill-smelling. It is also advantageous if worms pass with the evacuations as the crisis approaches.

3. Urine is best that has a very white sediment which is fine and uniform for the whole time until the condition reaches its crisis; this sign indicates that the disease will be safe and of short duration. If a disease remits subsequent to a sweat, and the urine is observed to be reddish-yellow with a white sediment, in such cases there is a recurrence of fever on the same day, and the person has a crisis free from danger in five days.

---

1 Del. Potter; cf. Ermerins and Preiser p. 29.
2 Littré.   3 ἡ νοῦσος later manuscripts: οἶνος V.

4. Τοῖσιν ἐλαχίστῳ χρόνῳ μέλλουσιν ὑγιάζεσθαι μέγιστα σημεῖα ἀπ' ἀρχῆς[4] γίνεται· ἀπονώτεροι γὰρ διατελοῦσι καὶ ἀκίνδυνοι, καὶ τὰς νύκτας κοιμέονται, καὶ τὰ ἄλλα σημεῖα προφαίνουσιν ἀσφαλέα.

5. Οἷς ἐν πυρετῷ μὴ θανατώδει κεφαλῆς ἄλγημα καὶ τὰ ἄλλα περιέστηκε[5] σημεῖα, χολὴ τούτων κρατεῖ.

6. Οἷς ἂν ἄρξηται ὁ πόνος τῇσι πρώτῃσιν ἡμέρῃσι, 278 τεταρταῖοί | τε μᾶλλον καὶ πεμπταῖοι πιέζονται· ἐς δὲ τὴν ἑβδόμην ἀπαλλάσσονται τοῦ πυρετοῦ.

7. Οἱ δὲ πυρετοὶ κρίνονται ἐν τῇσιν αὐτῇσιν ἡμέ-ρῃσι τὸν ἀριθμόν, ἐξ ὧν ἀπόλλυνται οἱ ἄνθρωποι καὶ ἐξ ὧν περιγίνονται· οἵ τε γὰρ εὐηθέστατοι τῶν πυρε-τῶν καὶ ἐπὶ σημείων ἀσφαλεστάτων τεταρταῖοι παύ-ονται ἢ πρόσθεν· οἵ τε φονικώτατοι καὶ ἐπὶ σημείων δεινοτάτων γινόμενοι τεταρταῖοι κτείνουσιν ἢ πρόσ-θεν· ἡ μὲν οὖν πρώτη ἔφοδος οὕτως τελευτᾷ. ἡ δ' ἑτέρη ἐς ἑπτὰ περιάγει, ἡ δὲ τρίτη ἐς τὴν ἑνδεκάτην, ἡ δὲ τετάρτη ἐς τὴν τεσσαρεσκαιδεκάτην, ἡ δὲ πέμπτη ἐς τὴν ἑπτακαιδεκάτην, ἡ δὲ ἕκτη ἐς τὴν εἰκοστήν· αὗται μὲν οὖν ἐπὶ τῶν ὀξυτάτων διὰ τεσσάρων ἐς τὰς εἴκοσι προσθέσεις. οὐ δύναται δ' ὅλαις ἡμέρῃσι οὐθὲν τούτων ἀριθμεῖσθαι ἀτρεκέως· οὐδὲ γὰρ οἱ ἐνιαυτοί τε καὶ μῆνες ὅλαις ἡμέρῃσι πεφύκασιν οὐδὲ ξυνεστήκασιν.

8. Ἐν τοῖσι καύσοισι τὰ ἀγαθὰ σημεῖα γινόμενα, οἷα ἐν τοῖσιν ὑγιεινοῖσι γέγραπται, μείονα μὲν ἐόντα ἐς τρίτην ἄνεσιν δηλοῖ, παχύτερα δὲ αὔριον, πάνυ παχέα δὲ αὐθημερόν.

9. Ἐν τοῖσι καύσοισιν ἢν ἑβδομαίῳ ὕστερον ἐπι-

4. In patients that are going to recover in the shortest time, the most significant signs appear from the beginning: these patients go through the disease with neither pains nor danger, nights they sleep, and they exhibit the other signs of safety.

5. Patients with a non-mortal fever whom pains in the head and other signs befall: these are dominated by bile.

6. Patients, in whom distress begins on the first days, are pressed even more on the fourth day and the fifth day. Towards the seventh day they are relieved of the fever.

7. Fevers come to a crisis on the same days, in number, as those after which people die or recover. For the most benign of fevers with the safest signs cease on the fourth day or before, while the most deadly of fevers with the most inauspicious signs kill the person on the fourth day or before. This is when the first access reaches its termination. The second access extends to the seventh day, the third to the eleventh day, the fourth to the fourteenth day, the fifth to the seventeenth day, and the sixth to the twentieth day: that is, these increments in the most acute (sc. of fevers) are by fours up to the number twenty. In fact, however, none of these periods can be numbered exactly in whole days, any more than the year or the months are composed or consist exactly of whole days.

8. If, in an ardent fever, favourable signs of the kind recorded in healthy persons are present, in a lesser quantity they point towards remission on the third day, more plentifully on the next day, and very plentifully on the same day.

9. If, in an ardent fever, jaundice supervenes later on

---

4 ἀ. ἀ. Preiser: ἅπαν V.
5 Aldina: -έστηκα V.

γένηται ἴκτερος, δῆλον ⟨ἀν⟩ίδρωτος·[6] τὸ γὰρ νόσημα
οὐ φιλέει ἐξιδροῦν, οὐδὲ ἄλλῃ ἀφίστασθαι οὐδαμῇ,
ἀλλὰ ὑγιὲς γίνεται.

10. Ἀνάγκη τοῦ θερμοῦ ἀπιόντος ἐφ᾽ ἑωυτὸ τὸ
ὑγρὸν ἑλκύσαντος, τῷ πυρετῷ κρίσιν γενέσθαι διὰ τὰ
οὖρα τὰ ἀποχωρέοντα ἢ καὶ τὰ διαχωρήματα κοιλίης,
ἢ αἵματος ἐκ τῶν ῥινῶν ῥύσιν, ἢ οὔρησιν πολλήν, ἢ
διαρροίην ἰσχυρήν, ⟨ἢ⟩[7] ἱδρῶτα, ἢ ἔμετον, γυναικὶ δὲ
καὶ ἐπιμηνίων ὁδόν· μάλιστα μὲν οὖν ταῦτα ποιέει
κρίσιν, ἢ ὅ τι ἂν τούτων ἐγγὺς γίνηται· ποιέει δὲ καὶ
ἕτερα[s][8] κρίσεις, ἧττον μὲν τούτων. |

11. Ἴκτερος δὲ ἢν ἑβδομαίῳ ἐπιγένηται ἢ ὕστερον
ἐν καύσῳ καὶ δυσχερείᾳ, σιάλου πολλοῦ ἀποχώρησις·
ἔν τε τοῖς καυσώδεσι πυρετοῖς καὶ τοῖς ἄλλοις, ἢν,
μηδενὸς τούτων τῶν σημείων γενομένου, ἀφίῃ ὁ πυρετός, ἀνάγκη τοιάσδε κρίσιας ἀντὶ τούτων γενέσθαι, ἢ
φυμάτων μεγάλων ἀπόστασιν, ἢ ὀδύνας ἰσχυρὰς ἀπὸ
τῆς ἀποστάσιος, ἢ τηκεδόνας τῶν ὑγρῶν ἐκ τοῦ θερμοῦ. κρίσιες δὲ καὶ ἀφέσιες τῶν καῦσον σημαινόντων,
μακροτέρα ἡ νοῦσος· τῶν δὲ ἰσχυρῶν, θάνατος ὡς ἐπὶ
τὸ πολύ· οἱ δὲ λοιποὶ ἀσφαλέες παύονται καῦσοι
ἑβδομαῖοι ἢ τεσσαρεσκαιδεκάταιοι. φιλέει δὲ καὶ ἐς
λιπυρίην περιίστασθαι, καὶ λαμβάνει μάλιστα τεσσαράκοντα ἡμέρας καὶ ἐξηπιαλοῦται. καὶ ἡ λιπυρίη
τῆς αὐτῆς ἡμέρης λαμβάνει τε καὶ μεθίησι· γίνεται δὲ
καὶ τῆς κεφαλῆς ὀδύνη. ἐὰν δὲ μὴ μεθίῃ αὐτὸν ἡ

---

6 Littré.   7 Linden.   8 Del. Littré.

the seventh day,[1] it is clear that it will be unaccompanied by sweats, for the condition will not tend to pass out with the sweat, nor indeed to include any other apostasis, but still it recovers.

10. It must follow that when heat recedes after having attracted moisture to itself, a crisis of the fever will occur through the excretion of urines, or also discharges of the cavity, or by a haemorrhage from the nostrils, or by copious urination, or by a severe diarrhoea, or by sweating, or by vomiting, or in a woman by the mentrual route. Now it is mainly these things that bring about a crisis, or something happening that is close to them; other things too provoke crises, but less than the ones named.

11. If jaundice supervenes on the seventh day or later in an ardent fever together with malaise, there is a secretion of copious saliva. In ardent fevers, as well in others, if none of these signs is present, but the fever remits, crises must necessarily take place in their stead, or an apostasis of large growths, or severe pains from the apostasis, or reductions of the moist components by heat. With crises and remissions of the signs indicating ardent fever, the disease will be longer; if the signs are severe, in most cases the outcome is death. Other ardent fevers cease safely in seven or fourteen days. An ardent fever also tends to change to an intermittent fever, and generally persists for forty days and becomes an ague: intermittent fever attacks and remits on the same day; headache also sets in. If intermittent fever

[1] Perhaps this should be emended to "on the seventh day *or later*"; cf. ch. 11 below.

λιπυρίη ἐν ταῖς τεσσαράκοντα ἡμέραις, ἀλλ᾽ ἄχθη καὶ ὀδύνη ἔχῃ τὴν κεφαλήν, καὶ φλυηρέῃ, ἐπικάθηρον αὐτόν. λήγοντος δὲ καύσου, ἂν ἐπιγένηται ἴκτερος, οὐ φιλέει ἔτι ἰδροῦν, οὐδ᾽ ἄλλῃ ἀφίστασθαι οὐδαμῇ, ἀλλ᾽ ὑγιὴς γίνεται.

12. Τριταῖος κρίνεται[9] ἐν ἑπτὰ περιόδοισιν ὡς ἐπὶ τὸ πολύ.

13. Ὁκόσοις ἐν ἀφορήτοις πυρετοῖς τῇ ἑβδόμῃ ἢ τῇ ἐνάτῃ ἢ τεσσαρεσκαιδεκάτῃ ἴκτεροι γίνονται, ἀγαθόν, ἐὰν μὴ τὸ δεξιὸν ὑποχόνδριον σκληρὸν γένηται· εἰ δὲ μή, ἐνδοιαστόν.

14. Τὰ ὀξέα νοσήματα κρίνεται ἐν τεσσαρεσκαίδεκα ἡμέρῃσιν ὡς ἐπὶ τὸ πολύ.

15. Ἱδρῶτες πυρεταίνοντι ἢν γίνωνται τριταίοις καὶ πεμπταίοις καὶ ἑβδομαίοις καὶ ἐναταίοις καὶ τεσσαρεσκαιδεκαταίοις καὶ μίῃ καὶ εἰκοσταίοις καὶ τριακοσταίοις, οὗτοι οἱ ἱδρῶτες νούσους κρίνουσιν· οἱ δὲ μὴ οὕτως γινόμενοι πόνους σημαίνουσιν.

16. Αἱ πεπάνσιες τῶν οὔρων κατὰ μικρὸν ἐκπεπαι-
282 νόμεναι, ἐν | τῇσι κρισίμοις ἐὰν πεπανθῶσι, λύουσι τὴν νοῦσον. παράδειγμα δεῖ τῶν οὔρων τὰ ἕλκεα ποιέεσθαι· τά τε γὰρ ἕλκεα, ἢν μὲν ἀνακαθαίρηται πύῳ λευκῷ, ταχείην θεραπείην δηλοῖ· ἐὰν δὲ μεταβάλλῃ ἐς τοὺς ἰχῶρας, κακοήθη γίνεται· τὸν αὐτὸν δὲ τρόπον καὶ τὰ οὖρα σημαίνει. ἐὰν ἐκ πόνου λεπτὰ γίνηται, ἀπὸ τῆς προφάσιος δεῖ λογίζεσθαι, ᾗ[10] τὸ νόσημα παρεγένετο, καὶ ταύτην ὁρᾶν ἐπεὶ παύεται·[11] ὡς ταύτης ὑπολειπομένης,[12] τῶν ἄλλων σημείων ἐπι-

does not release a patient in forty days, but he has heaviness and pain in his head, and talks nonsense, clean him. If jaundice comes on after an ardent fever ceases, the patient does not tend to sweat for a second time, nor indeed to experience any other apostasis; still, he recovers.

12. A tertian fever has its crisis within seven periods in most cases.

13. If, in patients with fevers that are hard to bear, on the seventh, the ninth or the fourteenth day jaundices supervene, the prognosis is good, unless the right hypochondrium becomes hard; otherwise the case is doubtful.

14. Acute diseases have their crises within fourteen days in most cases.

15. If, in patients with fevers, sweats occur on the third, fifth, seventh, ninth, fourteenth, twenty-first or thirtieth day, these sweats bring the disease to a crisis; sweats that occur on other days announce new sufferings.

16. The maturation of urines that proceeds a little at a time, if it occurs at the crises, resolves the disease. You should take ulcers as your model for urines: for if ulcers are mundified by white pus, it indicates a rapid recovery, whereas if there is an alteration to serous discharges, they become malignant—urines give indications in the same way. If they are thin due to the distress present, you should base your calculation on the cause from which the disease arose, and pay attention to when this ceases: as long as the cause remains, even if the other signs develop as they

---

9 Later manuscripts: γίνεται V.
10 Cornarius in marg.: ἤν V.
11 ἐ. π. Foes in note 16: ἐπιπαύεται V.
12 Littré: ἐπιλ. V.

γινομένων οἵων δεῖ, οὐκ εἶναι ἀπαλλαγὴν τῇ νούσῳ
οἰητέον.[13] ἐὰν ἀλγέῃ ἡ κεφαλή, καὶ ἀπὸ τούτου πυρετὸς ἐπιγένηται, <καὶ>[14] τούτου μὴ καταπαύσηται, μηδὲ τῆς ὀδύνης παυομένης, οὐ κρίσιμος ὁ πυρετός.

Κρίσεως μακρῆς [ἔτι][15] ἐπὶ τὸ ἄμεινον· πλεῖστα
ταῦτ᾽ ἐστι καὶ ἐπὶ τούτων ἅπερ ἐς ὑγίην ἰόντα.[16]

17. Ἐν τοῖσιν ὑποχονδρίοισιν οἰδήματα μαλθακὰ
καὶ ἀνώδυνα καὶ ὑπείκοντα ἐπεὶ[17] θιγγάνῃς αὐτοῦ,
χρονιωτέρας μὲν τὰς κρίσιας ποιέει, ἧσσον δὲ φοβε-
ρὰς τῶν ἐναντίων τουτοῖς φυμάτων· ὡσαύτως δὲ ἔχει
καὶ περὶ τῶν ἐν τῇ ἄλλῃ κοιλίῃ φυμάτων.

18. Οὖρον δὲ ἢν τὸ μὲν οὐρηθὲν μὴ καθαρὸν ᾖ, τὸ δ᾽
ὑπόστημα λευκόν τε καὶ λεῖον ἔχῃ, χρονιωτέρη ἡ
κρίσις, ἢ καὶ ἧσσον ἀσφαλὴς τοῦ βελτίστου οὔρου·
ἢν δέ ποτε ὑπέρυθρον <τὸ>[18] οὖρον καὶ τὸ ὑπόστημα
ὑπέρυθρον καὶ λεῖον, πολυχρονιώτερον μὲν τοῦτο τοῦ
προτέρου, σωτήριον δὲ κάρτα.

19. Ὁκόσα δὲ ποδαγρικὰ νοσήματα γίνεται, ταῦτα
ἐν τεσσαράκονθ᾽ ἡμέραις ἀφλέγμαντα καθίστανται.

Κρίσεως μακρᾶς ἐπὶ τὸ κάκιον· <καὶ>[19] ἐπὶ τούτων,
τὰ πλεῖστά ἐστιν [20.] ἅπερ ἐς θάνατον.[20]

Ἐν ἡμέρῃ καὶ νυκτὶ κρίνεται, ἅπερ ἀσθενεώσεως
σημεῖα, οἷον φαρμακοποσίης, κοιλίης ἐκταράξεως καὶ
284 ἄνω | καὶ κάτω, ἀσιτίης, καὶ τῶν ἄλλων τῶν τοιούτων·

should, no remission of the disease is to be thought of. If the head aches, and subsequently a fever comes on, and the condition does not cease even though the pain stops, fever does not bring a crisis.

Of a late crisis for the better: most signs in these are the same as those pointing towards health.

17. Swellings in the hypochondrium that are soft, painless, and that yield when you touch the person have their crises later, but are less to be feared than growths with contrary characteristics. The same also applies for growths in other parts of the cavity.

18. If the urines passed are not clean, and their sediment is fine and white, the crisis will be rather late, and less secure than in the case of the best urine. If urine is ever reddish with a fine, reddish sediment, this case will take a longer time than the one above, but is very safe.

19. Any gouty conditions that arise will settle down in forty days without the collection of phlegm.

Of a late crisis for the worse: most signs in these are (20.) the same as those which indicate death.

In a day and a night, signs indicating weakness have their crisis, such as after the drinking of a purgative medication, disturbance of the cavity either upwards or downwards, want of appetite, or other things of this sort. Now if

---

13 Later manuscripts: οἷον V.   14 Littré.
15 Del. Preiser.   16 Littré: ἐόντα V.
17 Later manuscripts: ἐπι- V.   18 Linden.
19 κ. κ. Potter: ἄμεινον V.; cf. Preiser p. 30.   20 Littré's chapter division separates what is certainly one sentence.

ἢν μὲν οὖν ἀπαλλάσσηται τούτων τὰ σημεῖα ἐν ἡμέρῃ καὶ νυκτί· εἰ δὲ μή, θανατώδη νομίζειν εἶναι.

21. Τῶν ἱδρώτων κάκιστοί εἰσιν οἱ ψυχροί τε καὶ περὶ τὸν αὐχένα γενόμενοι· οὗτοι γὰρ θανάτους καὶ μῆκος νούσων προσημαίνουσι.

22. Τὰ ποικίλα ὑποχωρήματα χρονιώτερα μὲν τῶν μελάνων καὶ τῶν ἄλλων θανασίμων ὑποχωρημάτων, οὐδὲν δὲ ἧσσον ὀλέθρια· ἔστι δὲ τοιάδε, ξυσματώδεα, χολώδεα, αἷμα, πρασοειδέα, μέλανα, καὶ τοτὲ μὲν ὁμοῦ πάντα διαχωρέει, τοτὲ δὲ κατὰ μέρος ἕκαστον.

23. Οὖρον δὲ ἐὰν τοτὲ μὲν καθαρὸν οὐρηθῇ, τοτὲ δὲ ὑπόστημα ἔχον λευκόν τε καὶ λεῖον, χρονιώτερα καὶ ἧσσον ἀσφαλῆ ταῦτ᾽ ἐστὶ τοῦ βελτίστου οὔρου. ἐὰν πυρρὸν καὶ λεπτὸν ᾖ τὸ οὖρον πολὺν χρόνον,[21] κίνδυνος μὴ οὐ δύνηται διαρκέσαι ὁ ἄνθρωπος, ἕως ἂν πεπανθῇ τὸ οὖρον· καὶ ἢν ἄλλως περιεσομένου σημεῖα ᾖ, προσδέχου τούτοις ἀπόστασιν παρεσομένην[22] ἐς τὰ κάτω τῶν φρενῶν χωρία.

24. Ἐν τοῖσι πυρετοῖσιν ἐὰν μεταβολὰς ἔχῃ τὸ οὖρον, χρόνον τε σημαίνει, καὶ ἀνάγκη τῷ ἀσθενέοντι μεταβάλλειν καὶ ἐπὶ τὰ χείρω καὶ ἐπὶ θάτερα.

25. Ἢν ἀρχόμενα οὖρα μὴ ὅμοια ᾖ, ἀλλὰ γένηται παχέα ἐκ λεπτῶν καὶ παντελῶς λεπτά, δύσκριτα καὶ ἀβέβαια τὰ τοιαῦτα.

26. Ψυχροὶ ἱδρῶτες ξὺν μὲν ὀξεῖ πυρετῷ θανάσιμοι, ξὺν δὲ πρηυτέρῳ μῆκος σημαίνουσι τῆς νούσου.

the signs alter from these in a day and a night, fine; if not, you should consider the condition to be fatal.

21. Of sweats the worst are cold sweats in the neck, for these presage death and long diseases.

22. Variegated stools signify longer diseases than do dark and other fatal stools, but they are no less deadly. They have the following characteristics: they contain shreds of flesh, bile, blood, material that is leek-green or dark, and sometimes all of these pass at once, sometimes each one at a time.

23. If urine is passed at one time clean, but at another time with a fine white sediment, these conditions are of longer duration and less secure than in the case of the best urine. If the urine is reddish and thin for a long time, there is a danger that the person will not be able to hold out until the urine becomes concocted. And if otherwise the signs are those of one who is going to survive, expect that in these cases an apostasis will occur in the regions beneath the diaphragm.

24. If in fevers the urine undergoes changes, this indicates chronicity, and the patient must of necessity take turns both for the worse and in the other direction.

25. If from the beginning the urines are not consistent, but they go from thin to thick and then become absolutely thin, such conditions have a difficulty in reaching their crisis and are unstable.

26. Cold sweats with acute fevers are deadly, whereas with a milder fever they indicate a long disease.

---

21 Froben: πολὺ χρόνου Aldina: πολυχρονον (sic) V.
22 H. Diller in Preiser: προεσ. V.

27. Καὶ ὅκου τοῦ σώματος θερμὸν ἢ ψυχρόν, ὅπου τοῦτο ἔνι, ἐνταῦθα ἡ νοῦσος.

28. Καὶ ὅτεῳ[23] ἐν ὅλῳ τῷ σώματι μεταβολαὶ ὀξεῖαι γίνονται, καὶ ἢν τὸ σῶμα ψύχηται, ἢ αὖθις θερμαίνηται, ἢ τὸ χρῶμα ἕτερον ἐξ ἑτέρου μεταβάλληται, μῆκος νούσου σημαίνουσι. |

286   29. Κἢν πυρέσσοντι ἱδρὼς ἐπιγένηται μὴ ἐκλείποντος τοῦ πυρετοῦ, κακόν· μηκύνει γὰρ ἡ νοῦσος καὶ ὑγρασίην σημαίνει.

30. Πυρέσσοντι ψυχροὶ ἱδρῶτες ἐπιγενόμενοι μακρὸν τὸν πυρετὸν σημαίνουσιν.

31. Ἱδρὼς πολὺς ἀκρήτως γινόμενος ὑγιαίνοντι νόσον σημαίνει, θέρεος μὲν μείω, ψυχρῶς[24] δὲ πλείω.

32. Ἐς τὸ αὐτὸ χωροῦντα, ἐὰν ἐάσῃς ξυστῆναι, ὑφίσταται ὁκοῖον ξύσματα, ἢν ὀλίγα, ὀλίγη ἡ νοῦσος, ἢν πολλά, πολλή· τούτοισι ξυμφέρει τὴν κοιλίην ἐπικλύζειν. ὁκόσοις δὲ ἐν τῇ κάτω ὑποχωρήσει χολῆς μελαίνης ὕπεστιν, ἢν πλεῖον, πλείων ἡ νοῦσος, ἢν ἐλάσσω, ἐλάσσων.

33. Ἐὰν αἱ φλέβες σφύζωσι, καὶ τὸ πρόσωπον ἐρρωμένον ᾖ, καὶ τὰ ὑποχόνδρια μὴ λαπαρά, ἀλλὰ ἐπηρμένα, χρονίη ἡ νοῦσος, καὶ ἄνευ σπασμοῦ οὐ λύεται, ἢ αἵματος πολλοῦ ἐκ τῶν ῥινῶν, ἢ ὀδύνης ἰσχυρῆς.

34. Καὶ οἱ παλμοὶ ἐν τῇσι χερσὶ πολυχρονίου πυρετοῦ σημεῖον.

27. Wherever in the body heat or cold is present, that is where the disease is.

28. When acute changes take place in the whole body, as for example if the body becomes cold or warms up again, or the colour of the skin alters its hue from one to another, these indicate a long disease.

29. If, in a person with a fever, a sweat comes on and the fever does not remit, it is bad; for the disease becomes long, and it indicates excessive moisture.

30. Cold sweats that come on in a person with fever indicate that the fever will be long.

31. A copious, violent sweat coming on in a healthy person indicates disease—in summer, less so; when it is cold, more so.

32. If you allow a person's excretions to run together and collect, they separate out something like shreds of flesh: if these are few, the disease is mild; if they are many, it is serious. In such patients, it benefits to give an enema. In persons whose stools contain dark bile, if it is to a greater degree, their disease is greater; if to a lesser degree, their disease is milder.

33. If the vessels throb and the face becomes healthy-looking, and the hypochondria are raised up rather than soft, the disease is long, and does not leave off without a convulsion, or a copious nose-bleed, or a severe pain.

34. Throbbing in the upper limbs is also the sign of a long fever.

---

23 H. Diller in Preiser: τουτέῳ V.

24 Potter: ψυχρὸς V: ψύξεος Aldina.

[Ἤ]²⁵ κρίσεως ξυντόμου ἐπὶ τὸ κάκιον· καὶ ἐπὶ τούτων τὰ πλεῖστα ἅπερ ἐς θάνατον.

35. Τοῖσιν ἐλαχίστῳ χρόνῳ μέλλουσιν ἀπόλλυ-
σθαι μέγιστα σημεῖα ἀπ᾽ ἀρχῆς γίνεται· δυσπνόητοι
γάρ εἰσι καὶ οὐ κοιμέονται τὰς νύκτας καὶ τὰ σημεῖα
προφαίνουσιν ἐπικίνδυνα.

36. Ξυνεχοῦς πυρετοῦ ἐὰν τεταρταῖος πονῆται καὶ
ἑβδομαῖος, καὶ μὴ κριθῇ ἑνδεκαταῖος, ὀλέθριος ὡς τὰ
πολλά.

37. Ὅσοι ὑπὸ τετάνου ἁλίσκονται, ἐν ταῖς τέσσαρ-
σιν ἡμέραις ἀπόλλυνται· ἢν δὲ ταύτας ἀποφύγωσιν,
ὑγιέες γίνονται.

38. Ἐν τοῖσι καύσοισιν, ἐὰν ἐπιγένηται ἴκτερος καὶ
λὺγξ πεμπταίῳ ἐόντι, θανατῶδες. |

39. [Ὑποστροφῆς γενομένης·]²⁶ ὑποστροφαὶ λαμβάνονται οἷς ἂν ἀπυρέτοισι γενομένοισιν ἀγρυπνίαι
ἐρρωμέναι προσγίνωνται, ἢ ὕπνοι ταραχώδεες, ἢ ἀρ-
ρωστίη²⁷ τοῦ σώματος, ἢ ἀλγήματα ἑνὸς ἑκάστου τῶν
μελέων, καὶ ὅσοις ἂν οἱ πυρετοὶ παύσωνται, μήτε
σημείων γενομένων λυτηρίων μήτ᾽ ἐν ἡμέρῃσι κρισί-
μῃσι· καὶ ἐάν, ἐκλελοιπότος τοῦ πυρετοῦ καὶ ἱδρῶτος
ἐπιγενομένου, πυρρὸν οὖρον οὐρήσῃ, λευκὴν ὑπόστασιν ἔχον, προσδέχου τούτοις ὑποστροφὴν πυρετοῦ
αὐθημερόν· αὗται δὲ αἱ ὑποστροφαὶ πεμπταῖαι κρίνονται ἀκίνδυνοι. καὶ ἤν, κρίσιος ἐγγενομένης, οὖρον
ἐρυθρὸν οὐρήσῃ ὑπόστασιν ἔχον ἐρυθρήν, καὶ τούτοις
ὑποσροφὴ γίνεται τοῦ πυρετοῦ αὐθημερόν, καὶ ὀλίγοι
ἐκ ταύτης σώζονται. ὅταν ὑποστρέφῃ ὁ καῦσος, τὰ

Of an early crisis for the worse: most signs in these are the same as those which indicate death.

35. In persons who are going to die in the least time the most significant signs are present from the start; for they have difficulty breathing, they cannot sleep nights, and they exhibit the most dangerous signs.

36. If in a continuous fever a patient suffers distress on the fourth and the seventh days, and he has no crisis on the eleventh, he generally dies.

37. Those befallen by tetanus die in four days; if they survive for that many, they recover.

38. In ardent fevers, if jaundice and hiccup come on on a person's fifth day, it is a fatal sign.

39. [A relapse occurring:] Relapses occur in patients who, after having become afebrile, suffer from severe insomnias or disturbed sleep or weakness in their body or pains in each individual limb, or in those whose fevers remit without the appearance of the signs that indicate delivery, and on days other than critical days. And if, when the fever has remitted and a sweat has supervened, the patient passes reddish-yellow urine with a white sediment, expect in such cases a relapse of the fever on the same day: these relapses have their crisis on the fifth day and are without danger. And if, when the crisis occurs, the person passes red urine with a red sediment, in these too a relapse of the fever on the same day occurs, and few escape it. When ar-

---

25 Del. Preiser.

26 Del. later manuscripts.

27 Linden after Foes' translation *corporis robur solvitur*: ῥώμη V.

πολλὰ καὶ ἐξιδροῖ, καὶ [ἢν] τόσας ἡμέρας λαμβάνει[28]
ὑποστρέψας ὅσας τὸ πρῶτον· ὑποτροπιάζει δὲ καὶ τρὶς
<ὁ>[29] πυρετός, ἢν μὴ περισσῇ ἡμέρῃ ἀφῇ ὑποτρο-
πιάσας. τὰ πολλὰ ἐάν, ἀπέπτων ἐόντων τῶν οὔρων,
καὶ τῶν ἄλλων σημείων μὴ κατὰ λόγον ἐόντων, ἡ
νοῦσος κρισίμη ἡμέρῃ ὑποτροπιάζει· ἔσται δὲ καὶ
ὑποστροφὴ[30] ἐν κρισίμῃ ἡμέρῃ, τούτων καταλειπο-
μένων τοιούτων.

40. Τὰ παρ' οὖς οἷς ἀμφὶ κρίσιν γενόμενα μὴ
ἐκπυήσῃ, τούτων ἀπαλλασσομένων, ὑποστροφὴ γίνεται κατὰ λόγον τῶν ὑποστροφῶν, ὁμοίᾳ περιόδῳ· ἐπὶ
τούτοις ἐλπὶς ἐς ἄρθρα ἀφίστασθαι, ἢ οὖρον παχύ,
οἷον τὸ λευκὸν ἐπὶ τοῖς κοπιώδεσι τεταρταίοισι, ῥύεται
τῆς ἀποστάσιος· ἐνίοις δὲ τούτων καὶ αἱμορραγίαι
γίνονται ἐκ τῶν ῥινέων, | καὶ πάνυ ταχὺ λύεται· καὶ
τούτοις πῦα ἀποχωρέουσιν ὑγιάζεται ἡ νοῦσος.[31]

41. Τοῖς μελαγχολικοῖς μετὰ φρενιτικῶν ἐχομένοις
αἱμορροΐδες ἐγγινόμεναι ἀγαθόν.

42. Ὅσοι μαίνονται ἢ αὐτόματοι <ἢ>[32] ἀπαλλασ-
σόμενοι ἐκ τῶν νούσων, τούτοις τὴν μανίην ὀδύνη ἐς
τοὺς πόδας εἰσελθοῦσα ἢ ἐς στῆθος, ἢ βὴξ ἰσχυρὴ
γενομένη λύει· ἐὰν δὲ τούτων μηδὲν γένηται, λυομένης
τῆς μανίης, στέρησις τοῦ ὀφθαλμοῦ γίνεται.

43. Ὁκόσοι τῇ γλώσσῃ παφλάζουσι τῶν χειλέων

290 (margin)

28 Potter: ἢν μέσας ἡμέρας λάβῃ V.
29 Linden.
30 H. Diller in Preiser: ὑποστρέφει V.

dent fever relapses, in most instances a sweat breaks out, and the ardent fever relapses for the same number of days as it lasted the first time. Fever can also recur for a third time, if in its relapse it did not go away on an uneven day. Generally, if the urines are unconcocted and the other signs are not in order, the disease relapses on a critical day. There will also be a relapse on a critical day if such signs continue.

40. Persons in whom at the time of their crisis there are swellings beside the ears which do not produce pus, after recovering, suffer a relapse according to the general pattern of relapses, and in a similar period. In such cases it is to be expected that there will be apostases to the joints, or thick urine (like the white urine on the fourth day in cases of weariness) that releases them from the apostasis. In some of these there are also haemorrhages from the nostrils, and complete recovery is swift. If these patients excrete pus with their stools, the disease gets better.

41. In persons suffering from melancholic conditions with phrenitis the occurrence of haemorrhoids is a good sign.

42. In persons who develop mania either spontaneously or with delivery from their diseases, a pain to the lower limbs or the chest, or the occurrence of a violent cough brings release from the mania. If none of these things occurs, but the mania is still resolved, there is loss of the sight.

43. Persons who stammer with their tongue and lose

---

31 καί πάνυ—νοῦσος Littré: ἤ τις τεταρταίοις οὐ λυτικὴ, καὶ τοῖς πύα ἀποχωρέοντα ὑγίαζειν νούσοις V.

32 Littré.

μὴ κρατέοντες, ἐὰν ταῦτα παύσηται, ἔμπυοι γίνονται,
ἢ ὀδύνη ἰσχυρὴ ἐν τοῖς κάτω χωρίοις λύει, ἢ κυφότης,
ἢ αἷμα πολὺ ἐκ τῶν ῥινῶν ῥυέν, ἢ μανίη.

44. Τοῦ μεγάλου νοσήματος ἐν ἔθει γενομένου
λύσις, [ὅσοις ἐν τοῖσι καύσοισιν][33] ἰσχίων ὀδύνη, ἢ
ὀφθαλμῶν διαστροφή, ἢ τύφλωσις, ἢ ὀρχίων οἴδή-
σεις, ἢ τιτθῶν ἄρσις.

45. Καῦσον λύει αἵματος ἐκ ῥινῶν ῥύσις.

46. Ἐν καύσῳ ἐὰν ἐπιλάβῃ ῥῖγος, φιλέει ἐξιδροῦν.

47. ‹Ὑπὸ› καύσου ἐχομένῳ,[34] ῥίγεος ἐπιγενομένου,
λύσις.

48. Ὅσοις ἐν τοῖσι καύσοισι τρόμοι ἐγγίνονται,
παρακοπὴ λύει.

49. Ὅσοις ἂν ἐν τοῖς πυρετοῖς τὰ ὦτα κωφωθῇ,
τούτοισι μὴ λυθέντος τοῦ πυρετοῦ μανῆναι ἀνάγκη·
λύει δὲ ἐκ τῶν ῥινῶν αἷμα ῥυέν, ἢ κοιλίη ἐκταρα-
χθεῖσα χολώδεα, ἢ δυσεντερίη ἐπιγενομένη, ἢ ὀδύνη
ἰσχίων ἢ γονάτων.

50. Ὅσοισι πυρετοῖσι ῥῖγος ἐπιγίνεται, ὁ πυρετὸς
λύεται. |

292   51. Ὅσοισιν ὀδύναι γίνονται ἐξαπίνης, τὸ ὑποχόνδριον ἐπῆρται ἄνω· καὶ ἐὰν περὶ τὴν νόθην πλευρὴν
‹καὶ› περὶ σκέλεα[35] αἱ ὀδύναι γίνονται, τούτοισι λύσις φλεβοτομίη καὶ κάθαρσις κάτω· οὐ γὰρ λαμβάνει
πυρετὸς ἰσχυρὸς ἀδυνατούντων τῶν χωρίων.

[33] Del. Littré as a repetition from ch.48 below.
[34] Froben, presumably after Aphorism 4, 58: Καύσῳ ἐπι-
γενομένῳ ἢ V.

command over their lips—if this ceases, they suppurate internally, or a powerful pain in the lower regions releases them, or becoming hunchbacked, or much blood flowing out of the nostrils, or mania.

44. Resolution of the serious disease (i.e. epilepsy) when it has become habitual [in persons with ardent fevers]: a pain in the hips, or strabismus of the eyes, or blindness, or swelling of the testicles, or enlargement of the breasts.

45. Ardent fever is relieved by a flux of blood from the nostrils.

46. If rigor comes on in ardent fever, the person is likely to break out in a sweat.

47. In a person suffering from ardent fever the occurrence of a rigor brings resolution.

48. In persons with ardent fevers who develop tremors, a delirium brings resolution.

49. Any persons with fevers whose hearing becomes indistinct must of necessity, unless the fever remits, develop mania. A flux of blood from the nostrils relieves them, or the cavity being stirred up by bilious stools, or the onset of dysentery, or pain in the hips or knees.

50. Whenever a rigor comes on in fevers, the fever is resolved.

51. In persons in whom pains suddenly set in, the hypochondrium swells up. And if the pains are located in the false ribs and the legs, resolution lies in phlebotomy and a downward cleaning of the cavity; for a strong fever does not persist if those parts lack strength.

---

35 Littré after Foes' note 48 ἢ περὶ σκέλεα: περισκελέες V.

52. Ὑπὸ ὕδρωπος ἐχομένῳ, κατὰ τὰς φλέβας ἐς τὴν κύστιν ἢ κοιλίην ὑδατώδους ῥυέντος, λύσις.[36]

53. Ἢν ὑπὸ λευκοῦ φλέγματος ἐχομένῳ διάρροια ἐπιγίνηται ἰσχυρή, λύσις.

54. Ὑπὸ διαρροίης ἐχομένῳ ἰσχυρῆς ἔμετος ἐπιγενόμενος ἀπὸ τοῦ αὐτομάτου λύσις.

55. Ὅσοι ὑπὸ διαρροίης πολὺν χρόνον λαμβάνονται ξὺν βηχί, οὐκ ἀπαλλάσσονται, ἐὰν μὴ ὀδύναι ἰσχυραὶ ἐν τοῖς ποσὶν ἐμπέσωσι· καὶ φιλέει[37] διαστροφὴ γίνεσθαι φύσιος, ἐπειδὰν μὴ διάρροια ἴῃ, ἢ κενὴ διαχώρησις πρὸς πᾶσαν λάβῃ· ἐπιγίνονται γὰρ φῦσαι ἔσωθεν[38] οὖσαι· δῆλον τοίνυν οὐκ ἔχουσιν οὐδὲν ὑγρόν, ὥστε προσφέρειν, εἰ δεῖ, σῖτα[39] ἀσφαλῶς τῷ οὕτως ἔχοντι.

56. Ἰλεοῦ ἐπιγενομένου οἶνον ψυχρὸν δίδου πίνειν πολὺν ἄκρατον κατ᾽ ὀλίγον, ἕως ὕπνος, ἢ σκελέων ὀδύνη γίνεται· λύει δὲ καὶ πυρετὸς ἢ δυσεντερίη.

57. Κεφαλὴν περιωδυνοῦντι καὶ νοσέοντι, πύου ῥέοντος ἢ κατὰ τὰ ὦτα ἢ κατὰ τὰς ῥῖνας, λύεται τὸ νόσημα.

58. Ὁκόσοισιν ὑγιαίνουσιν ἐξαπίνης ὀδύναι ἐγγίνονται ἐν ταῖς | κεφαλαῖς, καὶ παραχρῆμα ἄφωνοι γίνονται, καὶ ῥέγκουσιν, ἀπόλλυνται ἐν ἑπτὰ ἡμέραις, ἐὰν μὴ πυρετὸς ἐπιλάβῃ.

59. Κεφαλὴν περιωδυνέοντι, ὅ τι ἂν τῶν ἄνω χωρίων πονήσῃ, σικύην πρόσβαλλε· λύει ὀδύνη ἐς ἰσχία

---

[36] Later manuscripts: ῥύσις V.

52. In a person suffering from dropsy, the flux of watery material through the vessels into the bladder and the cavity brings resolution.

53. If violent diarrhoea supervenes in a person suffering from white phlegm, it brings resolution.

54. In a person suffering from violent diarrhoea, if vomiting spontaneously occurs it brings resolution.

55. Persons who suffer from diarrhoea for a long time and have a cough do not recover unless severe pains invade their lower limbs. Also there is likely to be a perversion of nature when the diarrhoea ceases or an empty straining occurs all day, for winds arise from within. Now it is clear that these patients lack moisture, so that to administer breads to a person in this state, if necessary, is safe.

56. If ileus occurs give much cold wine unmixed with water to drink, a little at a time, until sleep or a pain in the leg comes on. Fever or dysentery also bring resolution.

57. In a person who is ill and suffering great pain in his head, a flux of pus through either his ears or his nostrils resolves the disease.

58. Healthy persons who are suddenly befallen by pains in their heads, and who immediately lose their speech and breathe stertorously die in seven days, unless a fever supervenes.

59. In a person suffering great pain in his head, apply a blood-letting cup to any part in his upper regions where

---

37 Potter: εἰ βάλεται V: ἢ βούλεται Littré.

38 Littré: ἔξωθεν V.

39 Littré: εἰδήσει τὰ V.

καὶ γούνατα καὶ ἆσθμα, ὅ τι ἂν τούτων γίνηται.

60. Ὀφθαλμιῶντι ὑπὸ διαρροίης ἁλῶναι ἀγαθόν.

61. Ὑπὸ σπασμοῦ ἢ τετάνου ἐχομένῳ πυρετὸς ἐπιγενόμενος λύει τὸ νόσημα.

62. Ὑπὸ πυρετοῦ ἐχομένῳ σπασμὸς ἢν λάβῃ, παύεται ὁ πυρετὸς αὐθημερὸν ἢ τῇ ὑστεραίῃ ἢ τῇ τρίτῃ.

63. Ὁπόταν ξυντεταμένος τὰς χεῖρας καὶ τοὺς πόδας, μανίην ἐμποιέουσιν.

64. Ἢν αἱ φλέβες σφύζωσιν αἱ ἐν ταῖς χερσὶ καὶ τὸ πρόσωπον ἐρρωμένον ᾖ, καὶ τὰ ὑποχόνδρια μὴ μαλακά, ἀλλὰ ἐπηρμένα⁴⁰ ᾖ, χρονίη ἡ νοῦσος· ἄνευ τοῦ σπασμοῦ. . . .

---

40 Froben: ἠρημένα V.

the pain is present. Pain moving to the hips and knees brings a resolution, as does difficult breathing, of any of these that occurs.

60. For a person with ophthalmia to be seized by diarrhoea is a good sign.

61. In a person with a convulsion or tetanus for fever to supervene resolves the condition.

62. If a convulsion occurs in a person with a fever, the fever ceases on the same day, or the next day, or the third day.

63. When a person suffers spasms in his arms and legs, these provoke mania.

64. If the vessels in the upper limb throb and the face becomes healthy-looking, and the hypochondria are not soft, but raised up, the disease is long; without spasm . . . .

ΠΕΡΙ ΚΡΙΣΙΜΩΝ

1. Μέγα μέρος ἡγέομαι τῆς τέχνης εἶναι τὸ δύνασθαι κατασκοπέεσθαι περὶ τῶν γεγραμμένων ὀρθῶς· ὁ γὰρ γνοὺς καὶ χρεόμενος τούτοισιν οὐκ ἄν μοι δοκέοι μεγάλα σφάλλεσθαι κατὰ τὴν τέχνην. δεῖ δὴ καταμανθάνειν τὴν κατάστασιν τῶν ὡρῶν ἀκριβῶς καὶ τῶν νούσων· ἑκάστη ὅ τι [τὸ νόσημα][1] ἀγαθόν, ὅ τι καὶ κινδυνῶδες, ἢ ἐν τῇ καταστάσει, ἢ ἐν τῇ νούσῳ· μακρὸν ὅ τι νόσημα καὶ θανάσιμον, μακρὸν ὅ τι περιεστικόν· ὀξύ ὅ τι θανάσιμον, ὀξύ ὅ τι περιεστικόν. τάξιν τῶν κρισίμων ἐκ τούτων σκοπέεσθαι, καὶ τὸ προλέγειν ἐκ τούτων εὐπορέεται· ἔτι δὲ ἀπὸ τούτων ἔστιν οὕς, ὅτε καὶ ὡς δεῖ διαιτῆν.

2. Μέγιστον τοίνυν σημεῖον τοῖς μέλλουσι τῶν καμνόντων βιώσασθαι, ἐὰν μὴ παρὰ φύσιν ᾖ ὁ καῦσος· καὶ τἄλλα δὲ νοσήματα ὡσαύτως· οὐδὲν γὰρ δεινὸν τῶν κατὰ φύσιν γίνεται, οὐδὲ θανατῶδες. δεύτερον δέ, ἐὰν μὴ αὐτή γε ᾖ[2] ὥρη τῷ νοσήματι ξυμμαχήσῃ· ὡς γὰρ ἐπὶ τὸ πολὺ οὐ νικᾷ ἡ τοῦ ἀνθρώπου φύσις τὴν τοῦ ὅλου δύναμιν. ἔπειτα δέ, ἢν τὰ περὶ τὸ πρόσωπον ἰσχναίνηται, καὶ αἱ φλέβες αἱ ἐν τῇσι χερσὶ καὶ ἐν τοῖσι κανθοῖσι καὶ ἐπὶ τῇσιν ὀφρύῃσιν

---

[1] Del. Littré.   [2] αὐτή γε ἡ Littré: αὐτῇ τῇ Μ.

CRITICAL DAYS

1. A large part, I believe, of the medical art consists in being able to examine correctly its writings; for a person who knows and makes use of what has been written would not, in my opinion, be likely to go astray in the art. Indeed, you must understand precisely the constitutions of the seasons and diseases: on each (sc. day) what good factor there is, and what dangerous one, either in the constitution or in the disease; what condition is long and deadly, what long and tending to recovery; what condition is acute and deadly, what acute and tending to recovery. The order of the critical days is to be examined starting from these principles, and medical prediction as well derives from them; in which patients, when, and in what manner you must treat by regimen also follow from them.

2. Now the most important sign of patients that they will recover is if their ardent fever is not against nature; and in fact the same is true in other diseases as well: for nothing untoward occurs when they are according to nature, nor anything that leads to death. A second (sc. favourable) sign is if the season itself is not allied with the disease, for generally a person's nature does not overpower the force of the environment. And then, if swelling around the patient's face goes down and if the vessels in his arms, the corners of his eyes, and his eyebrows settle down, when

ἡσυχίην ἔχωσι, πρότερον μὴ ἡσυχάζουσαι. τοῦτο δέ,
ἢν ἡ φωνὴ [ᾖ]³ ἀσθενεστέρη καὶ λειοτέρη γίνηται, καὶ
τὸ πνεῦμα μανότερον καὶ λεπτότερον, ἐς τὴν ἐπιοῦσαν
ἡμέρην ἄνεσις τῆς νούσου. ταῦτα οὖν χρὴ σκοπεῖν
πρὸς τὰς κρίσιας, καὶ εἰ τὸ παρὰ δικροῦν τῆς γλώσσης ὥσπερ σιάλῳ λευκῷ ἐπαλείφεται· καὶ ἐν ἄκρῃ τῇ
γλώσσῃ ταὐτὸ τοῦτο γεγένηται, ἧσσον δέ· εἰ μὲν οὖν
σμικρὰ ταῦτα εἴη, ἐς τὴν τρίτην ἄνεσις τῆς νούσου·
ἢν δ' ἔτι παχύτερον, αὔριον· ἢν δ' ἔτι παχύτερον,
αὐθημερόν. τοῦτο δέ, [ὁκόταν]⁴ τῶν ὀφθαλμῶν τὰ
300 λευκὰ ἐν ἀρχῇ μὲν τῆς | νούσου ἀνάγκη μελαίνεσθαι,
ἐὰν ἰσχύῃ ἡ νοῦσος· ταῦτα οὖν καθαρὰ γινόμενα
τελείην ὑγείην δηλοῖ· ἀτρέμα μὲν βραδύτερον, σφόδρα δὲ γινόμενον, θᾶσσον.

3. Τὰ δ' ὀξέα γίνεται τῶν νοσημάτων ἀπὸ χολῆς,
ὁκόταν ἐπὶ τὸ ἧπαρ ἐπιρρυῇ, καὶ ἐς τὴν κεφαλὴν
καταστῇ. τάδε οὖν πάσχει· τὸ ἧπαρ οἰδέει καὶ ἀνα-
πτύσσεται πρὸς τὰς φρένας ὑπὸ τοῦ οἰδήματος, καὶ
εὐθὺς ἐς τὴν κεφαλὴν ὀδύνη ἐμπίπτει, μάλιστα δὲ ἐς
τοὺς κροτάφους· καὶ τοῖσί τε ὠσὶν οὐκ ὀξὺ ἀκούει,
πολλάκις δὲ καὶ τοῖσιν ὀφθαλμοῖσιν οὐχ ὁρῇ· καὶ
φρίκη καὶ πυρετὸς ἐπιλαμβάνει. ταῦτα μὲν κατ' ἀρχὰς
τοῦ νοσήματος αὐτῷ γίνεται διαλιμπάνοντα, τοτὲ μὲν
σφόδρα, τοτὲ δὲ ἧσσον· ὁκόσῳ δ' ἂν ὁ χρόνος τῆς
νούσου προΐῃ, ὅ τε πόνος πλείων ἐν τῷ σώματι. καὶ αἱ
κόραι σκίδνανται τῶν ὀφθαλμῶν, καὶ σκιαυγεῖ, καὶ ἢν
πορσφέρῃς τὸν δάκτυλον πρὸς τοὺς ὀφθαλμούς, οὐκ
αἰσθήσεται διὰ τὸ μὴ ὁρῆν. τούτῳ δ' ἂν γνοίης ὅτι

before they were not at peace. Another sign: if a patient's voice becomes weaker and softer, and his breath rarer and feebler, then remission of his disease will occur on the following day. Now the following things you must observe at the crises, and if the double root of the tongue is covered with a material resembling white saliva (this same also ocurs at the tip of the tongue, but to a lesser degree): now if the material is small in amount, the remission of the disease will take place on the third day; but if it is thicker, recovery will be on the next day, and if thicker yet, then on the same day. Another sign: at the beginning of a disease the whites of the eyes must necessarily become darker, if the disease is severe. If they become clean again, this indicates a complete recovery: if this happens gradually, the recovery is slower, if all at once, then quicker.

3. Acute diseases arise from bile, when bile collects in the liver, and also settles in the head. The patient suffers the following: his liver swells up and, by its swelling, expands against the diaphragm; pain immediately attacks the head, especially the temples; he does not hear clearly with his ears, and often he cannot see, either; shivering and fever set in. These things affect the patient at the beginning of the disease; they occur intermittently, sometimes more intensely, sometimes less so. The longer the disease goes on, the more pain there is in the body. The pupils of the eyes are dilated, the patient sees dimly, and if you bring your finger up to his eyes, he does not perceive it, because he cannot see: this is how you can tell that he does not

---

3 Del. Preiser.
4 Del. Littré after the Latin translation of *Sevens*.

οὐχ ὁρῇ, οὐ γὰρ σκαρδαμύσσει προσφερομένου τοῦ
δακτύλου. καὶ τὰς κροκύδας ἀφαιρέει ἀπὸ τῶν ἱμα-
τίων, ἤν περ ἴδῃ, δοκέων φθεῖρας εἶναι. καὶ ὁκόταν τὸ
ἧπαρ μᾶλλον ἀναπτυχθῇ πρὸς τὰς φρένας, παρα-
φρονέει· καὶ προφαίνεσθαί οἱ δοκέει πρὸ τῶν ὀφθαλμῶν ἑρπετὰ καὶ ἄλλα παντοδαπὰ θηρία, καὶ ὁπλίτας
μαχομένους, καὶ αὐτὸς αὐτοῖς δοκέει μάχεσθαι· καὶ
τοιαῦτα λέγει ὡς ὁρέων, καὶ ἐξέρχεται, καὶ ἀπειλεῖ, ἢν
μή τις αὐτὸν ἐῴη διεξιέναι· καὶ ἢν ἀναστῇ, οὐ δύναται
αἴρειν τὰ σκέλεα, ἀλλὰ πίπτει. οἱ δὲ πόδες αἰεὶ ψυχροὶ
γίνονται· καὶ ὁκόταν καθεύδῃ, ἀναΐσσει ἐκ τοῦ ὕπνου,
καὶ ἐνύπνια ὁρῇ φοβερά. τῷδε δὲ γινώσκομεν ὅτι ἀπὸ
ἐνυπνίων ἀναΐσσει καὶ φοβέεται· ὅταν ἔννοος γένηται,
ἀφηγεῖται τὰ ἐνύπνια τοιαῦτα ὁκοῖα καὶ τῷ σώματι
ἐποίεέ τε καὶ τῇ γλώσσῃ ἔλεγε. ταῦτα μὲν οὖν ὧδε
πάσχει. ἔστι δ᾽ ὅτε καὶ ἄφωνος γίνεται ὅλην τὴν
ἡμέρην καὶ τὴν νύκτα, ἀναπνέων πολὺ ἀθρόον πνεῦμα. ὅταν δὲ παύσηται παραφρονέων, εὐθὺς ἔννοος
γίνεται, καὶ ἢν ἐρωτᾷ τις αὐτόν, ὀρθῶς ἀποκρίνεται,
καὶ γινώσκει πάντα τὰ λεγόμενα· εἶτ᾽ αὖθις ὀλίγῳ
302 ὕστερον ἐν τοῖσιν αὐτοῖσιν ἄλγεσι κεῖται. αὕτη ἡ
νοῦσος προσπίπτει μάλιστα ἐν ἀποδημίῃ, καὶ ἤν πη
ἐρήμην ὁδὸν βαδίσῃ· λαμβάνει δὲ καὶ ἄλλως.

4. Τέτανοι δύο ἢ τρεῖς· ἢν μὲν ἐπὶ τρώματι γένηται,
πάσχει τάδε. αἱ γνάθοι πήγνυνται ὥσπερ ξύλα, καὶ τὸ
στόμα διοίγειν οὐ δύνανται, καὶ οἱ ὀφθαλμοὶ δακρύ-
ουσι θαμινὰ καὶ ἕλκονται, καὶ τὸ μετάφρενον πέπηγε,
καὶ τὰ σκέλεα οὐ δύνανται ξυγκάμπτειν, οὐδὲ τὰς

see: he does not blink when the finger is brought near. He removes pieces of wool from his blanket, if he sees them, believing they are lice. When his liver expands even more against the diaphragm, the patient becomes deranged; there seem to appear before his eyes reptiles and every other sort of beasts, and fighting soldiers, and he imagines himself to be fighting among them; he speaks out as if he is seeing such things, and he attacks and threatens, if someone will not allow him to go outside; if he does stand up, though, he cannot lift his legs, but falls. His feet are perpetually cold; when he goes to bed, he starts up out of his sleep on seeing fearful dreams. We know that his starting up and fear are due to dreams, from the following: when he comes to his senses, he reports having had dreams that correspond to the way he moved his body and spoke with his tongue. These things he suffers as described. Some times, he may also become speechless the whole day and night, taking frequent deep breaths. When his derangement ceases, he immediately regains his senses, and, if someone questions him, he answers correctly and understands everything that is said. But then, a little later, he labours again under the same distress. This disease usually attacks abroad, if a person is travelling a lonely road somewhere, although it does also occur under other circumstances.

4. Two or three tetanuses: if tetanus follows a wound, the patient suffers the following: his jaws are fixed as if they were wood, and he is unable to open his mouth; his eyes shed tears and look awry; his back is rigid, he cannot bend his legs, nor his arms and spine. When he is near death,

χεῖρας καὶ τὴν ῥάχιν· ὁκόταν δὲ θανατώδης ᾖ, τὸ
ποτὸν καὶ τὰ βρώματα, ἃ πρότερον ἐβεβρώκεεν, ἀνὰ
τὰς ῥῖνας ἔρχεται ἐνίοτε.

5. Ὁ δὲ ὀπισθότονος τὰ μὲν ἄλλα πάσχει διὰ
πλῆθος τὰ αὐτά, γίνεται δὲ ὁκόταν τοὺς ἐν τῷ αὐχένι
τένοντας τοὺς ὄπισθεν νοσήσῃ· νοσέει δὲ ἢ ἀπὸ συν-
άγχης, ἢ ὑπὸ σταφυλῆς, ἢ τῶν ἀμφιβραγχίων ἐμπύων γινομένων· ἐνίοισι δὲ καὶ ἀπὸ τῆς κεφαλῆς
πυρετῶν ἐπιγενομένων σπασμὸς ἐπιγίνεται· ἤδη δὲ
καὶ ὑπὸ τρωμάτων. οὗτος ἕλκεται ἐς τοὖπισθεν, καὶ
ὑπὸ τῆς ὀδύνης τὸ μετάφρενον πέπηγε καὶ τὰ στήθεα,
⟨καὶ⟩[5] οἰμώζει. οὗτος σπᾶται σφόδρα, ὥστε μόλις
κατέχουσιν οἱ παρεόντες, μὴ ἐκ τῆς κλίνης ἐκπίπτειν.

6. Ὁ δὲ τέτανος ἧσσον θανατώδης τῶν πρόσθεν.
γίνεται δὲ ἀπὸ τῶν αὐτῶν, καὶ σπᾶται πᾶν τὸ σῶμα
ὁμοίως.

7. Καῦσος δὲ τοῖσι προειρημένοισιν οὐχ ὁμοίως
γίνεται· φύσει γὰρ αὐτῇ ἠνάγκασται πῦρ ἔσεσθαι.[6]
δίψα μὲν οὖν πολλὴ ἔχει τὸν ἄνθρωπον καὶ πυρετὸς
σφοδρός. γλῶσσα δὲ ῥήγνυται τρηχυνομένη, καὶ ξηρὴ γίνεται, καὶ τὸ χρῶμα αὐτῆς τὸν μὲν πρῶτον
χρόνον ὠχρόν ἐστι, | οἷόν περ εἴωθε· προϊόντος δὲ τοῦ
χρόνου μελαίνεται, καὶ ἢν μὲν ἐν ἀρχῆσι μελαίνοιτο,
θάσσους αἱ κρίσιες εἰσίν· ἢν δ' ὕστερον, χρονιώτεραι.

8. Ἰσχιάδες δὲ ἀπὸ τῶνδε μάλιστα γίνονται τοῖσι
πολλοῖσιν, ἢν εἰληθῇ[7] ἐν ἡλίῳ πολὺν χρόνον, καὶ τὰ
ἰσχία διαθερμανθῇ, καὶ τὸ ὑγρὸν ἀναξηρανθῇ τὸ
ἐνεὸν ὑπὸ τοῦ καύματος ἐν τοῖσιν ἄρθροισιν. ὡς δ'

sometimes both the drink and the food that he has taken earlier come up through his nostrils.

5. The patient with opisthotonus suffers, on the whole, the same, but the disease arises when he is affected in the posterior tendons of the neck; his illness arises from angina, from staphylitis, or from a suppuration occuring in the parts about the tonsils; also, in some cases, such a convulsion originates from the head, when there are fevers; occasionally it also follows wounds. This patient is drawn backwards, and cries aloud from the pain in his back and chest; he is drawn so forcefully that the attendants can hardly prevent him from falling out of bed.

6. The following tetanus is less often mortal than the preceding ones. It arises from the same things, and the convulsion involves the whole body in a similar way.

7. Ardent fever does not arise in the same way as the conditions described above, since there is a necessity by its very nature for fever to arise. Great thirst befalls the person, as well as extreme fever. The tongue develops fissures and is rough, and it becomes dry; at first, it stays its normal colour, yellowish, but with time it turns dark. Now if the tongue turns dark at the beginning, the crises occur sooner, if later, the crises take longer.

8. The sciaticas generally arise from the following, in the majority of cases: if a person is exposed to the sun for a long time, and his hip-joints become heated, and the moisture present is dried up by the burning heat in the joints.

---

5 Linden.

6 αὐτῇ . . . πῦρ ἔσεσθαι Potter: ἅπαξ . . . πυρίασασθαι M.

7 Littré: ἔλθῃ M.

ἀναξηραίνεται καὶ πήγνυται, τόδε μέγα τεκμήριον· ὁ
γὰρ νοσέων ἐνστρέφεσθαι καὶ κινέειν τὰ ἄρθρα οὐ
δύναται ὑπὸ τῆς ἀλγηδόνος τῶν ἄρθρων καὶ τοῦ
συμπεπηγέναι τοὺς σπονδύλους. ἀλγέει δὲ μᾶλλον
τὴν ὀσφῦν καὶ τοὺς σπονδύλους τοὺς ἐκ πλαγίων τῶν
ἰσχίων καὶ τὰ γούνατα· ἵσταται δὲ ἡ ὀδύνη πλεῖστον
χρόνον ἐν τοῖσι βουβῶσιν, ἀλλὰ καὶ ἐν τοῖσιν ἰσχίοισιν, ὀξείη καὶ καυματώδης· καὶ ἤν τις αὐτὸν ἀνιστῇ
ἢ μετακινέῃ, οἰμώζει ὑπὸ τῆς ἀλγηδόνος, ὅσον δὴ
μέγιστον δύνηται· ἐνίοτε δὲ καὶ σπασμὸς ἐπιγίνεται
καὶ ῥῖγος καὶ πυρετός. γίνεται δὲ ἀπὸ χολῆς· γίνεται
δὲ καὶ ἀπὸ αἵματος· καὶ ὀδύναι παραπλήσιοι ἀπὸ
πάντων τῶν νοσημάτων, καὶ ῥῖγος καὶ πυρετὸς ἐνίοτε
ἐπιλαμβάνει βληχρός· ἀλλὰ χρὴ ὧδε μελετῆν.

9. Ἴκτερος δέ ἐστιν ὀξύς τε καὶ ταχέως ἀποκτείνων·
ἡ χροιὴ δὲ ὅλη σιδιοειδὴς σφόδρα ἐστί, χλωροτέρη ἢ
οἱ σαῦροι οἱ χλωρότεροι· παρόμοιος δὲ καὶ ὠχρός· καὶ
ἐν τῷ οὔρῳ ὑφίσταται οἷον ὀρόβιον πυρρόν, καὶ πυρετὸς καὶ φρίκη βληχρὴ ἔχει· ἐνίοτε δὲ καὶ τὸ ἱμάτιον
οὐκ ἀνέχεται ἔχων, ἀλλὰ δάκνεται καὶ ξύεται· τὰ
ἑωθινὰ ἄσιτος ἐών, τὰ ἔνδοθεν ἔπειτα μύζει τὰ
σπλάγχνα ὡς ἐπὶ τὸ πολύ. καὶ ὁκόταν ἀνιστῇ τις
αὐτὸν ἢ προσδιαλέγηται, οὐκ ἀνέχεται. οὗτος ὡς ἐπὶ
τὸ πολὺ θνήσκει ἐντὸς τεσσαρεσκαίδεκα ἡμερέων·
ταύτας δὲ διαφυγὼν ὑγιής.

10. Ἡ δὲ περιπνευμονίη τοιάδε ποιέει· πυρετός τε
306   ἰσχυρὸς ἴσχει, | καὶ πνεῦμα πυκνὸν καὶ θερμὸν ἀνα-
πνέει· καὶ ἀπορίη καὶ ἀδυναμίη καὶ ῥιπτασμὸς ἔχει,

The surest proof that the moisture is dried up and congealed is this: the patient cannot adduct and move his joints, because of the pain in them, and because the vertebrae have become fixed. He has pain especially in the loins, in the vertebrae that grow out of the oblique part of the hip-bone, and in the knees. Pain persists longest in the groins, but also in the hip-joints, and it is sharp and burning; if someone stands the patient up, or shifts him, he cries out at the top of his voice from the pain. Also, sometimes a convulsion supervenes, or chills and fever. This arises from bile, but also from blood; the pains from all of these diseases are similar; sometimes mild chills and fever are present. You must treat as follows.

9. Jaundice is both acute and rapidly fatal. The whole skin is very much the colour of pomegranate-peel, greener than quite green lizards, and similarly yellowish, too. In the urine a reddish sediment like vetch-meal precipitates; fever and mild shivering are present. Sometimes the patient will not even tolerate having his blanket on, but it scratches and irritates him. In the morning, before he has eaten, his inward parts then usually suffer tearing pains, and when anyone wakes him up or talks to him, he will not tolerate it. The patient generally dies within fourteen days; if he survives that many, he recovers.

10. In pneumonia the following happens: there is violent fever, the patient's breathing is rapid and hot; he is distraught, weak and restless, and beneath the shoulder-

καὶ ὀδύναι ὑπὸ τὴν ὠμοπλάτην καὶ ἐς τὴν κληῖδα καὶ
ἐς τὸν τιτθόν, καὶ βάρος ἐν τοῖσι στήθεσι, καὶ παραφροσύνη. ἔστι δ᾽ ὅτε ἀνώδυνός ἐστιν, ἕως ἂν ἄρξηται
βήσσειν, πολυχρονιωτέρη δ᾽ ἐκείνης καὶ χαλεπωτέρη.

Τὸ δὲ σίαλον λευκὸν καὶ ἀφρῶδες πτύει τὸ πρῶτον.
ἡ δὲ γλῶσσα ξανθή, προϊόντος δὲ τοῦ χρόνου μελαί-
νεται· ἢν μὲν οὖν ἐν ἀρχῇ μελαίνοιτο, θάσσους αἱ
ἀπαλλαγαί· ἢν δ᾽ ὕστερον, σχολαίτεραι· τελευτῶσι δὲ
καὶ ῥήγνυται ἡ γλῶσσα· κἢν προσθῇς τὸν δάκτυλον,
ἔχεται· τὴν δ᾽ ἀπαλλαγὴν τῆς νούσου σημαίνει ἡ
γλῶσσα, ἅπερ καὶ ἐν τῇ πλευρίτιδι ὁμοίως.

Ταῦτα δὲ πάσχει ἡμέρας τεσσαρεσκαίδεκα τὸ ἐλάχιστον, τὸ πλεῖστον δὲ εἴκοσι καὶ μίαν, καὶ βήσσει
τοῦτον τὸν χρόνον σφόδρα, καὶ καθαίρεται ἅμα τῇ
βηχὶ τὸ μὲν πρῶτον πολὺ καὶ ἀφρῶδες σίαλον, ἑβδόμῃ δὲ καὶ ὀγδόῃ, ὅταν ὁ πυρετὸς ἀκμάζῃ καὶ ὑγρὰ ᾖ ἡ
περιπλευμονίη, καὶ παχύτερον· εἰ δὲ μή, οὔ. ἐνάτῃ δὲ
καὶ δεκάτῃ ὑπόχλωρον καὶ ὕφαιμον· δωδεκάτῃ καὶ
μέχρι τεσσαρεσκαίδεκα πολὺ καὶ πυῶδες.

Ὧν ὑγραί εἰσιν αἱ φύσιες καὶ διαθέσιες τοῦ σώματος, καὶ ἡ νοῦσος ἰσχυρή· ὧν δὲ ἥ τε φύσις καὶ ἡ
στάσις τῆς νούσου ξηρή, ἧσσον οὗτοι.

11. Περὶ δὲ κρισίμων ἡμερῶν ἤδη μέν μοι καὶ
πρόσθεν λέλεκται· κρίνονται δὲ οἱ πυρετοὶ τεταρταῖοι,
ἑβδομαῖοι, ἑνδεκαταῖοι, τεσσαρεσκαιδεκαταῖοι, ἑπτα-
καιδεκαταῖοι, εἰκοστῇ καὶ πρώτῃ· ἐκ δὲ τούτων τῶν
ὀξέων τριακοσταῖοι, εἶτα τεσσαρακοσταῖοι, εἶτα ἑξ-
ηκοσταῖοι· ὅταν δὲ τούτους τοὺς ἀριθμοὺς ὑπερβάλλῃ,
χρόνια ἤδη γίνεται ἡ κατάστασις τῶν πυρετῶν.

blade he suffers pain that radiates toward the collar-bone and the nipple; he has a heaviness in his chest, and he is deranged. In some patients, there is no pain until they begin to cough; this pneumonia lasts longer than the one with pain from the beginning, and it is serverer.

The patient first expectorates white frothy sputum, and his tongue is yellow; as time passes, the tongue becomes dark. Now if it should become dark at the beginning, recovery is more rapid, but if it becomes dark later, recovery is slower; in the end, the tongue also develops fissures, and if you touch it with your finger, it is held fast. The tongue gives an indication of recovery in this disease just as in pleurisy.

The patient suffers these things for at least fourteen days, at most twenty-one; he coughs hard during this time, clearing with his cough first copious frothy sputum, and then on the seventh or eighth day, when the fever reaches its high point and the pneumonia is moist, a thicker sputum; not, however, if the pneumonia is not moist. On the ninth and tenth days, the sputum is somewhat yellow-green and charged with blood; on the twelfth to the fourteenth days, it is copious and purulent.

In patients whose natures and bodily propensities are moist, the disease is severe; in those whose nature and state of disease is dry, less so.

11. About critical days, I have already spoken before. Fevers have their crises on the fourth day, the seventh, the eleventh, the fourteenth, the seventeenth, and the twenty-first; and subsequent to these acute diseases, on the thirtieth, the fortieth, and then the sixtieth. When it goes beyond these numbers, the state of the fevers is already chronic.

SUPERFETATION

INTRODUCTION

This work, mentioned in no extant ancient writing, is the sole possible source of four terms in Galen's *Glossary*, suggesting that it was part of the Hippocratic Collection by the second century A.D.[1] Later, citations from *Superfetation* are to be found in Rhazes' *Continens* and the tenth century Spanish writer 'Arīb ibn Sa'īd's book on obstetrics.[2]

Superfetation is a loosely organized compilation of practical obstetrical knowledge, which takes its title from the subject matter of its first chapter; other topics handled include premature birth, still-birth, conception, abortion, and various therapeutic measures. Much of the material included in *Superfetation* also appears in other Hippocratic texts, such as *Diseases of Women I*, *Nature of Woman*, and most notably *Diseases of Women III* (= *Barrenness*), with which the following verbatim correspondences exist:

> 16: *Barrenness* 215
> 20: *Barrenness* 214
> 25: *Barrenness* 219

[1] Galen vol. 19, 80 ἀναχαίνεται; 19, 86 ἀρτίζωα; 19, 96 ἐκμιαίνεται; 19, 139 σκορδινᾶσθαι.

[2] M. Ullmann, *Die Medizin im Islam*, Leiden, 1970, pp. 31, 130, 139.

315

A detailed study of these passages has led Lienau[3] to conclude that *Diseases of Women III* is the source of the shared texts, and *Superfetation* the borrower.

The transmission of the treatise is exceptional among Hippocratic texts for a number of reasons. First, four short fragments of the Greek text of *Superfetation* are preserved along with others from *Diseases of Women I* in the sixth century Antinoopolis Papyrus 184.[4] Also, an anonymous Arabic translation of the entire text is contained in the thirteenth century manuscript Aya Sofya 3632.[5] These two early sources, which derive from the Greek tradition at a date anterior to the extant Greek manuscripts and their archetype (a), are each occasionally useful as evidence for the text. More significant, however, is the following: two parts of the twelfth century Greek manuscript Vaticanus Graecus 276 (Va, Vb) written by different scribes each contain a variant version of the treatise (fol. 119r-122v and

[3] On the relationships among these texts see Lienau pp. 37–42, 45–8.

[4] J. W. B. Barns and H. Zilliacus, *The Antinoopolis Papyri III*, London, 1967, pp. 130f.: Fragment 1(a) = *Superfetation* ch. 32; 2(a) = ch. 33–34; 2(b) = ch. 38–39; 3(a) = ch. 40–41.

[5] J. N. Mattock, *Kitāb Buqrāṭ fī Ḥabl 'alā ḥabl*, Cambridge, 1968. This translation, which the editor characterizes as "difficult, obscure, unfaithful to its original and written in poor and clumsy Arabic" (p. i), is of only limited value as evidence for the Greek original.

184v-187v) copied from a different branch of the tradition, so that we possess in fact three complete independent Greek witnesses to the text; albeit now contained in two manuscripts:

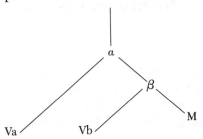

Superfetation is contained in all the collected Hippocratic editions and translations, and in 1963 was the subject of a Kiel dissertation subsequently published as:

Cay Lienau, *Hippokrates Über Nachempfängnis, Geburtshilfe und Schwangerschaftsleiden*, CMG I 2, 2, Berlin, 1973 (=Lienau).

It is upon this work that the present edition for the most part depends.

ΠΕΡΙ ΕΠΙΚΤΗΣΕΩΣ

1. Ὁκόταν ἐπικυΐσκηται γυνή, ἢν μὲν ἐν τῷ μέσῳ τῆς μήτρης τὸ πρῶτον ἔχῃ παιδίον καὶ τὸ ἐπικύημα, ⟨τὸ ἐπικύημα⟩[1] προεκπίπτει ὑπὸ τοῦ ἀρχαίου ἐξωθεύμενον· ἢν δ' ἐν τῷ κέρατι τῷ ἑτέρῳ τὸ ἐπικύημα ἔχῃ, ἐπιτίκτει ὕστερον οὐ γόνιμον, ἐπὴν χαλάσῃ ἡ μήτρη καὶ ὑγρανθῇ, τοῦ γονίμου ἀπολυθεῖσα. ἢν δὲ μὴ αὐτίκα ἀποχωρέῃ τὸ ἐπικύημα, ὀδύνας τε ἔχει καὶ ῥεῦμα δυσῶδες καὶ πυρετόν, καὶ οἰδεῖ τὸ πρόσωπον καὶ τὰς κνήμας καὶ τοὺς πόδας, καὶ † ἢν διάληται †[2] ἀσιτέει ἕως ἂν ἐκπέσῃ. ἐπικυΐσκονται δὲ τῶν γυναικῶν, ὧν ἂν ὁ στόμαχος μετὰ τὸ πρῶτον κύημα μὴ ξυμμεμύκῃ κάρτα ἐπιφαινομένων τῶν σημείων, ξυμμεμύκῃ δὲ μετά. τὰ δ' ἐπικυήματα τὰ ἐπιτικτόμενα ὕστερον, ἢν μήπω διάκρισιν ἔχῃ, ἀλλὰ σὰρξ ᾖ, οἰδέει μὲν οὔ, ἀλλὰ κατασήπεται μέχρι ἂν ἐξέλθῃ τῆς μήτρης.

2. Ἧιτινι ἂν ἐκδύῃ τὸ παιδίον τοῦ χορίου ἐν τῇσι

---

[1] Lienau.

[2] Several unsuccessful attempts have been made to solve this crux: e.g. si periat Cornarius; τὴν νηδὺν καὶ Littré; διαλύεται καὶ Ermerins.

SUPERFETATION

1. When a woman conceives superfetally, if she holds both fetuses in the central part of her uterus, the added fetus passes out first, being expelled by the original one. But if she holds the second fetus in one of the two horns (sc. of her uterus), she gives birth to it later, although still in an immature state, when her uterus has been slackened and moistened by the birth of the mature fetus. If the added fetus does not come out immediately (sc. after the first one), the woman has pain, an evil-smelling flux, and fever, she swells up in the face, calves and feet, and . . . she goes without food until the fetus passes out.[1] Superfetal conception occurs in women when the mouth of their uterus does not close after the first conception—so that significant (sc. menstrual) indications appear—but only later. Added fetuses born at a later date, if they have not yet reached the stage of differentiation but consist only of flesh, do not become oedematous, instead undergoing decomposition until they pass out of the uterus.

2. A woman, in whom the fetus has its membrane torn

[1] Cf. the case of superfetation in *Epidemics* V 11 (Loeb *Hippocrates* vol. 7, 160f.): "She swelled up greatly in the face and calves and feet and one thigh. She had no interest in food, but was very thirsty."

μήτρῃσι, πρὶν ἔξω ἄρξασθαι χωρέειν τὸ παιδίον,
δυστοκέει μᾶλλον καὶ ἐπικινδύνως, ἢν μὴ ἡ κεφαλὴ
ἡγέηται. ᾗτινι δὲ ἂν χωρέῃ ἔξω ξὺν τῷ χορίῳ τὸ
παιδίον καὶ προχωρῆσαν ἐς τὸν στόμαχον ἐκδύῃ τοῦ
χορίου ῥαγέντος, εὐτοκέει μᾶλλον· καὶ τὸ μὲν παιδίον
θύρηφι χωρέει, τὸ δὲ χόριον ἴσχεται καὶ ἀνασπᾷ
ἑωυτὸ καὶ μένει αὐτοῦ. |

478
3. Ὁκόταν ⟨μὴ⟩³ γόνιμον γένηται παιδίον, τούτου ἡ
σὰρξ ὑπερέχει τῶν ὀνύχων, οἱ δὲ ὄνυχες ἐλλείπουσι
τῶν χειρῶν καὶ τῶν ποδῶν.

4. Ὁκόταν γονίμου γενομένου τοῦ παιδίου χεὶρ
ὑπερέχῃ, πρῶτον ἀνωθέειν ὀπίσω μέχρι ἂν ἀπώσῃς·
καὶ ἢν ἀμφότεραι, ἀμφοτέρας ἀνῶσαι· καὶ ἢν τὸ
σκέλος ὑπερέχῃ, ἀνωθεῖν καὶ τοῦτο· ὁκόταν δὲ ἀμφότερα τὰ σκέλεα προφανέντα μείνῃ καὶ μηδετέρωσε
προχωρέῃ, πυριήματι χρῆσθαι, ἀπ' ὅτευ ὑγροτάτη
ἔσται ἡ ὑστέρη· ὀδμὴν δὲ ἐχέτω τὸ πυρίημα. καὶ
ὁκόταν ἡ μὲν κεφαλὴ προφανῇ, τὸ δ' ἄλλο σῶμα ἐνῇ,
καὶ ὧδε πυριῆν· καὶ ὁκόταν τὸ μὲν ἐν τῇσι μήτρῃσι
τοῦ σώματος ᾖ, τὸ δὲ ἐν τοῖσιν αἰδοίοισι, τὸ δὲ ἔξω
τῶν αἰδοίων οἰδέῃ καὶ μένῃ, καὶ οὕτως πυριῆν. καὶ ἢν
μὲν ἀπὸ τῆς πυριήσιος ἀποχωρήσῃ· εἰ δὲ μή, περι-
αλείψας τὸ στόμα τῆς μήτρης ἐλατηρίῳ παχεῖ, διεὶς
ὕδατι, ὅκως ὠδῖνα ἐμποιήσῃ, καὶ ἀπὸ τῶν ποτῶν καὶ
ἐδεστῶν ὠδῖνα ἐμποιέειν· καὶ τὰ αἰδοῖα αὐτὰ ἀλείφειν
κηρωτῇ, ἤν σοι δοκέῃ ξηροτέρη εἶναι τοῦ καιροῦ.

³ Lienau after Cornarius' *non vitalis*.

off in the uterus before it begins to move outward, generally has a difficult and dangerous birth, unless its head leads the way. A woman, in whom the fetus moves outward with its membrane present and only has it torn off when it reaches the mouth of the uterus, tends to have an easy birth: the fetus moves forward towards the exit, but the membrane is held back, retracts on itself, and remains in place.

3. When a child is born prematurely, flesh covers its nails, and the nails of its hands and feet are rudimentary.

4. When at the birth of a fetus that has reached maturity one arm comes out, first push it back up until you succeed in replacing it; if both arms come out, push them both back; if one leg comes out, push it up, too. When both legs come into view and remain there, but there is no progress on either side, employ a vapour-bath by which the uterus will be moistened: let the vapour-bath be fragrant. When the fetus' head comes into view, but the rest of its body is still inside, use a vapour-bath in the same way. When one part of the body is in the uterus, another part is at the external genitals, and a third part which is already outside the genitals swells and remains fixed, use a vapour-bath in the same way, too. Now if from the vapour-bath the fetus moves out, fine. But if not, anoint the mouth of the uterus with thickened squirting-cucumber juice and soak it with water in order to induce birth pangs, and also employ drinks and foods to the same end. The external genitals themselves you should anoint with a salve, if you think they are drier than they should be.

5. Ὁκόταν παιδίου τῆς κεφαλῆς προφανείσης ἐκ τοῦ στομάχου μηκέτι ἐθέλῃ τὸ ἄλλο σῶμα προχωρεῖν, τὸ δὲ παιδίον τεθνήκῃ, τοὺς δακτύλους βρέξας ὕδατι, μεταξὺ τοῦ στομάχου καὶ τῆς κεφαλῆς παραβύσας τὸν δάκτυλον, ἐν κύκλῳ περιάγειν· εἶθ᾽ ὑποβαλὼν τὸν δάκτυλον ὑπὸ τὸ γένειον, διαβύσας ἐς τὸ στόμα, ἔξω ἕλκειν.

6. Ὁκόταν τὸ μὲν ἄλλο σῶμα θύρηφι ᾖ τῶν αἰδοίων, ἡ δὲ κεφαλὴ εἴσω, ἐπὶ πόδας φερομένου τοῦ ἐμβρύου, ἐπὴν περιαγάγῃς τὸν δάκτυλον ἐν κύκλῳ, ἀμφοτέρας τὰς χεῖρας παρεὶς μεταξὺ τοῦ στομάχου καὶ τῆς κεφαλῆς, βρέξας ὕδατι, ἐξελκύσαι. ἢν δὲ τοῦ 480 στομάχου | ἔξω ᾖ, τῶν δὲ αἰδοίων εἴσω, παρεὶς τὰς χεῖρας, περιλαβὼν τὴν κεφαλήν, ἐξέλκειν.

7. Ἢν δὲ τὸ ἔμβρυον ἔνδον μένῃ τετελευτηκὸς καὶ μὴ δύνηται μήτε αὐτόματον μήτε διὰ φαρμάκων ἐκπεσεῖν κατὰ φύσιν, χρίσας τὴν χεῖρα κηρωτῇ, ἥτις ὀλισθητικὴ μάλιστα, ἔπειτα ἐνείρας ἐς τὴν μήτρην, διελεῖν τοὺς ὤμους ἀπὸ τοῦ τραχήλου, ἐπερείσαντα τῷ μεγάλῳ δακτύλῳ· ἔχειν δὲ χρὴ πρὸς τὰ τοιαῦτα καὶ ὄνυχα ἐπὶ τοῦ μεγάλου δακτύλου· διελόντα δὲ ἐξενεγκεῖν τὰς χεῖρας· ἔπειτα πάλιν ἐνείραντα τὴν κοιλίην ἀνασχίσαι καὶ ἀνασχίσαντα ἡσυχῇ ἐξελεῖν τὰ ἐντοσθίδια, ἔπειτα ἐξελόντα ξυντρῖψαι τὰ πλευρία, ὅκως ξυμπεσὸν τὸ σωμάτιον εὐσταλέστερον γένηται καὶ ῥᾷον ἐξίῃ, μὴ ὀγκῶδες ἐόν.

5. When, with the fetus' head already in view outside the mouth of the uterus, the rest of its body is not yet willing to move forward, and the fetus is dead, moisten your fingers with water, insert one finger between the mouth of the uterus and the fetus' head, and move it around in a circle. Then work your finger under the fetus' chin, force it into its mouth, and pull the fetus out.

6. When the rest of a fetus' body is at the opening of the mother's external genitals, but the head is still inside—the embryo is presenting feet first—after you have moved your finger around it in a circle, wet your hands with water and pass the two of them between the mouth of the uterus and the fetus' head, and pull the fetus out. If the head is out of the mouth of the uterus, but still inside the external genitals, pass your hands in, grasp the head, and pull the fetus out.

7. If a fetus remains inside after it has died, and cannot be brought out in the natural way either spontaneously or with the help of medicines, anoint your hand with a cerate that is especially slippery, then insert it into the uterus, and separate the shoulders from the neck by pressing with your thumb—in order to achieve this you must have a "claw"[2] on your thumb; separate and remove the arms. Next reinsert your hand, cut open the abdomen (sc. of the fetus), and after carefully doing this extract the inner parts; then, when you have extracted these, crush the ribs in order that the little body, in collapsing, will become more manageable and pass out more easily, since it has no bulk.

[2] A surgical instrument consisting of a scalpel blade mounted on a ring. See J. S. Milne, *Surgical Instuments in Greek and Roman Times*, Oxford, 1907, p. 31 and plate VII 1.

8. Τὸ δὲ χόριον, ἢν μὴ ῥηϊδίως ἐκπίπτῃ, μάλιστα μὲν ἐᾶν πρὸς τὸ ἔμβρυον προσκρέμασθαι, καὶ τὴν λεχὼ προσκαθῆσθαι ὥσπερ ἐπὶ λασάνου· ἔστω δὲ κατεσκευασμένον ὑψηλόν τι, ἵνα τὸ ἔμβρυον ἐκκρεμάμενον ξυνεπισπᾶται τῷ βάρει ἔξω· ἡσυχῇ δὲ τοῦτο ποιέειν καὶ μὴ πρὸς βίην, ὅκως μὴ ἀποσπασθέν τι παρὰ φύσιν φλεγμονὴν ἐμποιήσῃ. ὑποκεῖσθαι οὖν δεῖ τῷ ἐμβρύῳ εἴρια ὡς ὀγκωδέστατα, νεόξαντα, ἵνα κατὰ μικρὸν ἐνδιδῷ, καὶ[4] ἀσκία δύο ἐζευγμένα ὕδατος μεστά· ἐπάνω δὲ τῶν ἀσκίων εἴρια, ἐπάνω δὲ τῶν εἰρίων τὸ ἔμβρυον· ἔπειτα τρυπῆσαι ἑκάτερον τῶν ἀσκίων ῥαφίῳ, ὅκως ῥυῇ κατὰ σμικρὸν τὸ ὕδωρ· ἐκρέοντος δὲ τοῦ ὕδατος ἐνδιδόασιν οἱ ἀσκοί· ἐνδιδόντων δὲ ἐπισπᾶται τὸ ἔμβρυον τὸν ὀμφαλόν, ὁ δὲ ὀμφαλὸς τὸ χόριον.

482 Ἢν δὲ μὴ δύνηται καθῆσθαι ἐπὶ τοῦ λασάνου, | ἐπ' ἀνακλίτου δίφρου τετρυπημένου καθήσθω. ἢν δὲ ἀσθενέῃ τὸ ξύνολον καθῆσθαι, τὴν κλίνην ὡς ὀρθοτάτην ἀπὸ τῶν πρὸς τῆς κεφαλῆς μερῶν ἀείραντας, ἵνα κάτω ῥέπῃ ὡς μάλιστα καὶ ξυνεπισπᾶται κάτω τὸ βάρος, ὑπὸ δὲ τὰς μασχάλας δῆσαι τὴν λεχὼ ἔξωθεν τῶν ἱματίων πρὸς τῇ κλίνῃ τῇ ταινίῃ ἢ ἱμάντι πλατεῖ καὶ μαλθακῷ, ὅκως μὴ ὀρθῆς ἐούσης τῆς κλίνης κάτω τὸ σῶμα φέρηται. τὸν αὐτὸν δὲ τρόπον, κἂν ἀπορραγῇ ὁ ὀμφαλὸς κἂν ἀποτάμῃ τις πρὸ τοῦ καιροῦ, βάρεα ξύμμετρα ἐκκρεμαννύντα τὴν ἐξαγωγὴν τοῦ χορίου

___

4 Potter: ἢ codd.

8. The placenta, if it does not come out easily, you had best leave attached to the child, and have the parturient sit on something like a night-stool: let some high construction be devised so that the suspended child, by its weight, pulls the placenta out. Do this gently and without violence in order to prevent anything from being unnaturally stretched and thereby provoking an inflammation. Now under the child should be placed some very bulky masses of newly-carded wool, in order that they will give way a little at a time, and two linked skin bags filled with water: on top of the bags, the wool, and on top of the wool, the child. Then puncture each of the bags with a small needle, so that the water slowly runs out. As this is taking place, the bags gradually collapse, and, as they do, the child will pull on the umbilical cord and the umbilical cord will pull on the placenta.[3]

If the woman is unable to sit on a night stool, she should sit on a reclining chair that has an opening below. If she is too weak to sit at all, raise the head end of her bed very high in order that she is as inclined as much as possible, and the weight (sc. of the child) pulls downward; bind the parturient to her bed with straps or a soft, flat cloth under her axillae on the outside over her clothes, so that even though the bed is upright her body will not be pulled down. If the umbilical cord is torn or someone has removed it from the child too soon, attach weights sufficient to make the placenta come out, in the same manner, for

[3] Grimm comments (vol. 4, 632): "Diese Art, die Nachgeburt zu holen ist zwar sehr sinnreich; aber wohl nie versucht worden."

ποιέεσθαι· βελτίστη γὰρ αὕτη θεραπείη τῶν τοιῶνδε
γίνεται καὶ ἥκιστα βλάπτει.

9. Ἢν δέ τινι ἐναποθάνῃ τὸ παιδίον ἐν τῇ μήτρῃ
καὶ μὴ ἐξέλθῃ ὑγρῆς ἐούσης τῆς μήτρης καὶ [μὴ]⁵
ἐχούσης ἔτι τὸ ὑποκείμενον, ἀλλὰ ξηρανθῇ ἡ μήτρη,
πρῶτον μὲν οἰδίσκεται αὐτό· ἔπειτα σήπεται καὶ τήκε-
ται τὰς σάρκας καὶ ῥέει θύρηφι· ὕστατα δὲ τὰ ὀστέα
χωρέει· καὶ ῥόος ἐνίοτε λαμβάνει, ἢν μὴ πρότερον
ἀποθάνῃ.

10. Ὁκόταν τὸ παιδίον ἐντεθνήκῃ, καὶ τοῖσιν ἄλ-
λοισι ξυντεκμαίρεσθαι σημείοισι, καὶ κελεύειν τοτὲ
μὲν ἐπὶ δεξιὰ κεῖσθαι τοτὲ δὲ ἐπ᾿ ἀριστερὰ μετα-
βάλλειν· μεταπίπτει γὰρ ἐν τῇ μήτρῃ τὸ παιδίον
ὁποτέρωθι ἂν καὶ ἡ γυνή, ὥσπερ λίθος ἢ ἄλλο τι, ἢν
τεθνήκῃ, καὶ τὸ ἦτρον ψυχρὸν ἔχει· ἢν δὲ ζώῃ, καὶ τὸ
ἦτρον θερμὸν ἔχει καὶ ἡ μὲν γαστὴρ ὅλη μεταπίπτει
τεταμένη μετὰ τοῦ ἄλλου σώματος, ἐν αὐτῇ δὲ οὐδὲν
μεταπίπτει χωρὶς τοῦ ἄλλου σώματος.

11. Ἥτινι ἂν ὠδινούσῃ πρὸ τοῦ παιδίου ῥόος
484 αἱματώδης | γένηται πολλὸς ἀνώδυνος, κίνδυνος
τεθνηκὸς ἀπολυθῆναι ἢ οὐ βιώσιμον.

12. Κυουσέων τῶν γυναικῶν ὁ στόμαχος τῶν πλείστων ἐγγὺς τυγχάνει ἐὼν πρὸ τοῦ τόκου.

13. Κυέουσα γυνή, ἢν μὴ λαγνεύηται, ῥηΐτερον
ἀπολυθήσεται τοῦ τόκου.

⁵ Del. Lienau after Littré's translation.

this therapy is the best one in such cases, and causes the least harm.

9. If a fetus dies in a woman's uterus and does not pass out as long as the uterus is moist and still retains its (sc. other) contents, but the uterus dries up, then the dead fetus first swells up with fluid; then it decomposes and dissolves its tissues, and these flow out; finally the bones pass out. Sometimes a flux also befalls such a women, if she has not already died.

10. When the fetus has died inside a woman, this is confirmed both by other signs and by the following test: order her first to lie on her right side and then to move on to her left side: if the fetus is dead, it will fall in the uterus whichever direction the woman turns, just as a stone or any other object would fall; coldness also occupies her lower abdomen. But if the fetus is alive, then her lower abdomen is warm, and the whole belly is held tight and moves the same way the rest of the body does, and nothing in it falls in any direction on its own apart from the rest of the body.

11. If, in a woman suffering birth pangs, there is a copious, painless, bloody flux before the fetus is expelled, there is a danger that she will give birth to a dead child or to one that is not likely to survive.

12. In pregnant women the mouth of the uterus is in most instances near (sc. the exterior) just before they give birth.

13. A pregnant woman will give birth more easily if she does not have intercourse.

14. Ἡ τὰ δίδυμα κυέουσα [οὐ]⁶ τίκτει τῇ αὐτῇ ὡς καὶ ξυνέλαβεν· ἔχει δὲ ἑνὶ χορίῳ τὰ παιδία ἀμφότερα.

15. Γυναικὶ δυστοκεούσῃ, ἢν τὸ παιδίον ἐν τῇσι γονῇσιν ἐνέχηται καὶ μὴ εὐπόρως ἐξέλθῃ, ἀλλὰ ξὺν πόνῳ καὶ μηχανῇσιν ἰητροῦ, ταῦτα τὰ παιδία ἀρτί-ζωα. τούτων οὐ χρὴ τὸν ὀμφαλὸν ἀποτάμνειν πρὶν οὐρήσῃ ἢ πτάρῃ ἢ φωνήσῃ, ἀλλὰ ἐᾶν, προσχωρῆσαι δὲ τὴν γυναῖκα ὡς ἐγγυτάτω τοῦ παιδίου, καὶ ἢν διψῇ, πινέτω μελίκρητον. καὶ ἢν ὁ ὀμφαλὸς ἐμφυσῆται ὥσπερ στόμαχος, [καὶ]⁷ κινηθήσεται ἢ πταρεῖ τε τὸ παιδίον καὶ φωνὴν ῥήξει, καὶ τότε ἀποτάμνειν ἀνα-πνέοντος τοῦ παιδίου· ἢν δὲ μὴ φυσῆται ὁ ὀμφαλὸς μηδὲ κινῆται χρόνου ἐγγενομένου, οὐ βιώσεται.

16. Κυέουσαν γυναῖκα, ἢν μὴ ἄλλως γινώσκῃς, οἱ ὀφθαλμοὶ εἱλκυσμένοι καὶ κοιλότεροι γίνονται, καὶ τὰ λευκὰ τῶν ὀφθαλμῶν οὐκ ἔχει τὴν φύσιν τῆς λευκό-τητος, ἀλλὰ πελιδνότερα.

17. Ἦν τις ἐπίτοκος ἐοῦσα, κοίλη δὲ τοὺς ὀφθαλ-μούς, καὶ ὑποιδέῃ τὸ πρόσωπον καὶ ὅλη αὐτή, καὶ τοὺς πόδας οἰδέῃ, καὶ ὥσπερ ὑπὸ φλέγματος λευκοῦ ἐχομένη φαίνεται, καὶ τὰ ὦτα ⟨ἔχῃ⟩⁸ λευκὰ | καὶ τὴν ῥῖνα ἄκρην λευκὴν καὶ τὰ χείλεα πελιδνά, αὗται ἢ τεθνηκότα τίκτουσιν ἢ ζῶντα πονηρὰ καὶ οὐ βιώσιμα καὶ ἄναιμα ὥσπερ νοσηλὰ ἐόντα,⁹ ἢ προέτεκον οὐ γόνιμα. ταύτῃσι τὸ αἷμα ἐξυδατώθη. δεῖ οὖν αὐτῇσι

486

⁶ Del. Lienau after Calvus.
⁷ Del. Littré after Calvus.

14. A woman pregnant with twins gives birth to them both on the same day, just as she conceived them on the same day: both fetuses are contained in one membrane.

15. In a woman that is having difficulty giving birth, if the fetus is held back in the parts of generation and does not pass out readily, but only with great effort and the application of medical contrivances, such children are not likely to survive. You should not remove their umbilical cord before they pass urine or sneeze or give voice, instead leaving it connected; also have the woman move very near to the child, and if she is thirsty give her melicrat to drink. If the umbilical cord puffs up with air like a pouch, the child will move or sneeze and give voice; then you should remove the cord while the child is taking a breath. If, after a certain time, the umbilical cord does not puff up with air nor the child move, it will not survive.

16. That a woman is pregnant, if you do not recognize it otherwise: her eyes are compressed and become more hollow than usual, and their whites do not have the natural whiteness, but are more livid.

17. If a woman approaching childbirth has a hollowness around her eyes, swells up in her face and her whole body, has oedema in her feet, appears to be suffering from white phlegm, becomes white in her ears and at the point of her nose, and livid in her lips, such women give birth to children that are dead or that live only with difficulty and are unlikely to survive, and that lack blood to a morbid degree; or they have given birth before to children that were incapable of life. In such a woman the blood has become wa-

---

8 H. Diller in Lienau.

9 *ν. ἐ.* Lienau: *νοσηλέοντα* M Vb: *νόθα ὄντα* Va.

μετὰ τὸν τόκον τὰ εὐώδεα προστιθέναι, καὶ πίνειν τὰ
εὐώδεα, καὶ σιτίοισιν ἀνατρέφειν, καὶ πρῶτον τοῦ
προσώπου ἡ ῥὶς ἔνσημος γίνεται ἄκρη καὶ χρῶμα
λαμβάνει.

18. Ἤν τις κυΐσκομένη γῆν ἐπιθυμέῃ ἐσθίειν ἢ
ἄνθρακας καὶ ἐσθίῃ, ἐπὶ τῆς κεφαλῆς τοῦ παιδίου
φαίνεται, ὁκόταν γένηται, σημεῖον ἀπὸ τοιούτων.

19. Γυναικὶ χρὴ γινώσκειν τῶν μαζῶν ὁκότερος
μέζων αὐτῇ, κεῖθι γὰρ τὸ ἔμβρυον· ὁμοίως δὲ καὶ τῶν
ὀφθαλμῶν· ἔσται γὰρ μέζων καὶ λαμπρότερος τὸ πᾶν
εἴσω τοῦ βλεφάρου, ὁκοτέρωθι καὶ μαζὸς μέζων.

20. Ἧιτινι ἂν ἀπὸ προσθέτων μὴ λίην ἰσχυρῶν
ὀδύναι ἐς τὰ ἄρθρα ἀφίκωνται καὶ βρυγμὸς ἔχῃ καὶ
σκορδινέηται καὶ χασμᾶται, ἐλπὶς ταύτην κυῆσαι
μᾶλλον ἢ ἥτις ἂν τῶν τοιούτων μηδὲν πάσχῃ.

21. Γυνὴ ἥτις παχεῖα παρὰ φύσιν ἐγένετο καὶ
πίειρα καὶ φλέγμα|τος ἐπλήσθη, οὐ κυΐσκεται τούτου
τοῦ χρόνου· ἥτις δὲ φύσει τοιαύτη ἐστί, κυΐσκεται
τούτων ἕνεκεν, ἢν μή τι ἄλλο κωλύῃ αὐτήν.

22. Τῶν γυναικῶν τῇσι πλείστῃσιν ὁκόταν τὰ ἐπιμήνια μέλλῃ φαίνεσθαι, ὁ στόμαχος αὐτῶν ἑωυτὸν
ἀνέσπακε μᾶλλον ἢ ἄλλοτε.

23. Γυνὴ ἥτις ἀρικύμων ἐοῦσα πέπαυται κυΐσκομένη, φλεβοτομείσθω δὶς τοῦ ἐνιαυτοῦ ἀπὸ τῶν χειρῶν καὶ τῶν σκελέων.

24. Ἧιτινι ὀδύναι ἐν ἰσχίῳ ἢ ἐν κεφαλῇ ἢ ἐν χερσὶν
ἢ που ἄλλοθι τοῦ σώματος, ὅταν δὲ κύῃ ἐκλείπουσιν,
ὁκόταν δ' ἀπολυθῇ ἀπὸ τῶν μητρέων ἔνεισι, ξυμφέρει

tery. Thus, after they give birth you should administer fragrant suppositories, have them take fragrant drinks, and build them up with nourishing foods. The first facial sign (sc. that you are succeeding) is for the point of the nose to take on colour.

18. If a pregnant woman wishes to eat earth or coal, and she does so, a mark will appear on the head of the child at birth as a result.

19. You should observe in a woman which of her breasts is larger, for the fetus is on that side. In the same way, learn from her eyes, too: for the eye will be altogether larger and brighter within the eyelid on the same side that the breast is larger.

20. If a woman, after receiving moderately strong suppositories, suffers pains in her joints, chattering of her teeth, and she stretches and yawns, she is more likely to be pregnant than one who does not experience any of these signs.

21. A woman who has become stout in an unnatural way, has added fat, and has filled up with plegm, does not become pregnant at that time; but one who is naturally like this does become pregnant as a result of these things, unless some other factor prevents her.

22. In most women, at the time that the menses are about to appear the mouth of the uterus draws itself upward more than at other times.

23. Let any woman who was once prolific, but who has ceased becoming pregnant, be phlebotomized twice a year from the arms and legs.

24. If pains in the hip, the head, the arms, or any other part of the body disappear when a woman becomes pregnant, but then return after she has given birth from her

τὰ εὐώδεα καὶ πίνειν καὶ προστίθεσθαι πρὸς τὸ στόμα
τῆς μήτρης.

25. Ὅταν τινὰ ἰητρεύῃς γυναῖκα κυήσιος ἕνεκα,
ὁκόταν δοκέῃ κεκαθάρθαι καὶ τὸ στόμα καλῶς ἔχῃ
τῆς μήτρης, λουσάσθω καὶ σμησάσθω τὴν κεφαλὴν
καὶ μὴ ἀλειψάσθω μηδενί· ἔπειτα ὀθόνιον ἄνοδμον
περιθεῖσα περὶ αὐτὰς τὰς τρίχας πεπλυμένον, κεκρυ-
φάλῳ πεπλυμένῳ ἢ μηδενὸς ὄζοντι καταδησάσθω τὸ
ὀθόνιον, ἐπιθεῖσα πρῶτον· ἔπειτα ἀναπανέσθω προσ-
θεμένη χαλβάνην πρὸς τὸν στόμαχον ἑψήσασα καὶ
μαλθάξασα πρὸς πῦρ καὶ μὴ ἥλιον· ἔπειτα πρωὶ
ἀποδυσαμένη τὸν κεκρύφαλον μετὰ τοῦ ὀθονίου,
ὀσφρανθῆναι παρεχέτω τινὶ τὴν κορυφήν· καὶ ὀζέσει,
ἢν καλῶς ἔχῃ τῆς καθάρσιος· εἰ δὲ μή, οὐκ ὀζέσει·
ἄσιτος δὲ ταῦτα ποιείτω. καὶ ἢν μὴ τεκοῦσα ᾖ,
οὐδέποτε ὀζέσει, οὔτε καθαιρομένη οὔτ' ἄλλως· οὐδ'
ἢν κυεούσῃ προσθῇ<ς>,[10] οὐδ' οὕτως ὀζέσει. ἥτις δὲ |
490 κυΐσκεται θαμέως καὶ ἀρικύμων ἐστὶ καὶ ὑγιαίνει, ἢν
προσθῇς μηδὲ καθήρας, ὀζέσει αὐτῆς ἡ κορυφή, ἄλλο
δὲ οὐδέν.

26. Ὅταν δὲ δοκέῃ καλῶς ἔχειν πάντα καὶ δέῃ
παρὰ τὸν ἄνδρα ἰέναι, ἡ μὲν γυνὴ ἄσιτος ἔστω, ὁ δὲ
ἀνὴρ ἀθώρηκτος, ψυχρῷ λελουμένος καὶ εὐωχημένος
σῖτα ξύμφορα. καὶ ἢν γνῷ ξυλλαβοῦσα τὴν γονήν, μὴ
ἔλθῃ τοῦ πρώτου χρόνου πρὸς τὸν ἄνδρα, ἀλλ' ἡσυ-
χαζέτω. γνώσεται δέ, ἢν μὲν ὁ ἀνὴρ φῇ ἀφεικέναι, ἡ

---

[10] Lienau.

uterus, it benefits her to take fragrant drinks and to receive such a suppository against the mouth of her uterus.

25. When you are treating a woman to promote pregnancy, after she has been well cleaned and the mouth of her uterus is in a good state, have her bathe, wipe her head clean, but not anoint it with anything. Then she should place an odourless, newly washed, linen cloth over her hair, and bind it down, when it is once set in place, with an odourless, newly washed hairnet. Next, apply a suppository of all-heal juice, which has been warmed and softened by being exposed to a fire but not to the sun, against the mouth of her uterus, and have her retire for the night. Then, early next morning after unbinding the hairnet and the linen cloth, let someone smell the crown of her head: if the cleaning has occurred as it should, he will smell it (i.e. the all-heal juice), but if not, he will not smell it. (The woman should do these things in the fasting state.) If a woman is barren, she will never smell in this way, whether or not she is cleaned, nor indeed under any other circumstances; nor if you give this suppository to a woman that is pregnant will she smell in this way. But any woman that becomes pregnant often, is prolific, and is healthy, if given such a pessary even when she has not been cleaned beforehand, will smell from the crown of her head, not, however, otherwise.

26. When a woman seems to be the right way in everything and she is ready to approach her husband, let her be in the fasting state, and let him be sober, newly bathed in cold water, and well nourished on appropriate foods. If the woman knows that she has taken up the seed, let her not again approach her husband at first, but keep herself quiet; she will know this if her husband says he has ejaculated,

δὲ γυνὴ ἀγνοῇ ὑπὸ ξηρότητος. ἢν δὲ ἀποδῷ πάλιν ἡ
μήτρη τὴν γονὴν ἐς τὰ αἰδοῖα καὶ γένηται ὑγρή, αὖθις
μισγέσθω μέχρι ἂν ξυλλάβῃ.

27. Γυνὴ ἥτις κυΐσκεται μέν, διαφθείρει δὲ δίμηνα
τὰ παιδία ἀκριβῶς ἐς τὸν αὐτὸν χρόνον καὶ μήτε
πρότερον[11] μήτε ὕστερον, καὶ τοῦτο πάθῃ δὶς ἢ τρὶς
κατὰ ταὐτὰ καὶ πλεονάκις, ἤν τε τρίμηνα ἤν τε τε-
τράμηνα ἤν τε πλείονα χρόνον γεγονότα διαφθείρῃ
κατὰ τὸν αὐτὸν τρόπον τοῦτον, ταύτης αἱ μῆτραι οὐκ
ἐπιδιδόασιν ἐπὶ τὸ μέζον, τοῦ παιδίου αὐξανομένου
καὶ ὑπερβάλλοντος [τοῦ][12] ἐκ τοῦ διμήνου ἢ τριμήνου
ἢ ὁπηλίκου ἄν ποτε ᾖ· ἀλλὰ τὸ μὲν αὐξάνεται, αἱ δὲ
μῆτραι οὐκέτι εἰσὶν ἱκαναί, ἀλλὰ κατὰ τοῦτο δια-
φθείρεται ἐς τὸν αὐτὸν χρόνον. ταύτῃ χρὴ κλύσαι[13]
τὴν μήτρην[14] ὡς μάλιστα προσθέτοισι φαρμάκοισι
τοισίδε· τῆς σικύης τὴν ἐντεριώνην | κόψαντα δια-
σῆσαι· ἔπειτα ἐν μέλιτι ἑφθῷ μῖξαι ὀλίγον πλείονι τῷ
μέλιτι καὶ σιλφίου ὀλίγον· τὸ δὲ μέλι κάθεφθον ἔστω.
τοῦτο περιπλάσαι περὶ μήλην, τὸ πάχος ποιέοντα
ὁκόσον παραδέξεται ὁ στόμαχος· προστιθέναι δὲ καὶ
πρὸς τὸ στόμα τῆς μήτρης καὶ ὦσαι ὅκως ἂν περήσῃ
ἐς τὸ εἴσω τῆς μήτρης· ὅταν δ᾽ ἀποτακῇ τὸ φάρμακον,
ἐξελεῖν τὴν μήλην. καὶ τὸ ἐλατήριον ὧδε ποιέων προστιθέναι, καὶ τῆς κολοκυνθίδος τῆς ἀγρίης ὡσαύτως.
καὶ ἐσθιέτω τοῦτον τὸν χρόνον σκόροδα ὡς πλεῖστα

492

---

11 IR: πρῶτον codd.   12 Del. Littré.
13 Littré: κυῆσαι ποιῆσαι codd.

even if she does not conclude it herself from being dry.[4] If her uterus returns the seed to her external genitals, and these become wet, the woman should have intercourse again until she takes the seed up.

27. If a woman becomes pregnant, but her fetuses are always aborted at the same time after two months, neither earlier nor later, and she experiences this two or three or even many times in just the same way, or if the conceptus is aborted this same way after three or four months or even a longer time, then her uterus is not expanding enough, while the fetus increases and goes beyond it after these two or three months, or whenever it is. As the fetus increases, the uterus is no longer large enough to contain it, so that abortion then regularly occurs for this reason at the same time. In such a case you must wash out the uterus thoroughly by applying the following suppository medications. Pound the pulp of a bottle-gourd and pass it through a sieve; then mix a small amount of this into a larger amount of boiled honey—the honey should be thoroughly boiled—and also mix in a little silphium; smear this on a spatula, making its consistency such that the mouth of the uterus will admit it. Apply this to the mouth of the uterus, and press it into the interior; after the medication melts, withdraw the spatula. Also prepare squirting-cucumber juice in the same way and apply it as a suppository; also a suppository made from wild gourd in the same way. Have the patient eat a large amount of garlic at this time, a stalk

---

[4] I.e. if her uterus has taken the seed up, leaving her external genitals dry.

---

[14] Add. δεῖ οὖν ἀποποιῆσαι καὶ φυσῆσαι M Vb.

ΠΕΡΙ ΕΠΙΚΤΗΣΕΩΣ

καὶ καυλὸν σιλφίου καὶ ὅτι φῦσαν ἐμποιέει ἐν τῇ
κοιλίῃ. προστιθέσθω δὲ τὸ πρόσθετον διὰ τρίτης
ἡμέρης ἕως ἂν δοκέῃ καλῶς ἔχειν, καὶ πλῆθος ὅκως
ἂν προσίηται· τὰς δὲ μεταξὺ μαλθακτηρίοισι χρῆσθαι. ἐπειδὰν δὲ καταστῇ τοῖσι μαλθακτηρίοισι τὸ
στόμα τῆς μήτρης, μετὰ τὸ ἐπιφανῆναι ἐπιμείνασαν,
ὁκόταν δὲ ξηρὴ ᾖ, μίγνυσθαι.

28. Ἧιτινι ἂν ἡ μήτρη ἔμπυος γένηται ἢ μετὰ τὸν
τόκον ἢ ἐκ διαφθορῆς ἢ ἄλλως πως, καὶ μὴ ἐν τεύχει
ἑτέρῳ καὶ χιτῶνι τὸ πῦον ὥσπερ ἐπὶ φύματος ᾖ,
ξυμφέρει ταύτῃ μήλην ὑπαλειπτρίδα καθιέναι ἐς τὸ
στόμα τῆς μήτρης· ἧσσον γὰρ ⟨ἂν⟩ δέοι[15] καύσιος, εἰ
χωρήσειε πρὸς τὴν μήλην· ἔπειτα κάμπας τὰς ἀπὸ
τῆς τιθυμαλλίδος ξυλλέξαι, αἵπερ δὴ κέντρα ἔχουσιν·
ἔπειτα δὲ ταύτας ἀποτάμνειν ἡσυχῇ, ὅπως ἂν ἡ φορ-
βὴ μὴ ἐκρυῇ· ἔπειτα ξηραίνειν αὐτὰς ἐν ἡλίῳ· καὶ τοὺς
σκώληκας δὲ τοὺς κοπρίνους ὡσαύτως ξηραίνειν ἐν τῷ
ἡλίῳ· ἔπειτα δὲ λειοτριβέειν· καὶ τῆς μὲν κάμπης δύο
ὀβολοὺς σταθμῷ Αἰγιναίους, τῶν δὲ σκωλήκων δι-
πλάσιον, καὶ ἀννήσου ὀλίγον παραμῖξαι, ἢ τῶν |
τοιουτοτρόπων τινός· κακῶδες γὰρ γίνεται· ταῦτα δὲ
λεῖα τρίψας, δίες οἴνῳ λευκῷ εὐώδει· καὶ ἐπὴν πίῃ,
βάρος ἐπιγίνεται καὶ νάρκη ἐμπίπτει ἐν τῇ γαστρί· ἢν
οὖν ἐπιγένηται, μελίκρητον ἐπιπινέτω ὀλίγον.

29. Κυήσιος δὲ καὶ παιδοποιήσιος ἥτις δεῖται καὶ
ἄτεκνος ἐοῦσα καὶ ἤδη κυήσασα, ἐοῦσα δὲ τεκνοῦσ-
σα, ὅταν ὁ στόμαχος ξηρὸς ᾖ ἀκρόπλοος καὶ ξυμμε-

494

336

of silphium, and whatever will produce air in the cavity. She should apply the suppository every other day until she seems to have recovered, and as much as she will accept; on the days between the applications, employ softening agents. After the mouth of her uterus is repaired by the softening agents, the woman should wait for the appearance of her menses, and, when these dry up, have intercourse.

28. If a woman's uterus is suppurating after she has given birth, or as the result of an abortion, or in some other way, and the pus is not contained in another recepticle and a tunic the way it is in a new growth, it benefits her to introduce an ointment spatula into the mouth of her uterus: for if the pus moves towards the spatula, she will have less need of a cauterization. Then collect caterpillars from the spurge plant, the ones with a sting, and then carefully cut these off in such a way that their juice does not run out; then dry them in the sun. Also dry some of the worms that form in excrement the same way in the sun, and then grind them fine. Mix together two obols weight by the Aeginetan standard of the caterpillars, twice that measure of the worms, and a small amount of anise or something like it, since the mixture has an evil smell. Grind this fine and dissolve it in fragrant white wine: when the patient drinks it, she feels a heaviness, and a numbness invades her belly: now, if this comes on, give her a little melicrat to drink.

29. A woman who wishes to become pregnant and to bear children, whether she is childless, or has already been pregnant, or has children: when the mouth of her uterus is

---

15 ἂν δέοι Lienau after the Arabic translation: δοκέοι codd: δεήσεται Froben.

μύκη καὶ μὴ ὀρθὸς ᾖ, ἀλλὰ πρὸς τὸ ἰσχίον ἀπεστραμ-
μένος τὸ ἕτερον ἢ ἐς τὸν ἀρχὸν κεκύφῃ ἢ ἀνεσπάκῃ
ἑωυτὸν ἢ τὸ χεῖλος ἐπιβάλλῃ τοῦ στομάχου ὁκοθενοῦν
ἐπ᾽ ἑωυτῷ ἢ τρηχὺς καὶ πεπωρωμένος ᾖ—σκληρὸς δὲ
γίνεται καὶ ἀπὸ ξυμμύσιος καὶ ἀπὸ πωρώσιος—, ταύτῃσι τὰ ἐπιμήνια οὐ φαίνεται ἢ ἐλάσσονα καὶ κακίονα
τοῦ δέοντος, καὶ διὰ πλείονος χρόνου ἐπιφαίνεται.
ἔστι δὲ ἐν ᾗσι τὰ ἐπιμήνια καὶ κατὰ ὑγιείην τοῦ
σώματος καὶ τῶν μητρέων τὴν ἔξοδον εὑρίσκεται, καὶ
κατά γε τὸ ξύμφυτον καὶ δίκαιον, καὶ ὑπὸ θερμότητος
καὶ ὑγρότητος τῶν ἐπιμηνίων, [καὶ]¹⁶ τοῦ στομάχου
μὴ κάρτα βεβλαμμένου· τὴν δὲ γονὴν οὐ δέχεται κατὰ
τὴν βλάβην, ἥτις ἂν κωλύῃ ἀπὸ τοῦ στομάχου μὴ
καλῶς ἔχοντος τοῦ δέχεσθαι. ταύτην χρὴ πυριήσαντα
τὸ σῶμα ὅλον πιεῖν δοῦναι φάρμακον καὶ κάθαρσιν
ποιήσασθαι τοῦ σώματος πρῶτον, ἤν τε ἄνω καὶ κάτω
δέηται ἤν τε ἄνω μοῦνον· καὶ ἢν μὲν ἄνω διδῷς τὸ
φάρμακον, μὴ πυριῆν πρότερον τῆς καθάρσιος· πυρι-
496 ήσας δὲ μεταπῖσαι | κάτω· ἢν δὲ μὴ δοκέῃ δεῖσθαι
ἀνωτερικοῦ, προπυριήσας κάτω πῖσαι. ὅταν δὲ δοκέῃ
καλῶς ἔχειν καθάρσιος τὸ σῶμα, μετὰ τοῦτο πυριῆν
τὰς μήτρας, ἐγκαθίζοντα αὐτὰς πυκνὰ ⟨ἐν ὅτῳ⟩¹⁷ ἂν
δοκέῃ ξυμφέρειν· ἐπιβάλλειν δὲ ἐς τὸ πυρίημα κυπαρίσσου ῥινήματα καὶ δάφνης φύλλα κόψας· καὶ λούειν
πολλῷ καὶ θερμῷ. ὅταν δὲ νεόλουτος καὶ νεοπυρίητος

¹⁶ Littré.
¹⁷ Littré.

dry on its surface, has closed together, and is not straight, but has turned toward one or other of her hips or has bent toward the rectum, or it has drawn itself up, or the lip of the mouth has folded over on itself at some point, or the mouth of the uterus is rough and petrified—it becomes hard from closing together and being petrified—in such women the menses do not appear, or if they do, then less and worse than usual, and they reappear at longer intervals. In some cases the menses still find their way out, on account of the healthiness of the body and the uterus, and because it is natural and right, and also due to the heat and moistness of the menses, as long as the mouth of the uterus is not too damaged. The seed, however, she does not receive, due to the impairment which results from the mouth of her uterus being disordered and prevents its reception. To this woman apply a vapour-bath all over her body, then give her a medication first to carry out a cleaning of her body, either both upwards and downwards if she requires it, or just upwards. If you give a medication to clean upwards, do not apply a vapour-bath before the cleaning takes place; then apply the vapour-bath, and afterwards give a potion to act downwards. If the patient does not seem to require a medication to clean upwards, first apply a vapour-bath and afterwards give a potion to act downwards. After the body seems to be in an appropriate, clean state, apply a vapour-bath to the uterus by having the patients repeatedly sit on any kind of stool that will have the desired effect: into the vapour-bath drop filings of cypress wood and pounded bay leaves. Bathe the patient in copious hot water, and immediately after the bath and the

ἦ, ἀνευρύνειν τὸ στόμα τῆς μήτρης μήλῃ τινὶ[18] κασσι-
τερίνῃ, καὶ ἀνορθοῦν ὅπῃ ἂν δέηται, ἢ μολιβδίνῃ,
ἀρξάμενος ἐκ λεπτῆς, εἶτα παχυτέρῃ, ἢν παραδέχη-
ται, ἕως ἂν δοκέῃ καλῶς ἔχειν· βάπτειν δὲ τὰς μήλας
ἐν ἑνὶ τῶν μαλθακτηρίων διειμένῳ, ὅτι ἂν δοκέῃ ξυμφέρειν, ὑγρὸν ποιήσας· τὰς δὲ μήλας ποιέειν ὄπισθεν
κοίλας, εἶτα περὶ ξύλοισι μακροτέροισιν ἁρμόσαι, καὶ
οὕτως χρῆσθαι. τὸν δὲ χρόνον τοῦτον πινέτω καθ-
εψοῦσα ἐν οἴνῳ λευκῷ εὐώδει ὅτι ἡδίστῳ γλυκεῖ δαῖδα
ὡς πιοτάτην κατασχίσασα λεπτὰ καὶ σελίνου καρπὸν
κόψασα καὶ κυμίνου Αἰθιοπικοῦ καρπὸν καὶ λιβανωτὸν ὡς κάλλιστον· τούτου πινέτω νῆστις, ὁκόσον ἂν
δοκέῃ μέτριον εἶναι πλῆθος, ἡμέρας ὁκόσας ἂν δοκέῃ
ἅλις ἔχειν· καὶ ἐσθιέτω σκυλάκια ἑφθὰ καὶ πουλύποδα
ἐν οἴνῳ ἑφθὸν γλυκεῖ· καὶ τοῦ ζωμοῦ πινέτω, καὶ
κράμβην ἑφθήν, καὶ οἶνον λευκὸν ἐπιπινέτω, καὶ μὴ
διψήτω· καὶ λουέσθω θερμῷ δὶς τῆς ἡμέρης· σιτίων δὲ
ἀπεχέσθω τοῦτον τὸν χρόνον.

Μετὰ δὲ τοῦτον, ἢν μὲν χωρέῃ κατὰ τὸν στόμαχον
καὶ φαίνηταί τι ἔξω καθάρσιος, πίνειν τε ἔτι τοῦ
πόματος ἡμέρην μίαν καὶ δύο, καὶ τῇσι μήλῃσι παύεσθαι χρεόμενον, καὶ πειρῆσθαι καθαίρειν τὰς μήτρας
προσθέτοισι φαρμάκοισιν. ᾗτινι τοῦ στομάχου ὀρθοῦ
καὶ μαλθακοῦ καὶ ὑγιέος καὶ καλῶς ἔχοντος καὶ ἐν τῷ
δέοντι κειμένου τὰ ἐπιμήνια μὴ φαίνεται πάμπαν ἢ
ἐλάσσονα καὶ διὰ πλείονος χρόνου καὶ μὴ ὑγιεινά,
τὴν νοῦσον ἀνευρών, ἥντινα ἔχουσιν αἱ μῆτραι, ἤν τε
καὶ τὸ σῶμα ξυμβάλληταί τι, ἐξευρὼν τὸ αἴτιον, ἀπ'

vapour-bath widen the mouth of her uterus with a tin or lead spatula, and straighten it wherever this is necessary, beginning with a thin one and then moving to a broader one if the uterus will admit it, until it appears to be as it should. Dip the spatulas into a solution of one of the softening agents that seems likely to have a beneficial effect, making this fluid. Make the spatulas hollow from behind, and then attach them around longer pieces of wood, and employ them thus. At this time have the patient drink very pleasant, sweet, fragrant, white wine, boiled down, to which are added very oily resinous wood cut fine, pounded celery seed, the seed of Ethiopian cumin, and the best grade of frankincense: have her drink, in the fasting state, as much of this as seems fitting, and for as many days as seem necessary. Also have her eat the meat of puppy and polypus boiled in sweet wine, drink the sauce from this, take some boiled cabbage, and after that drink white wine. She should prevent thirst, and bathe twice a day in warm water. She should avoid cereals during this period.

After this, if there is movement through the mouth of the uterus and a cleaning appears outside, you should continue her drink for a day or two longer, discontinue the use of the spatulas, and attempt to clear out her uterus with medicated suppositories. If the mouth of the uterus is straight, soft, healthy, in a good state, and lying where it should, but the menses do not appear at all, or in a decreased amount, or at longer intervals than usual, or in an unhealthy way: after investigating what disease the uterus is suffering from and whether or not the body is a contributing factor, and discovering the cause of her not becom-

---

18 μ. τ. H. Diller in Lienau: $\tau\hat{\eta}$ $\mu\acute{\eta}\lambda\eta$ $\tau\hat{\eta}$ codd.

ὅτευ οὐ κυΐσκεται καὶ ταῦτα οὕτως ἔχει, τὴν ἴησιν
ποιέεσθαι προσφέρων, ἢν δέχηται, θεραπείην, ἀρχόμενος ἐξ ἰσχυρῶν, ὅκως ἂν δοκέῃ καιρὸς εἶναι, τελευτᾶν δὲ ἐς μαλθακώτερα, ἕως ἂν δοκέῃ καλῶς ἔχειν
καθάρσιος ἡ μήτρη καὶ ὁ στόμαχος καθεστηκέναι
ὀρθῶς ἔχων καὶ ἐν τῷ ἐξαρκέοντι κείμενος.

Ἢν δὲ ἀπὸ τοῦ ποτοῦ καὶ φαρμάκου μὴ προ-
χωρήσῃ μηδὲ πινούσης χρόνον τὸν μέτριον, τούτου δὲ
παύεσθαι τοῦ πόματος· ὅταν δὲ καλῶς ἔχῃ τοῦ ἀπὸ
τῶν μηλέων ἔργου, μαλθάξαι τὸ στόμα τοῦ στομάχου,
καὶ ποιέειν ὅκως ἀναχάνηται ἐς ὁδὸν τῷ προσθέτῳ
ἀπό τε θυμιητῶν [καὶ]¹⁹ φαρμάκων καὶ μαλθακτηρίων.
ὅταν δὲ δοκέῃ καλῶς ἔχειν μαλθάξιος καὶ θυμιήσιος,
προστιθεὶς φάρμακον κάθαρσιν ποιέεσθαι τῆς μήτρης ἕως ἂν δοκέῃ καλῶς ἔχειν, ἀρχόμενος ἐκ μαλ-
θακῶν ἐπὶ ἰσχυρότερα, τελευτᾶν δὲ αὖθις ἐς μαλθακὰ
εὐώδεα· τῶν γὰρ ἰσχυρῶν φαρμάκων τὰ πλεῖστα ἑλ-
κοῖ τὸν στόμαχον καὶ δάκνει· ἔπειτα τόν τε στόμαχον
καθιστάναι ὀρθὸν καὶ ὑγιέα καὶ καλῶς ἔχοντα πρὸς
τὴν δέξιν τῆς γονῆς, καὶ τὴν μήτρην ξηρὴν ποιέειν.

Ἢν γυνὴ δοκέῃ ὑπὸ πιμελῆς τὰς μήτρας βε-
βλάφθαι ἐς τὴν κύησιν, λεπτύνειν ὡς μάλιστα καὶ
ἰσχναίνειν πρὸς τοῖσιν ἄλλοισιν.

30. Ὥρη δ' ἐαρινὴ ἀρίστη κυήσιος· ὁ δὲ ἀνὴρ μὴ
μεθυσκέσθω, | μηδ' οἶνον λευκὸν πινέτω, ἀλλ' ὅστις
ἰσχυρότατος καὶ ἀκρητέστατος· καὶ σιτία <σιτεί-
σθω>²⁰ ὡς ἰσχυρότατα· καὶ μὴ θερμολουτείτω· ἰσχυ-

ing pregnant and this being so, carry out her cure by administering a treatment, if she will accept it, beginning with strong agents at a time that seems to be appropriate, and ending with milder ones, until her uterus seems to be in a good state of cleaning and to have its mouth settled in a proper condition and lying in a satisfactory position.

If from the potion and the medication there is no progress, even after the patient has been drinking it for a reasonable time, discontinue the potion. But when the procedure with the spatulas has been effective, soften the mouth of the uterus and induce it, with fumigating medications and softening agents, to gape open and form a passage for a suppository; when the softening and fumigation seem to have been successful, apply medicated suppositories to clean the uterus, and continue until they seem to have had the desired effect: begin with mild agents, proceed to more powerful ones, and return at the end to mild fragrant ones, for most of the powerful medications ulcerate the mouth of the uterus and irritate it. Then set the mouth in a straight position, make it healthy and such as it should be to receive the seed, and dry the uterus.

If a woman seems to be prevented by fat in her uterus from conceiving, thin her down as much as possible and reduce her swelling by other means.

30. Spring is the best time for becoming pregnant. Let the man be sober, avoid white wine, but drink very potent wine unmixed with water; he should also eat the most potent foods, avoid the hot bath, make himself strong and

---

19 Del. Lienau.
20 Lienau.

έτω καὶ ὑγιαινέτω καὶ σιτίων ἀπεχέσθω τῶν μὴ ξυμφερόντων τῷ πρήγματι.

31. Ὅταν βούληται ἄρσεν φυτεύειν, τῶν ἐπιμηνίων ἀπολήγοντων ἢ ἐκλελοιπότων μίσγεσθαι· καὶ ὠθέειν ὡς μάλιστα ἕως ἂν ἐκμιαίνηται·[21] ὅταν δὲ θῆλυ βούληται γενέσθαι, ὅταν πλεῖστα ἐπιμήνια ἴῃ τῇ γυναικί, καὶ ἔτι δ᾽ ἐόντων, τὸν δὲ ὄρχιν τὸν δεξιὸν ἀποδῆσαι ὡς ἂν μάλιστα καὶ ἀνέχεσθαι δύνηται· ἔπὴν δὲ ἄρσεν βούληται φυτεύειν, τὸν ἀριστερὸν ἀποδῆσαι.

32. Στόμαχος μήτρης· ἀπὸ μὲν θυμιημάτων ξυμμεμυκὼς ἀναχάσκει, ἀπὸ δὲ τῶν μαλθακτηρίων μαλθάσσεται. θυμιῆν δὲ λωτοῦ φλοιόν, σπέρμα, δάφνης φύλλα χλωρὰ μᾶλλον κεκομμένα, λιβανωτόν, σμύρναν, ἀρτεμισίης καρπὸν ἢ φύλλα, καὶ ἄννησον κόψας ἢ στέαρ καὶ κηρὸν καὶ θεῖον καὶ κυπαρίσσου σπέρμα, πευκεδάνου ῥίζαν, μυρσίνης κόψας φύλλα χλωρά, κάστορος ὄρχιν, ὀνίδας ἄρσενος ὄνου, σκόροδα, στύρακα, ὑὸς στέαρ· κἂν ἀπεστραμμένον ᾖ τὸ στόμα, τούτοισι θυμιῆν· ἀναχάσκει μὲν οὖν οὕτως καὶ στρέφεται.

Μαλθάσσειν τε ἀπὸ τούτων τὸ στόμα τῆς μήτρης· σανδαράκην, στέαρ αἰγός, ὀπὸν συκέης, ὀπὸν σιλφίου, κυκλαμίνου χυλόν, θαψίην, ὀπὸν τιθυμάλλου, καρδαμώμου καρπόν, ποιήν ἢ καλεῖται πέπλος, κάστορος ὄρχιν, κράδης ὄξος, λίνου καρπόν, λίτρον, ἄρου ῥίζαν, σταφίδα ἀγρίην, καλαμίνθης φύλλα χλωρά, στρουθίου καρπόν, σκίλλης τὸ ἐκ τοῦ μέσου.

healthy, and stay away from foods that do not contribute to the matter.

31. When he wishes to beget a male child, let him have intercourse when his wife's menses are ceasing or have stopped, and he should push very hard until he ejaculates. When he wishes to beget a female child, he should have intercourse when his wife's menses are still present and flowing in their greatest amount, and also he should bind up his right testicle as tightly as he can stand. When he wishes to beget a male child, bind up the left testicle.

32. Mouth of the uterus: fumigations make it open up when it has been closed, emollient agents soften it. Fumigate it with lotus bark and seeds, green laurel leaves well pounded, frankincense, myrrh, seeds or leaves of wormwood, pounded anise or fat, wax, sulphur and seeds of cypress, root of sulphurwort, pounded green leaves of myrtle, castoreum, excrement of a male ass, garlic, storax, and lard. Also fumigate the mouth of the uterus with these agents if it is turned aside, for it will open up in this way and turn back.

Soften the mouth of the uterus with the following: red arsenic, goat's fat, fig juice, silphium juice, cyclamen juice, thapsia, spurge juice, cress seeds, the plant called *peplos* (i.e. wartweed), castoreum, sour fig-juice,[5] linseed, soda, cuckoo-pint root, wild raisin, green leaves of catmint, soapwort seed, and the pulp of squill.

---

[5] Lienau comments: "vielleicht der scharfe Saft aus der Rinde und den Ästen des [Feigen]baumes."

---

[21] Littré: -μαίνηται codd., cf. Galen's *Glossary* (vol. 19, 96).

33. Φάρμακα μαλθακτήρια καὶ πρὸς τὸ ἰσχυρὴν κάθαρσιν γενέσθαι ἀνακινῆσαι· θαψίης ῥίζαν, μυελὸν βοός, χήνειον στέαρ, | ῥόδινον· ταῦτα τρίψασα, ἀναζέσασα προστιθέσθω ἡμέρας τέσσαρας· καὶ πινέτω πράσου χυλόν, καὶ οἶνον γλυκὺν λευκόν· καὶ ῥητίνην, καὶ ἔλαιον χλιερόν, καὶ κύμινον, λίτρον, μέλι ἐν ῥυπαρῷ εἰρίῳ, ⟨ᾧ⟩[22] χρήσθω ἡμέρας τέσσαρας, πίνουσα σελίνου καρπόν, καὶ λιβανωτοῦ πυρῆνας πέντε, καὶ κύμινον Αἰθιοπικὸν ἐν οἴνῳ λευκῷ ἀκρήτῳ γλυκεῖ· καὶ λουέσθω δὶς τῆς ἡμέρης.

Σμύρνα, λίβανος, βοὸς χολή, ῥητίνη τερεβινθίνη ἢ νέτωπον· τούτων ἴσον ἑκάστου μίξασα προστιθέσθω ἐν εἰρίῳ καθαρῷ ἢ ῥάκει λεπτῷ· βάψασα δὲ τὸ ῥάκος ἐν μύρῳ λευκῷ Αἰγυπτίῳ εὐώδει καὶ ἀποδήσασα λίνῳ, λουσαμένη προστιθέσθω. καὶ πώλυπον φλεύσασα ἐσθιέτω, καὶ πινέτω σελίνου καρπὸν καὶ ἀσπαράγου, καὶ οἶνον λευκὸν τρὶς τῆς ἡμέρης νῆστις ἐοῦσα. σμύρνα, κασίη, λιβανωτός, κιννάμωμον, νέτωπον· τούτων ἑκάστου ἴσον ἐν εἰρίῳ ἢ βαλάνους ποιέουσα προστιθέσθω. κολοκύνθης ἀγρίης, κύμινον πεφρυγμένον, ἀνήθου καρπόν, κυπαρίσσου ῥίζαν, ταῦτα τρίψας λεῖα, μέλιτι ἑφθῷ φυρήσας, βαλάνους ποιέων, δίδου προστίθεσθαι. καὶ πινέτω γλυκυσίδης ῥίζαν, σελίνου καρπόν, ὀπὸν σιλφίου, οἶνον. τὸ βόλβιον δὲ καὶ αὐτὸ καθαίρει προστιθέμενον. καὶ σμύρναν πρώτην, ἄνθος ὀλίγον ἐν οἴνῳ λευκῷ εὐώδει προστίθει.

Φάρμακα πρόσθετα μήτρην καθῆραι· λαβὼν ἄνθος χαλκοῦ καὶ λίτρου τρίτον μέρος, μέλιτι ἑφθῷ φυρή-

33. Medications that soften and initiate powerful cleaning: root of thapsia, beef marrow, goose fat, oil of roses: grind these, boil them up, and apply as a suppository for four days; also have the woman drink leek juice and sweet white wine. Also resin, warm olive oil, cumin, soda, and honey in greasy wool, which she should employ (sc. in the form of a suppository) for four days, while drinking cress seed, five grains of frankincense, and Ethiopian cumin in sweet white wine unmixed with water. She should bathe twice daily.

Myrrh, frankincense, bull's gall, turpentine-resin or oil of bitter almonds: mix together an equal amount of each of these and apply as a suppository on clean wool or a thin cloth: first soak the cloth in fragrant white Egyptian unguent, and after the woman has bathed tie it on with a piece of linen and apply it; also have her eat a polypus she has seared, and drink celery and asparagus seeds in white wine, three times a day in the fasting state. Myrrh, cassia, frankincense, cinnamon, oil of bitter almonds: prepare an equal amount of each of these in wool or formed into a pessary, and have the woman apply it. Some wild gourd, roasted cumin, dill seeds, cypress root: grind these fine, mix into boiled honey, form into pessaries, and give them to the woman to apply; she should also drink peony root, celery seeds, silphium juice and wine. A small onion applied alone as a suppository cleans, too. Also apply as suppository myrrh of the first grade and a little flower of copper, in fragrant white wine.

Medicated suppositories to clean the uterus: take flower of copper and one third as much soda, mix them into

---

22 H. Diller in Lienau.

σας, βαλάνους ποιήσας, ὁκόσαι ἂν δοκέωσι μέτριαι
εἶναι μέγεθος καὶ πάχος, οὕτως προστίθει πρὸς τὸ
στόμα τῆς μήτρης.

Ἢν δὲ βούλῃ ἰσχυρότερον εἶναι, ἐλατήριον παρα-
μῖξαι καὶ τὸ ἄνθος μοῦνον, καὶ οὕτως ποιέων δίδου
προστίθεσθαι· καὶ κράδης παραμιγνύναι [πρόσθε-
τον][23] φλοιὸν ξύων καὶ λεῖον τρίβων, ὅταν τὸ στόμα
504 δοκέῃ ξηρό|τερον εἶναι τῆς μήτρης, τὸ ἥμισυ ὡσαύ-
τως.

Πρόσθετον ἕτερον· τρίψας ἐλατήριον καὶ ἄνθος
χαλκοῦ λεῖον, δύο μοίρας ἄνθους, ἐλατηρίου μίαν,
ταῦτα διέσθαι· κυκλαμίνου τρίψας, ταύτῃ ἀναμῖξαι,
ὅκως ἂν δοκέῃ καιρὸς εἶναι· καὶ ποιησαμένην πρόσθετα προστίθεσθαι ἐν εἰρίῳ.

Πρόσθετον λευκὰ καθαῖρον· ἀρτεμισίην ποίην, λίτρον, κυκλάμινον ἡμίξηρον, κύμινον. ἕτερον τὰ αὐτὰ
καθαίρει· ἀρτεμισίην ποίην χλωρὴν τρίψας καὶ σμύρνης τρίτον μέρος, οἴνου ἀναμίξας εὐώδεος, εἰρίῳ λευκῷ περιελίξας αὐτό, ἐν οἴνῳ βρέξας δὸς προστίθεσθαι.

Ὅταν χολῶσιν αἱ μῆτραι, λίτρον, σικύης ἐντε-
ριώνην, κυκλάμινον ἡμίξηρον ἐν εἰρίῳ.

Πρόσθετον παντοῖα καθαῖρον· σταφίδα ἀγρίην
τρίψας χλωρήν, περιπλάσας, ἀρτεμισίης ποίης τρίψας τὰ φύλλα, οἴνῳ[24] πλάσας καὶ ξηρήνας ἐν σκιῇ,
ἀπὸ τούτου ποίει πρόσθετα καὶ δίδου προστίθεσθαι.
ἄνθος μιγνὺς χαλκοῦ ἢ στυπτηρίην Αἰγυπτίην διεὶς
τῇ κυκλαμίνῳ ὥσπερ τὰ πρότερα, ἐν μέλιτι ἑφθῷ

boiled honey, form this into pessaries of the required length and breadth, and apply them in the same way to the mouth of the uterus.

If you want something more forceful, mix together squirting-cucumber and flower of copper alone, make this up in the same way, and give it to the patient to apply as a suppository; also mix in in the same way bark of a fig-tree shredded and ground fine, to half the amount, when the mouth of the uterus seems to be quite dry.

Another suppository: grind squirting-cucumber and flower of copper fine (two parts of the flower to one of the squirting-cucumber) and liquify this; into this mix ground cyclamen in the amount that seems correct: have the woman make suppositories of this and apply them in wool.

Suppository to clean the white flux: wormwood plant, soda, half-dry cyclamen, and cumin. Another, to clean the same: grind green wormwood plant and one third as much myrrh, mix this with fragrant wine, wind white wool around it, soak in wine, and give to the patient to apply as a suppository.

When the uterus suffers from bile: soda, pulp of a bottle-gourd, and half-dry cyclamen in wool.

A suppository that cleans all sorts of matters: grind green stavesacre and knead it, grind leaves of the wormwood plant, knead this with wine and dry it in the shade: from this form suppositories and give them to the woman to apply. Mix flower of copper or Egyptian alum, dissolve it in cyclamen as above, form it to a suppository in boiled

---

23 Del. Littré.
24 Potter after the Arabic translation: φλοιο- codd.

πλάσας ἢ ἐν ἰσχάδι ποιήσας, καὶ σμύρναν ὀλίγην.
ἕτερον· τὴν κυκλάμινον τρίψας τὴν λευκήν, οἴνῳ εὐώδει παραμίξας, ἐν ῥάκει δήσας ὡς λεπτοτάτῳ λίνῳ
καθαρῷ, δίδου προστίθεσθαι. ἄλλο· κυκλάμινος ἡμί-
ξηρος, λίτρον, κανθαρίδες, στέαρ, σανδαράκης.

34. Παρθένῳ ὁπόταν ὡραία μὴ γένηται, χολᾷ καὶ
πυρεταίνει καὶ ὀδυνᾶται <καὶ>[25] διψῇ καὶ πεινῇ καὶ
ἐξεμεῖ καὶ μαίνεται καὶ πάλιν σωφρονέει· κινέονται αἱ
μῆτραι. καὶ ὁκόταν μὲν πρὸς τὰ σπλάγχνα τράπων-
506   ται, ἐξεμεῖ καὶ πυρέσσει καὶ παιραφρονεῖ ὅταν δ'
ἀπολίπωσι, πεινῇ καὶ διψῇ καὶ ἠπίαλος ἔχει. χρὴ
ταύτῃσιν ἀρνακίδας προστιθέναι θερμὰς πρὸς τὴν
γαστέρα, καὶ ὑποκάπνιζε ἐς αὐτὰ τὰ αἰδοῖα ὅτι μάλιστα ἐπ' ἀμφορέως αὐχένι καθίσας· σμύρνης ὅσον
κύαμον, λιβανωτὸν δὶς ὅσον, τοσοῦτον μίξας καὶ ζέας
ἐρηριγμένας, ὁμοῦ θυμιῆν, καὶ ἐπὶ τὸ πῦρ ἐπιβάλλων,
νῆστις ὡς μάλιστα, καὶ λούειν πολλῷ θερμῷ.

35. Πρόσθετον· Αἰγυπτίην στυπτηρίην μαλθακὴν
εἰρίῳ περιειλήσασα προστιθέσθω. πρόσθετον· ἀρτεμισίην τρίψασα, ἐν οἴνῳ λευκῷ δεύσασα προστιθέσθω.

36. Νεοτόκῳ γυναικί· ῥόδινον, σμύρναν, κηρὸν
μίξαντα ἐν εἰρίῳ δοῦναι προστίθεσθαι· ὅταν προπέ-
σωσιν[26] αἱ μῆτραι, τὰ ξηρὰ καὶ στρυφνὰ προσφέρειν
καὶ πίνειν καὶ προσέχειν.

37. Καθαρτήριον· σῦκον μέλαν, σκόροδον, λίτρον,
κύμινον, ταῦτα τρίψας λεῖα, ἐν εἰρίῳ δοῦναι προσ-

honey or apply it to a dried fig, and add a little myrrh. Another: grind white cyclamen, mix it with fragrant wine, bind this in a piece of very fine, clean linen cloth, and give it to be applied as a suppository. Another: half-dry cyclamen, soda, blister-beetles, fat, and red arsenic.

34. When the first menstruation fails to appear in a young woman, but she suffers from bile, fever, pain, thirst and hunger, she vomits, and she is out of her senses but then returns to them, her uterus is in motion: when it turns towards the viscera, she vomits, has a fever, and is deranged; when this ceases, she has hunger, thirst, and nightmares. In these patients one must apply warm goat-skins to their belly, and fumigate them from below directly into the external genitals, if possible by having them sit on the neck of an amphora: myrrh to the amount of a bean, twice that amount of frankincense, mix these with bruised spelt, fumigate together, and cast onto a fire. She should do this, as far as possible, in the fasting state, and then wash herself with copious hot water.

35. Suppository: bind mild Egyptian alum in wool and apply it. Suppository: grind wormwood, soak it in white wine, apply.

36. Suppository for a recent parturient: mix oil of roses, myrrh and wax in wool, and give it to the woman to apply. When her uterus prolapses, administer dry and astringent medications for her to drink and to apply to her uterus.

37. Cleaning suppository: grind a black fig, garlic, soda and cumin fine, and give this to apply in wool. Another one:

---

25 Lienau.
26 Ermerins: προσπέσ. codd.

τίθεσθαι. ἕτερον· σηπίης ὄστρακον κόψας λεῖον, οἴνῳ δεύσας, ἐν λαγωοῦ θριξὶ καὶ εἰρίῳ προστίθεσθαι.

38. Ἢν μετὰ τόκον ὑστέρας ἀλγέῃ, πτισάνην καὶ πράσα καὶ στέαρ αἴγειον ἑψήσασα, ῥοφείτω τοῦτο ὡς ὀλίγιστον.

39. Πρόσθετον· λίτρον, κύμινον, σύκου τὸ ἴσον. καθαρτήριον πρόσθετον καὶ μαλθακτήριον· νέτωπον, ῥόδινον μύρον, χηνὸς ἄλειφα, ἐς ὀθόνια λεπτά. |

508   40. Ἢν τὰ ἐπιμήνια πολλὰ γίνηται, γλυκυσίδης κόκκους τοὺς μέλανας δὶς ἑπτὰ πίνειν ἐν οἴνῳ δυσὶ κυάθοις.

41. Ἢν ἐξίωσιν αἱ μῆτραι πυκνὰ βρέξας ὕδατι χλιερῷ τὰς μήτρας, ὑπτίην ἀνακλίνας, μίξας σίδιον, κηκῖδα, ῥοῦν τὴν ἐρυθρήν, ἐν οἴνῳ λευκῷ διατρίψας, τούτῳ χρίσας, ἐντιθέναι· ἔπειτα πῖσαι δάφνης φύλλα ἐν οἴνῳ αὐστηρῷ.

42. Ὅταν γυνὴ κυέουσα ῥέηται, ὀνίδα ξηρὴν καὶ μίλτον καὶ ὄστρακον σηπίης, ταῦτα τρίψας λεῖα, ἐν ῥάκει ἀποδήσας, προστιθέναι.

43. Τὸ ὕστερον, ἢν μὴ καθαρθῇ· λεβηρίδος[27] ἐν οἴνῳ λευκῷ ὅσον κυάθῳ τρίψας ὀβολὸν Ἀττικόν, διδόναι πίνειν, καὶ καθαρεῖται.

---

[27] Potter (cf. *Diseases of Women II* 78): λεάναι codd.

pound the bone of a cuttle-fish fine, dissolve it in wine, and apply it in a hare's fur or in wool.

38. If, after giving birth, a woman suffers pain in her uterus, boil peeled barley, leeks and goat's fat, and have her drink a very little of this at a time.

39. Suppository: soda, cumin, and the same amount of fig. A cleaning and softening suppository: oil of bitter almonds, oil of roses, and goose grease on a soft linen cloth.

40. If the menses become excessive: drink fourteen black pennyroyal seeds in two cyathoi of wine.

41. If the uterus prolapses repeatedly, sprinkle it with warm water, and have the woman recline on her back; mix pomegranate peel, oak-gall, and red sumach, grind these together into white wine, anoint the uterus with this, and then replace it. After that give laurel leaves in astringent wine to drink.

42. When a pregnant woman has a flux, grind dry ass's excrement, red ochre, and cuttle-fish bone fine, bind it in a rag, and apply it as a suppository.

43. If the placenta is not cleaned out: grind serpent's skin to the amount of an Attic obol in a cyathos of white wine and give it to drink; it will be cleaned out.

GIRLS

INTRODUCTION

Whether this fragment stems from the treatise(s) the author of *Diseases of Women I* refers to as his "Diseases of Girls" and "About the Girl"[1] cannot be determined, since no textual coincidence exists between his references and our *Girls*. The only secure ancient testimony for the work's existence is the appearance of one rare word in Galen's *Glossary*.[2]

Girls is present in the collected Hippocratic editions and translations, and received some individual attention from Renaissance scholars;[3] recently it appeared first in a partial English translation[4] and then as a complete edition and translation:

> R. Flemming and A. E. Hanson, "Hippocrates' *Peri Partheniôn (Diseases of Young Girls)*: Text and Translation," in *Early Science and Medicine* 3 (1998), 241–52.

The present edition is based on a reading of the independent manuscripts M and V from microfilm.

---

1 Littré vol. 8, 10 and 98.

2 Galen vol. 19, 153 φονᾷ.

3 See the works cited at Littré vol. 8, 465.

4 M. R. Lefkowitz and M. B. Fant, *Women's Life in Greece and Rome*, 2nd edn., Baltimore, 1992, pp. 242f.

ΠΕΡΙ ΠΑΡΘΕΝΙΩΝ

1. Ἀρχή μοι τῆς ξυνθέσιος ⟨ἀπὸ⟩[1] τῶν αἰειγενέων[2] ἰητρικῆς· οὐ γὰρ δυνατὸν τῶν νοσημάτων τὴν φύσιν γνῶναι, ἥπερ ἐστὶ τῆς τέχνης ἐξευρεῖν, ἢν μὴ γνῷ τὴν ἐν τῷ ἀμερεῖ κατὰ τὴν ἀρχήν, ἐξ ἧς διεκρίθη.

Πρῶτον περὶ τῆς ἱερῆς νούσου καλεομένης, καὶ περὶ τῶν ἀποπληκτικῶν, καὶ περὶ τῶν δειμάτων, ὁκόσα φοβεῦνται ἰσχυρῶς ἄνθρωποι, ὥστε παραφρονέειν καὶ ὁρῆν δοκέειν δαίμονάς τινας ἐφ' ἑωυτῶν δυσμενέας, ὁκότε μὲν νυκτός, ὁκότε δὲ ἡμέρης, ὁκότε δὲ ἀμφοτέρῃσι τῇσιν ὥρῃσιν. ἔπειτα ἀπὸ τῆς τοιαύτης ὄψιος πολλοὶ ἤδη ἀπηγχονίσθησαν, πλέονες δὲ γυναῖκες ἢ ἄνδρες· ἀθυμοτέρη γὰρ καὶ λυπηροτέρη[3] ἡ φύσις ἡ γυναικείη. αἱ δὲ παρθένοι, ὁκόσῃσιν ὥρη γάμου, παρανδρούμεναι, τοῦτο μᾶλλον πάσχουσιν ἅμα τῇ καθόδῳ τῶν ἐπιμηνίων, πρότερον οὐ μάλα ταῦτα κακοπαθέουσαι. ὕστερον γὰρ τὸ αἷμα ξυλλείβεται ἐς τὰς μήτρας, ὡς ἀπορρευσόμενον· ὁκόταν οὖν τὸ στόμα τῆς ἐξόδου μὴ ᾖ ἀνεστομωμένον, τὸ δὲ αἷμα πλέον ἐπιρρέῃ διά τε σιτία καὶ τὴν αὔξησιν τοῦ

---

[1] Ermerins.   [2] M: νεηγεν. V.
[3] V: ὀλιγωτέρη M.

GIRLS[1]

1. The beginning point of my composition is from what is eternal[2] in medicine; for it is not possible to know the nature of diseases, which it is the task of medicine to discover, unless one knows nature in its indivisibility, from the beginning point out of which it proceeded.[3]

First, concerning the sacred disease, as it is called, and persons who are paralysed, and the terrors by which people are so thoroughly frightened that they become deranged and think they see malevolent spirits, sometimes by night, sometimes by day, and sometimes at both hours. From such a vision many persons have hanged themselves — actually more women than men, for womanly nature is more fainthearted and sorrowful. When young women in the season of marriage remain without a husband, they suffer, in particular at the time of the downward passage of their menses, this evil to which before they were not very subject. For at this later time in their life, blood collects in the uterus, destined to run out, but when the mouth of the exit does not open up, more blood keeps being added from food and the growth of the body, and then, left with no-

---

[1] Literally the title means "On Girlish Matters."
[2] With V's reading: "new."   [3] Cf. *Fleshes* 1–2 for a proem which shares certain thoughts and vocabulary with this one.

σώματος, τηνικαῦτα οὐκ ἔχον τὸ αἷμα ἔκρουν ἀναΐσσει ὑπὸ πλήθους ἐς τὴν καρδίην καὶ ἐς τὴν διάφραξιν.
ὁκόταν οὖν ταῦτα πληρωθέωσιν, ἐμωρώθη ἡ καρδίη,
εἶτ' ἐκ τῆς μωρώσιος νάρκη, εἶτ' ἐκ τῆς νάρκης παρά-
νοια ἔλαβεν. ὥσπερ ὁκόταν καθημένου πολὺν χρόνον
τὸ ἐκ τῶν ἰσχίων καὶ μηρῶν αἷμα ἀποπιεχθὲν[4] ἐς τὰς
468   κνήμας καὶ τοὺς πόδας | νάρκην παράσχῃ. ὑπὸ δὲ τῆς
νάρκης ἀκρατέες οἱ πόδες ἐς ὁδοιπορίην γίνονται, ἔστ'
ἂν ἀναχωρήσῃ τὸ αἷμα ἐς ἑωυτό· ἀναχωρέει δὲ τάχιστα, ὁκόταν ἀναστὰς ἐν ὕδατι ψυχρῷ τέγγῃ τὸ[5] ἄνω
τῶν σφυρῶν. αὕτη μὲν οὖν ἡ νάρκη εὐήνιος, ταχὺ γὰρ
παλιρροεῖ διὰ τὴν ἰθύτητα[6] τῶν φλεβῶν, καὶ ὁ τόπος
τοῦ σώματος οὐκ ἐπίκαιρος. ἐκ δὲ τῆς καρδίης καὶ τῶν
φρενῶν βραδέως παλιρροεῖ· ἐπικάρσιαι γὰρ αἱ φλέβες καὶ ὁ τόπος ἐπίκαιρος ἔς τε παραφροσύνην καὶ
μανίην ἕτοιμος. ὁπόταν γὰρ πληρωθέωσι ταῦτα τὰ
μέρεα, καὶ φρίκη ξὺν πυρετῷ ἀναΐσσει πλανήτης.[7]
ἐχόντων δὲ τούτων ὧδε, ὑπὸ μὲν τῆς ὀξυφλεγμασίης
μαίνεται, ὑπὸ δὲ τῆς σηπεδόνος φονᾷ, ὑπὸ δὲ τοῦ
ζοφεροῦ φοβέεται καὶ δέδοικεν, ὑπὸ δὲ τῆς περὶ τὴν
καρδίην πιέξιος ἀγχόνας κραίνουσιν, ὑπὸ δὲ τῆς
κακίης τοῦ αἵματος ἀλύων καὶ ἀδημονέων ὁ θυμὸς
κακὸν ἐφέλκεται. ἕτερον δὲ καὶ φοβερὰ ὀνομάζει· καὶ
κελεύουσιν ἄλλεσθαι καὶ καταπίπτειν ἐς φρέατα ἢ
ἄγχεσθαι, ἅτε[8] ἀμείνονά τε ἐόντα καὶ χρείην[9] ἔχοντα

---

⁴ M: ἀναπιεχ. V.   ⁵ Littré: τεγγέτω (δὲ) MV.
⁶ M: παχύτητα V.   ⁷ Potter: -ίτας M: -ήτες V.

where to flow out, the blood springs up in its excess to the heart and the diaphragm. Now when these parts are filled, the heart becomes stupefied, then from the stupefaction numb, and finally from the numbness these women become deranged. It is like when, in a person who sits for a long time, the blood is pressed out of his hips and thighs into his lower legs and feet, and this provokes numbness. As a result of the numbness, the feet lose their capacity to walk, until the blood moves back into its natural place: it moves back soonest when the person stands up and immerses his legs above the ankles in cold water. Now this numbness is tractable, for it goes away quickly on account of the straightness[4] of the vessels, and furthermore those places in the body are not critical. But from the heart and the diaphragm the blood recedes only slowly, since the vessels there are transverse and those places are critical and can bring about derangement and raging. For when these parts are filled, a transient shivering with fever arises. When the situation is such, from the acute inflammation the woman rages, from the putrefaction she becomes murderous, from the darkness she is frightened and afraid, from the compression around their heart they are desirous of throttling themselves, and from the bad state of the blood the mind, being distraught and dismayed, tempts them to evil. She names strange and frightful things, and these urge the women to take a leap and to throw themselves down wells, or to hang themselves, as being better

4 With V's reading: "wideness."

---

8 Froben: καὶ MV.
9 Littré: χροιὴν M χρονίην V.

παντοίην. ὁκότε δὲ ἄνευ φαντασμάτων, ἡδονή τις ἀφ'
ἧς ἐρᾷ τοῦ θανάτου ὥσπερ τινὸς ἀγαθοῦ. φρονεούσης
δὲ τῆς ἀνθρώπου, τῇ Ἀρτέμιδι αἱ γυναῖκες ἄλλα τε
πολλὰ καὶ τὰ ἱμάτια τὰ πολυτελέστατα καθιεροῦσι
τῶν γυναικείων, κελευόντων τῶν μάντεων ἐξαπατε-
ώμεναι. ἡ δὲ τῆσδε ἀπαλλαγή, ὁκόταν μὴ ἐμποδίζῃ
τι[10] τοῦ αἵματος τὴν ἀπόρρυσιν. κελεύω δὴ τὰς παρ-
θένους, ὁκόταν τι τοιοῦτο πάσχωσιν, ὡς τάχιστα
ξυνοικῆσαι ἀνδράσιν· ἢν γὰρ κυήσωσιν, ὑγιέες γίνον-
470   ται. εἰ δὲ μή, ἢ εὐθέως[11] ἅμα τῇ ἥβῃ ἢ | ὀλίγον
ὕστερον ἁλώσεται, εἴπερ μὴ ἑτέρη[12] νούσῳ. τῶν δὲ
ἠνδρωμένων γυναικῶν στεῖραι ταῦτα πάσχουσιν.

[10] Potter: om. τι M: ἐμποδίζηται V.
[11] Potter: ἢ αὐτίκα Littré: ἐηἀ*τέων M: αὐτέων ἢ V.
[12] μ. ἑ. Littré: μητὴρ MV.

and in every way advantageous. When there are no visions, there is a pleasure from which the woman loves death as some kind of good. When there is a return to the senses, women dedicate many different things to Artemis, including the most costly cloaks of the female sort, being deceived by the bidding of seers. Release from this disease comes when nothing prevents the discharge of blood. I urge young women suffering from a condition of this kind to cohabit with men as soon as they can: for if they become pregnant, they recover. If not, then either at once in puberty or a little later she will be seized by this disease, if not by another one. Among married women, some barren ones suffer these things.

EXCISION OF THE FETUS

INTRODUCTION

This short collection of obstetrical notes unknown in the extant ancient literature is the source of one word in Galen's *Glossary*,[1] evidence that it was part of the Hippocratic Collection by the second century A.D.

The text of this treatise is unique in that it appeared twice in the manuscript M, once under the title *Excision of the Fetus* on fol. 297r-298r (M[I]), and once under the title *Excision of the Child* on lost leaves once situated between the leaves now numbered 408 and 409 (M[II]). The present edition is based on three independent witnesses: M[I], M[II] reconstructed from a copy in the manuscript Parisinus Gr. 2140 (I[II]) made before M[II] was lost, and V.[2]

Excision of the Fetus appears in the collected Hippocratic editions and translations, but has never been the subject of a special study.

The present edition is based on collations of the three independent manuscripts M, I, and V from microfilms.

---

[1] Galen vol.19, 107 ἰχθύην.

[2] See A. Anastassiou, "Zur Frage der Struktur des Hippokratescodex Marcianus Venetus 269 (M)," in M. D. Grmek, *Hippocratica. Actes du Colloque hippocratique de PARIS*, Paris, 1980, pp. 24–31.

ΠΕΡΙ ΕΓΚΑΤΑΤΟΜΗΣ
ΕΜΒΡΥΟΥ[1]

1. Περὶ δὲ τῶν μὴ κατὰ τρόπον κυΐσκομένων, ἀλλ᾽ ἐγκατατεμνομένων οὕτως·[2] πρῶτον μὲν ἐπὶ τὴν γυναῖκα σινδόνα ἐπιβαλών, κατάζωσον ἀνώτερον τοῦ μαζοῦ, καὶ τὴν κεφαλὴν κατακαλύψαι χρὴ τῇ σινδόνι, ὅπως μὴ ὁρῶσα φοβῆται ὅ τι ἂν ποιέῃς. ἢν οὖν ἐξίσχῃ τὴν χεῖρα τὸ ἔμβρυον παραπλάγιον παραπεσόν, ἐπιλαβόμενος τῆς χειρὸς προάγειν ἔξω ὡς μάλιστα, καὶ παραδεῖραι τὸν βραχίονα, καὶ ἀποψιλώσας τὸ ὀστέον· ἰχθύην περιδῆσαι[3] περὶ τοὺς δακτύλους τῆς χειρὸς τοὺς δύο, ὅκως μὴ ἀπολισθάνοι ἡ σάρξ, μετὰ δὲ ταῦτα τὸν ὦμον περισάρκισον καὶ ἄφελε κατὰ τὸ ἄρθρον. ἔπειτα τὴν κεφαλὴν κατὰ φύσιν παρώσας, ὑπάγειν ἔξω τὴν κεφαλὴν τοῦ ἐμβρύου· τῷ δὲ δακτύλῳ τὸ ἔμβρυον εἴσω ἀπῶσαι, ἢ μαχαιρίῳ διὰ τῶν πλευρέων ἢ διὰ τῆς κληῗδος, ὅκως τὴν φῦσαν ἀφῇ καὶ ξυμπέσῃ τὸ ἔμβρυον καὶ ἡ ἔξοδος αὐτῷ εὐπετεστέρη ᾖ. τὴν δὲ κεφαλήν, ἢν μὲν δύνησαι

---

[1] M[I]: ΠΑΙΔΙΟΥ I[II] V.   [2] Περὶ—οὕτως M[I]: ἐγκατατομὴν παιδίου ποιήσεις οὕτως I[II]: ἐγκατατομὴ παιδίου V.
[3] Potter: -δήσας codd.

EXCISION OF THE FETUS[1]

1. Concerning pregnancies that do not proceed in the normal way, but which are cut to pieces inside (sc. the uterus), the matter is as follows. First place a cloth over the woman, girding it above each breast, and also you must cover her head with a cloth, so that she will not see what you are doing and become frightened. Now, if the fetus falls sideways and one arm comes out, take hold of the arm and, drawing it as far out as possible, excoriate the upper arm and strip its bone bare; bind a fish-skin around two fingers of the hand so that the flesh will not slip away, and after that make an incision all around the shoulder and separate it at the joint. Next replace the fetus's head in its natural position, and then draw the fetus downward; with your finger cave the fetus's body in, by using a blade through the ribs or the collar bone, so that the body will expel air and collapse, which makes its passage to the outside easier. If you are able to bring out the head in the natural

---

[1] This is the traditional title; more accurate would be: "Cutting the fetus to pieces inside (sc. the uterus)."

κατὰ φύσιν ἔξω ὦσαι· εἰ δὲ μή, ξυμφλάσαι, καὶ οὕτως
ὑπεξάγειν ἔξω τὸ ἔμβρυον. ἔπειτα τῷ θερμῷ πολλῷ
καταχέας καὶ ἀλείψας ἐλαίῳ, κατακεῖσθαι κελεύειν
514 ἐπαλλάξασαν[4] τὼ πόδε, | καὶ μεταπῖσαι οἶνον γλυκὺν
κεράσας εὐζωρότερον καὶ λευκόν, καὶ ῥητίνην, μέλιτι
διατρίψας, μίξας τῷ οἴνῳ, δοῦναι πιεῖν. τὰ δὲ ἄλλα
θεραπεύειν ὡς λεχώ,[5] κατὰ τὰ εἰρημένα.

2. Ὅταν δὲ τικτούσῃ γυναικὶ πλάγιον παραπέσῃ τὸ
ἔμβρυον, γίνεται δὲ ὁπόταν στρέφηται τὸ τοιόνδε· ὁ
ὀμφαλὸς περὶ τὸν τράχηλον περιελίσσεται καὶ ἐπί-
σχει τὴν ἔξοδον τοῦ ἐμβρύου, καὶ ἐς τὸ ἰσχίον ἐπεμ-
βάλλει τὴν κεφαλήν, καὶ ἡ χεὶρ ὡς ἐπὶ τὸ πολὺ ἔξω
γίνεται. ἢν μὲν οὖν ἤδη τεθνηκὸς ἔξω γένηται, τοῦτο
προσημαίνει· ᾗσι δὲ μὴ ἔξω ἡ χεὶρ τοῦ ἐμβρύου, ὡς
ἐπὶ τὸ πολὺ ζῇ τὸ ἔμβρυον· κίνδυνος καὶ οὕτως.

3. Ἔνιαι δὲ καὶ τὰ λόχια πρὸ τοῦ ἐμβρύου ἀφιᾶσιν,
ὥστε ἀναγκαῖον τὴν ὠδῖνα ξηράν τε εἶναι καὶ ἐπί-
πονον· ὅσαι δὲ τὰ λοχία μὴ προκαθαίρονται, ῥᾷον
ἀπαλλάττουσι ἐν τῷ τόκῳ.

4. Ἀνασείειν δὲ δεῖ ὧδε· σίνδονα ὑποστορέσαντα,[6]
ἀνακλῖναι τὴν γυναῖκα, καὶ ἑτέρην ἐπιβαλεῖν ὅκως ἂν
τὸ αἰδοῖον κεκρυμμένον ᾖ, καὶ περικαλύψαι περὶ
ἑκάτερον τὸ σκέλος τὴν σίνδονα καὶ περὶ ἑκάτερον τὸ
516 γυῖον. γυναῖκας δὲ δύο λαβέσθαι τοῦ σκέλεος ἑκα|τέ-
ρου, καὶ τῆς χειρὸς ἑκατέρης ἑτέρας γυναῖκας δύο·
ἔπειτα διασείειν λαβούσας ἐγκρατῶς, μὴ ἔλασσον ἢ

---

4 Froben: -άξαντα Μ¹ V· -άξασας Ι¹¹.

way, fine; if not, crush it to pieces, and in this way draw the fetus down and out. Then pour copious warm water over the woman and anoint her with olive oil; command her to lie down and cross her legs; after that have her drink sweet white wine hardly diluted with water; and grind resin into honey, mix this with wine, and give it to her to drink. Otherwise treat her as you would any parturient, according to what has been said.

2. When the fetus falls sideways in a woman who is giving birth, this happens when the fetus gets turned in the following way: the umbilical cord becomes wrapped around its neck, holds back its movement to the outside, and dashes its head against the hip, so that usually an arm comes out. Now, if the arm comes out after the fetus has already died, this is normally an indication of the death; in women in whom the fetus' arm does not come out, the fetus is generally alive, although even in this case there is danger.

3. Some women expel their waters before the fetus, so that of necessity their birth pangs must be dry and difficult. Others who are not cleaned prematurely of their waters give birth more easily.

4. A parturient should be shaken in the following way. Spread a cloth beneath the woman, lean her back, and place another cloth over her so that her genitals are hidden; put the cover around each leg and each limb. Have two women each take hold of one leg, and two others each take hold of one arm; then they should hold the parturient

---

5 Littré: λέγω codd.
6 Littré: -έσασαν codd.

δεκάκις. ἔπειτα δὲ ἐς κλίνην ἀνακλῖναι τὴν γυναῖκα
ἐπὶ κεφαλήν· τὰ δὲ σκέλεα ἄνω ἔχειν, καὶ τὰς γυναῖκας πάσας λαβέσθαι τοῖν σκελέοιν, ἀφείσας τὰς
χεῖρας. ἔπειτα σείειν τὰς γυναῖκας ἐπὶ τοὺς ὤμους
πολλάκις, ἀναβολὰς ἐπὶ τὴν κλίνην, ὅκως ἐς τὴν
εὐρυχωρίην ἐπανασεισθὲν τὸ ἔμβρυον στραφῇ καὶ
δύνηται ἐπὶ φύσιν ἰέναι. καὶ ἤν ἔχῃς δίκταμνον Κρη-
τικόν, μεταπῖσαι· εἰ δὲ μή, κάστορος ἐνεψῆσαι τῷ
ἰσοχόῳ.

5. Ἢν δὲ αἱ ὑστέραι ἔξω χωρέωσιν, ἤν τε ἐκ πόνου,
ἤν τε ἐκ τόκου, ἤν μὲν οὖν παραλάβῃς νέας, ἄξιον
ἐπιχειρεῖν· εἰ δὲ μή, ἐῆν. ποιέειν δὲ χρὴ ὧδε· ἐπιταμὼν
τὸν ὑμένα τῆς ὑστέρης κατὰ φύσιν καὶ κατὰ πλάγιον,
τρῖψαι ὀθονίῳ ὡς φλεγμαίνῃ, κᾆτα[7] ἀλείψας φώκης
ἐλαίῳ ἢ πίσσῃ, καταπλάσας ἅμα κυτίνοισι· καὶ μαλ-
θακοὺς σπόγγους οἴνῳ ῥήνας, προσθείς, ἀναδῆσαι ἐκ
518 τῶν | ὤμων. καὶ ἀνακείσθω ὡς ἀνωτάτω τὰ σκέλεα
ἔχουσα, ἐσθιέτω δὲ σιτία μέτρια.

---

[7] Littré after a conjecture in I¹: κατ- M¹ I¹¹ V.

and shake her forcibly at least ten times. Then slant the woman on her bed towards her head, with her legs higher: all the women should release her arms and grasp her by her legs. Then have the women shake her several times towards her shoulders, after setting her on a bed, so that the fetus will be shaken upward into the open space, turned, and made ready to pass forward in the natural way. Also, if you have Cretan dittany available, (sc. have the woman) drink it afterwards; if not, boil crocuses to an equal amount.

5. If the uterus moves outside, either from a labour or a birth, and you take the women on when this has just happened, such cases are worth attempting; otherwise, decline them. You must do as follows: make incisions in the membrane of the uterus, both straight and crosswise, rub them with a piece of linen so that they will become inflamed, and then anoint them with seal oil or pitch, applying a plaster together with flowers of pomegranate.[2] Sprinkle soft sponges with wine and apply them as suppositories, and bind the uterus from over the woman's shoulders. Have her recline with her legs as high as possible, and let her eat moderate dishes.

[2] No reference is made to the actual replacement of the prolapsed uterus, which presumably occurs at this point.

SIGHT

INTRODUCTION

The only ancient reference to a treatise of this title is found in Galen's *Commentary on Hippocrates' Epidemics II*, a work lost in the original but preserved in Arabic translation:

> They (i.e. Numesianus and Pelops) are of the opinion that blue eyes indicate a warm temperament, as is stated in the book written on *Sight* which is attributed to Hippocrates.[1]

Another, unnamed work on eye diseases is announced in *Affections* 5. However, confirmative evidence that one or both of these allusions are to the *Sight* transmitted in the Hippocratic manuscripts is lacking.

This treatise has the appearance of being the fragmentary remains of a textbook of ophthalmology, including chapters on cataracts, trachoma, conjunctivitis, amaurosis, etc. Much attention is paid to treatment, in particular cautery and cleaning.

Sight was most recently edited in:

R. Joly, *Hippocrate, . . . De la Vision . . .* , Budé XIII, Paris, 1978. (=Joly)

[1] F. Pfaff, *Galeni In Hippocratis Epidemiarum libr. II Comm.* V, CMG V 10,1, Leipzig and Berlin, 1934, pp. 349f.

E. M. Craik, *Two Hippocratic Treatises*, On Sight . . . ,
 Leiden, 2006. (=Craik)

Particularly useful for the interpretation of the work are
comments in:

J. Hirschberg, *Geschichte der Augenheilkunde I. Alter-
 thum*, Leipzig, 1899, pp. 61–143. (=Hirschberg)

E. Craik, "The Hippocratic Treatise *Peri Opsios (De
 videndi acie, On the Organ of Sight)*," in Eijk, pp.
 191–207.

The present edition is based on a collation of the sole
independent witness, M, from microfilm.

ΠΕΡΙ ΟΨΙΟΣ

IX 152
Littré

1. Αἱ ὄψιες αἱ διεφθαρμέναι, αὐτόματοι μὲν κυανί-
τιδες γινόμεναι, ἐξαπίνης γίνονται, καὶ ἐπειδὰν γένωνται, οὐκ ἔστιν ἴησις τοιαύτη. αἱ δὲ θαλασσοειδέες
γινόμεναι, κατὰ μικρὸν ἐν πολλῷ χρόνῳ διαφθεί-
ρονται, καὶ πολλάκις ὁ ἕτερος ὀφθαλμὸς ἐν πολλῷ
χρόνῳ ὕστερον διεφθάρη. τούτου δὲ χρὴ καθαίρειν
τὴν κεφαλὴν καὶ καίειν τὰς φλέβας· κἢν ἀρχόμενος
πάθῃ ταῦτα, ἵσταται τὸ κακὸν καὶ οὐ χωρέει ἐπὶ τὸ
φλαυρότερον. αἱ δὲ μεταξὺ τῆς τε κυανίτιδος καὶ τῆς
θαλασσοειδέος, ἢν μὲν νέῳ ἐόντι γένωνται, πρεσβυτέρῳ γινομένῳ καθίστανται· ἢν δὲ πρεσβυτέρῳ ἐόντι
γίνωνται, ἐτέων ἑπτὰ βέλτιον ὁρῇ καὶ τὰ μεγάλα πάνυ
καὶ λαμπρά, καὶ ἀπὸ πρόσθεν ὁρῇ μέν, σαφέως δὲ οὔ,
καὶ ὅ τι ἂν πάνυ πρὸς αὐτὸν[1] τὸν ὀφθαλμὸν προσθῇ,
καὶ τοῦτο, ἄλλως δὲ οὐδέν. ξυμφέρει δὲ τούτῳ καῦσις
καὶ κάθαρσις τῆς κεφαλῆς· αἷμα δὲ τούτοισιν οὐ
ξυμφέρει ἀφιέναι, οὔτε τῇ κυανίτιδι, οὔτε τῇ θαλασ-
σοειδεῖ.

2. Τὸ ὄμμα ἐν τοῖσιν ὀφθαλμοῖσι, τῆς ὄψιος ὑγιέος
ἐούσης τῶν νεωτέρων ἀνθρώπων, ἤν τε θήλεια ᾖ, ἤν τ᾽
154 ἄρσην, οὐκ ἂν | ὠφελοίης ποιέων οὐδέν, ἕως ἂν αὔξη-

SIGHT

1. Pupils that spontaneously become deep blue when they are damaged become this way suddenly, and once they do, there is no cure for them. But pupils that become aquamarine in colour are damaged a little at a time over a long period, and often the opposite eye too becomes damaged after a long time. Such a patient must be treated by cleaning his head and cauterizing his vessels; if he receives this treatment at the beginning, the evil comes to a halt and does not progress to a worse state. Pupils that take on a colour between deep blue and aquamarine: if this happens in a young person, when he becomes older the pupils settle down. If it happens in someone older, for seven years he sees quite well objects that are both very large and bright, and what is before him he still sees, although not clearly, and whatever is placed very near to the eye itself, this too; but otherwise he sees nothing. Cautery and cleaning of the head benefit such a patient, but blood-letting is not useful in such cases, neither for the deep blue pupil nor for the aquamarine,

2. The vision in the eyes—as long as the pupil is healthy in young persons, whether female or male—you have no possible means of helping as long as the body is still grow-

---

1 Ermerins: ἑωυτὸν M.

ται τὸ σῶμα ἔτι. ὅταν δὲ μηκέτι αὐξάνηται, αὐτὼ τὼ
ὀφθαλμὼ σκεψάμενος τὰ βλέφαρα λεπτύνειν, ξύων,
ἢν δοκέη προσδέεσθαι, καὶ ἐπικαίων ἔνδοθεν μὴ δια-
φανέσιν.

3. . . . Ἔπειτα ἀναδήσας, τὰ σκέλεα ἐκτείνας,
δίφρον ὑποθεὶς ἀφ' οὗ στηρίζεται τῇσι χερσί· μέσον
δέ τις ἐχέτω. ἔπειτα διασημήνασθαι τὰς νωτιαίας
φλέβας, σκοπεῖν δὲ ὄπισθεν. ἔπειτα καίειν παχέσι
σιδηρίοισι καὶ ἡσυχίῃ διαθερμαίνειν, ὅκως ἂν μὴ
ῥαγῇ αἷμα καίοντι· προαφιέναι δὲ τοῦ αἵματος, ἢν
δοκέῃ καιρὸς εἶναι. καίειν δὲ πρὸς τὸ ὀστέον ὄπισθεν.
ἔπειτα ἐνθεὶς σπόγγον ἠλαιωμένον ἐγκατακαίειν,
πλὴν τοῦ πάνυ πρὸς αὐτῷ τῷ ὀστέῳ· ἢν δὲ προσ-
δέχηται τῷ καυστηρίῳ τὸ σπόγγιον, ἕτερον λιπαρώ-
τερον ἐνθεὶς ἐγκατακαίειν. ἔπειτα τοῦ ἄρου μέλιτι
δεύων, ἐντιθέναι τῇσιν ἐσχάρῃσιν. ὅταν δὲ φλέβα
παρακαύσῃς ἢ διακαύσῃς, ἐπειδὰν ἐκπέσῃ ἡ ἐσχάρη,
156 ὁμοίως τέταται ἡ φλὲψ καὶ πεφύσηται καὶ | πλήρης
φαίνεται, καὶ σφύζει ὅτε κάτωθεν τὸ ἐπιρρέον· ἢν δὲ
διακεκαυμένος ᾖ [ὁ]² κάτωθεν, ταῦτα πάντα ἧσσον
πάσχει. διακαίειν δὲ χρὴ αὖθις, ἢν μὴ τὸ πρῶτον
διακαύσῃς· τά τε σπόγγια χρὴ ἰσχυρῶς ἐγκατακαίειν,
πρὸς τῆς ῥεούσης φλεβὸς μᾶλλον. αἱ ἐσχάραι αἱ
μᾶλλον ὀπτηθεῖσαι τάχει ἐκπίπτουσιν. αἱ καιόμεναι
οὐλαὶ πρὸς τὸ ὀστέον καλλίονες γίνονται. ἐπειδὰν δὲ
τὰ ἕλκεα ὑγιέα γένωνται, αὖθις ἀναφυσῶνται καὶ
ἐπαίρονται, καὶ ἐρυθραί εἰσι παρὰ τὸ ἄλλο, καὶ ὥσπερ
ἀναρραγησόμεναι³ φαίνονται, ἕως ἂν χρόνος ἐπιγένη-

ing. When it stops growing, protect the two eyes themselves and thin the eyelids by scarification, if it seems necessary, and cauterize them inside with irons that are not
red-hot.

3. . . . Then, binding the patient and extending his legs,
place a stool under him on which he is held steady by the
arms: let someone hold him by the middle. Then mark the
vessels at his spine, and examine him from behind. Then
cauterize him with wide irons, heating through slowly in
order that no haemorrhage occurs while you are cauterizing; draw off some blood first, if the time seems propitious.
Burn down to the bone of the spine. Then, inserting a
sponge soaked in olive oil, cauterize right down, while
sparing the part very close to the bone itself; if the sponge
adheres to the cautery iron, insert another one with more
oil, and cauterize down again. Then smear some arum with
honey and place it on the eschars. If you cauterize near
or through a vessel, when the eschar sloughs off the vessel
will remain just as stretched, puffed up, and seemingly
full; and it pulsates when there is an afflux from below; if
the cautery is performed lower down on the back, such a
patient suffers all these things less. You must cauterize
again if you do not succeed the first time. The sponges you
must burn thoroughly, especially near the bleeding vessel.
Eschars that are well singed slough off rapidly. Cautery
scars that are near the bone take on a better appearance.
But when these lesions have healed, the vessels puff up
again and swell, are redder than the surrounding tissue,
and, until some time has passed, look as if they are about to
haemorrhage. And whether it is the head that is cauterized

---

² Del. Joly.　　　³ Ermerins: ἀναιρησόμεναι M.

ται· καὶ κεφαλῆς καυθείσης καὶ στήθεος, ὁμοίως δὲ
καὶ πάντι τῷ σώματι ὅκου ἂν κανθῇ.

4. Ὅταν δὲ ξύῃς βλέφαρα ὀφθαλμοῦ, ξύειν <εἶτα
καίειν>⁴ εἰρίῳ Μιλησίῳ, οὔλῳ, καθαρῷ, περὶ ἄτρακτον
περιειλέων, αὐτὴν τὴν στεφάνην τοῦ ὀφθαλμοῦ φυλασσόμενος, μὴ διακαύσῃς πρὸς τὸν χόνδρον. σημεῖον δὲ ὅταν ἀποχρῇ⁵ τῆς ξύσιος, οὐκ ἔτι λαμπρὸν
αἷμα ἔρχεται, ἀλλὰ ἰχὼρ αἱματώδης ἢ ὑδατώδης. τότε
δὲ χρή τινι τῶν ὑγρῶν φαρμάκων, ὅκου ἄνθος ἐστὶ
χαλκοῦ, τούτῳ ἀνατρῖψαι. ὕστερον δὲ [τὸ]⁶ τῆς ξύσιος
καὶ [τὸ]⁶ τῆς καύσιος, ὅταν αἱ ἐσχάραι ἐκπέσωσι καὶ
κεκαθαρμένα ᾖ τὰ ἕλκεα καὶ βλαστάνῃ, τάμνειν το-
μὴν διὰ τοῦ βρέγματος. ὅταν δὲ τὸ αἷμα ἀπορρυῇ,
χρὴ διαχρίειν τῷ ἐναίμῳ φαρμάκῳ. ὕστερον δὲ τούτου
ἔργον καὶ πάντων τὴν κεφαλὴν καθῆραι.

5. Τὰ βλέφαρα τὰ παχύτερα τῆς φύσιος, τὸ κάτω
ἀποταμὼν | τὴν σάρκα ὡς ἂν⁷ εὐμαρέστατα δύνῃ,
ὕστερον [δὲ]⁸ τὸ βλέφαρον ἐπικαῦσαι μὴ διαφανέσι,
φυλασσόμενος τὴν φύσιν τῶν τριχῶν, ἢ τῷ ἄνθει
ὀπτῷ λεπτῷ προστεῖλαι. ὅταν δὲ ἀποπέσῃ ἡ ἐσχάρη,
ἰητρεύειν τὰ λοιπά.

6. Ὁκόταν δὲ βλέφαρα ψωριᾷ καὶ ξυσμὸς ἔχῃ·
ἄνθος χαλκοῦ βώλιον πρὸς ἀκόνην τρίψας, ἔπειτα τὸ
βλέφαρον ἀποτρίψας αὐτοῦ, καὶ τότε τὴν φολίδα τοῦ
χαλκοῦ τρίβειν ὡς λεπτοτάτην· ἔπειτα χυλὸν ὄμφακος
διηθημένον παραχέας καὶ τρίψας λεῖον· τὸ δὲ λοιπὸν
ἐν χαλκῷ ἐρυθρῷ παραχέων, κατ᾿ ὀλίγον ἀνατρίβειν,

or the chest, it is the same in every part of the body where cautery is performed.

4. When you scarify the lids of the eye, scarify ⟨and then cauterize⟩ them with compact, clean Milesian wool wound around a wooden spindle, avoiding the rim itself of the eye in order not to burn right through to the cartilage. Here is a sign by which you can tell when you should stop your scarification: pure blood no longer flows out, but a bloody or watery serum. Then you should rub the patient with one of the moist medications consisting of flower of copper. Later, after the scarification and the cautery, when the eschars slough off and the lesions become clean and begin to regenerate, make an incision through the bregma; when blood flows out, anoint with a styptic medication. After this it is a matter of cleaning the head in all these cases.

5. Eyelids thicker than normal: excise beneath the tissue as effectively as you can, and later cauterize the lid with irons that are not red-hot, avoiding the roots of the hairs; or apply fine, burnt flower (sc. of copper). When the eschar sloughs off, heal what remains.

6. When the eyelids become scabby, and itchiness is present: grind a lump of flower of copper against a whetstone, next rub off the eyelid with it, and then grind some scale of copper as fine as you can. Then add strained juice of unripe grapes, grind fine, and pour what is left into a red copper vessel, continuing to grind it a little at a time until it

---

4 Add. J. Sichel in Littré.

5 A. Anastassiou in *Gnomon* 52 (1980), p. 314: -όχρη M.

6 Del. Ermerins.   7 ὡς ἂν Joly: ὁκόσην M.

8 Del. Ermerins.

ἕως ἂν πάχος γένηται ὡς μυσσωτός· ἔπειτα, ἐπειδὰν
ξηρανθῇ, τρίψας λεῖον χρῆσθαι.

7. Νυκτάλωπος φάρμακον πινέτω ἐλατήριον, καὶ
κεφαλὴν καθαιρέσθω, καὶ κατάξας τὸν αὐχένα ὡς
μάλιστα, πιέσας πλεῖστον χρόνον. ἐπανιεὶς δὲ διδόναι
ἐν μέλιτι βάπτων ἧπαρ βοὸς ὠμὸν καταπιεῖν μέγιστον ὡς ἂν δύνηται, ἓν ἢ δύο.

8. Ἤν τινι οἱ ὀφθαλμοὶ ὑγέες ἐόντες διαφθείροιεν
τὴν ὄψιν, τούτῳ χρὴ ταμόντα κατὰ τὸ βρέγμα, ἐπανα-
δείραντα, ἐκπρίσαντα τὸ ὀστέον, ἀφελόντα τὸν ὕδρωπα, ἰῆσθαι· καὶ οὕτως ὑγιέες γίνονται.

9. Ὀφθαλμίης τῆς ἐπετείου[9] καὶ ἐπιδημίου συμ-
160 φέρει κάθαρ|σις κεφαλῆς καὶ τῆς κάτω κοιλίης κάθαρ-
σις· καὶ εἰ ἔχοι τὸ σῶμα, αἵματος ἀφαίρεσις συμφέρει
πρὸς ἔνια τῶν τοιούτων ἀλγημάτων, καὶ σικύαι κατὰ
τὰς φλέβας. σῖτος ὀλίγος ἄρτος, καὶ ὕδατος πόσις.
κατακεῖσθαι δὲ ἐν σκότῳ, ἀπό τε καπνοῦ καὶ πυρὸς
καὶ τῶν ἄλλων λαμπρῶν, πλαγίως,[10] ἄλλοτε ἐπὶ τὰ
δεξιά, ἄλλοτε ἐπ᾽ ἀριστερά. μὴ τέγγειν τὴν κεφαλήν,
οὐ γὰρ συμφέρει. κατάπλασμα ὀδύνης μὴ ἐνεούσης,
ἀλλ᾽ ὡς ῥεύματος ἐπέχοντος, οὐ συμφέρει. οἰδημάτων
ἀνωδύνων καὶ μετὰ τὰ δριμέα φάρμακα τῆς ὀδύνης
ἐναλειφόμενα, ἐπειδὰν ἥ τε ὀδύνη παύσηται καὶ δια-
χωρισθῇ μετὰ τὴν ἐσάλειψιν τοῦ φαρμάκου, τότε
συμφέρει καταπλάσσειν τῶν καταπλασμάτων ὅ τι ἂν

---

[9] Later manuscripts: ἐπαιτίου M.
[10] Cornarius in marg.: πλαγίων M.

has the thickness of *mussotos*.[1] Then, when this becomes dry, grind it fine and apply.

7. As a medication for nyctalopia[2] let the patient drink squirting-cucumber juice, have his head cleaned, and reduce his neck as much as possible, compressing it for a very long time. When remission occurs give him raw bull's liver dipped in honey, and have him drink down as much as he can, one or two.

8. If someone's eyes, though in a healthy state, impair his vision, you must incise him at the bregma, scrape off the skin, saw out the bone, and by removing the swelling heal him; in this way he will recover.

9. In annual and epidemic ophthalmia cleaning the head and the lower cavity helps. And if the patient's body will tolerate it, it is beneficial to draw blood for some of these pains and to apply bleeding-cups to the vessels. As food give bread, a small amount, and as drink water. Have the patient lie down in the shade—away from smoke, fire, and other bright things—on his side, sometimes on the right, sometimes on the left. Do not moisten his head, since this would do no good; poultices are of no use when no pain is present, but there are fluxes. When there are swellings unaccompanied by pain, and when there has been an anointment of pungent medications against pain —when the pain ceases and goes away after anointment with this medication, then it helps to apply poultices that

---

[1] "A savoury dish of cheese, honey, garlic, etc.," H. G. Liddell and R. Scott, *A Greek-English Lexicon*, 9th edn., Oxford, 1940.

[2] For a discussion of nyctalopia in the Hippocratic writings see Hirschberg pp. 98–104.

δοκέῃ ξυμφέρειν. οὐδὲ διαβλέπειν ξυμφέρει πολὺν χρόνον, δάκρυον γὰρ προκαλέεται, οὐ δυνάμενος πο-νέειν ὁ ὀφθαλμὸς πρὸς τὰ λαμπρά· οὐδὲ ξυμμύειν πολὺν χρόνον, ἢν ἔχῃ ῥεῦμα θερμὸν μάλιστα· θερ-μαίνει γὰρ τὸ δάκρυον ἰσχόμενον. ῥεύματος δὲ μὴ ἔχοντος, μετὰ τοῦ ξηροῦ τὴν ὑπάλειψιν ξυμφέρει ποιέ-εσθαι.

are likely to be beneficial. It is not good for the person to stare for a long time, as this brings forth tears, since such an eye cannot tolerate bright objects. Nor should he close his eyes for a long time, especially if he has a hot flux: for tears that are held back cause warming. If there is no flux, it helps to apply ointments together with a dry medication.

INDEX

INDEX